New Organic Chemistry

New Organic Chemistry

by H. L. HEYS M.A. (CANTAB.)

Formerly Senior Science Master,
Liverpool Collegiate School

SECOND EDITION

HARRAP LONDON

First published in Great Britain 1957
as *A New Organic Chemistry for Schools and Colleges*
by GEORGE G. HARRAP & CO. LTD
182–184 High Holborn, London WC1V 7AX

This edition © *H. L. Heys* 1973

ISBN 0 245 50693 4

The cover design shows part of the large
scale poly(propene) model at the Philips Science
Centre 'Evoluon', Eindhoven, Holland, and is
reproduced by kind permission of Paul Brierley.

Text set in 10/11 pt. Monotype Old Style, printed by photolitho-
graphy, and bound in Great Britain at The Pitman Press, Bath

Preface to the First Edition

This book is intended to meet the requirements in organic chemistry of students preparing for the Advanced and Scholarship Levels of the General Certificate of Education. It is also suitable for candidates for Open Scholarships at Oxford and Cambridge and for first-year university students who wish to improve their knowledge of general elementary organic chemistry.

The chief aim of the author in writing the book has been to modernize the subject-matter of organic chemistry. During the last twenty or thirty years profound changes have taken place in the subject. A new branch of chemical industry based on petroleum has arisen, and from it has come a variety of new products, many of which are now well known in everyday life. Fresh reagents have appeared in the laboratory, and both here and in industry new techniques have been introduced. The most important of these modern developments have been incorporated in the present work. Unfortunately, considerations of space have prevented the inclusion of the application of the electronic theory of valency to organic compounds, but it is hoped to deal with this aspect of the subject in a small supplementary book.

In general the arrangement of the subject-matter follows conventional lines, although departures from the logical sequence have been made in some cases on the grounds of expediency. Thus, as a result of many years' teaching experience, the author has found it advisable for pupils to study qualitative and quantitative analysis before methods of purification. Not only does this help to consolidate the interest of the pupil in what is, for him or her, a new branch of chemistry, but the knowledge of inorganic analysis previously gained can be used as a stepping-stone to further study. Similarly it is preferable to deal with halogen derivatives of the paraffins, not immediately after the parent hydrocarbons, but at a later stage, when the student is in a better position to tackle the more complicated reactions involved.

The author has tried to keep in mind the difference in mental capacity between the beginner and the more advanced student. Particular attention has been paid to the grading of the subject-matter, and, as far as possible, more difficult parts have been placed at the ends of chapters. Where this has not been feasible more difficult material has been printed in smaller type to indicate that its study may be postponed until a later stage.

Examiners frequently criticize candidates for their ignorance of the conditions under which organic reactions occur. The author has endeavoured to remedy this defect by a careful description of reaction conditions. In addition, he has included a large number of small-scale experiments, which can be used for demonstration purposes or for individual practical work. Experience has shown that these experiments are of great assistance in helping pupils to become familiar with the compounds which they are studying.

The choice of subject-matter has not been determined, however, by examination requirements alone. Organic chemistry has considerable cultural value in the many contacts which it makes with human experience. Unfortunately, lack of time often compels the teacher to neglect this aspect of the subject. The author has endeavoured to foster the natural interest of young students by giving brief accounts of plastics, detergents, insecticides, synthetic fibres, silicones, and many other substances of everyday importance. It is hoped that pupils will be sufficiently interested to read these accounts on their own initiative and thus acquire a broader outlook on science in general.

The author wishes to express his gratitude to Mr E. R. Oldman, M.Sc., for reading the manuscript of the book in proof and making many valuable suggestions for its improvement. He is also indebted to the following bodies for permission to reproduce questions from their examination papers: the Oxford and Cambridge Schools Examination Board; the Senate of the University of London; the Welsh Joint Education Committee; and the Joint Matriculation Board of the Universities of Manchester, Liverpool, Leeds, Sheffield, and Birmingham.

H. L. H.

Preface to the Second Edition

In this edition a number of alterations have been made, including simplification of the previous title, *A New Organic Chemistry for Schools and Colleges*. The most important change has been to modernize the approach to the study of the subject. From the outset students are now expected to acquire an understanding of the electronic structures of organic molecules and some of the mechanisms by which they react. A new chapter early in the book explains the production of ionic character in covalent bonds by inductive and mesomeric effects, the influence of this character on properties, and the general features of organic reactions and reaction mechanisms. The principles are then applied throughout the remainder of the book.

An important innovation is the incorporation of systematic naming of compounds (organic and inorganic) as recommended in *Chemical Nomenclature, Symbols and Terminology*, published by the Association for Science Education in 1972. Trivial or common names in current use, however, have been included alongside systematic names sufficiently often to avoid confusion and to facilitate the change-over to modern nomenclature.

The units used are generally those of SI (*Système International d'Unités*), although a few exceptions (e.g. grams, atmospheres) have been made where traditional units are likely to continue in use because of their convenience.

Another new feature in this edition is the inclusion of 'short answer' questions at the end of each chapter, with answers at the end of the book. The object of these is to encourage students to quiz themselves on their grasp of factual knowledge and understanding of basic principles.

The author wishes to thank the following bodies for permission to reproduce additional questions from their GCE examination papers at A level and S level.

University of Cambridge Local Examinations Syndicate
Oxford and Cambridge Schools Examination Board
Southern Universities' Joint Board for School Examinations

The author is indebted to Unilever Ltd for permission to reproduce the infrared absorption spectrum of methyl octadecanoate (Fig. 3-3) from *The Physics of Chemical Structure*, a Unilever Educational Booklet. His thanks are also due to Professor E. G. Cox FRS and the Royal Society for allowing the use of an electron density map of the benzene molecule (Fig. 3-5).

Finally the author wishes to express his gratitude to Miss V. C. Coghlan, B. Eng., for her most valuable assistance in the preparation of the manuscript for the new format adopted for the book.

H. L. H.

Contents

part one

GENERAL PRINCIPLES

1

A Brief History of Organic Chemistry —Features of Organic Compounds

A BRIEF HISTORY OF ORGANIC CHEMISTRY

Period of the 'Vital Force' Theory

Organic chemistry developed as a separate branch of chemistry during the early part of the last century. Stimulated by Priestley's discovery of oxygen (1774) and Lavoisier's researches on combustion (1773–78), chemistry as a whole was expanding rapidly at this time. Many chemists devoted their attention to the compounds, such as sugar and gelatine, produced naturally in plant and animal 'organisms,' isolating these compounds and analysing them to find their chemical composition. The term 'organic' was first applied to such compounds by the famous Swedish chemist Berzelius. Compounds like salt and chalk, which came from mineral sources, were distinguished by calling them 'inorganic.'

At this early period it was generally believed that 'organic' substances could be made only in living cells by means of a mysterious *vital force*. It was also thought that the ordinary laws of chemistry, such as the law of definite proportions, did not hold for these compounds. In the words of Berzelius, "In living Nature the elements appear to obey laws which are quite different from those in non-living Nature." This error probably arose from the difficulty of preparing the so-called organic substances in a state sufficiently pure for accurate analysis.

The 'vital force' theory held sway in organic chemistry for many years. It was abandoned only when experiments showed that organic compounds could be made in the laboratory from inanimate materials. Thus, in 1828 a German chemist, Wöhler, discovered a method of making the compound urea without the aid of plants or animals. Urea was regarded as a typical organic substance, the only method known for obtaining it being by evaporation of urine. Wöhler found that when he evaporated an aqueous solution of ammonium cyanate the atoms rearranged themselves to form urea. Since the ammonium cyanate was

$$NH_4CNO \rightarrow CO(NH_2)_2$$

ammonium   urea
cyanate

readily obtained from ordinary laboratory materials no 'vital force' could be involved in Wöhler's method of preparing urea. Wöhler well realized the revolutionary nature of his discovery. In a letter to Berzelius he proudly announced, "I can prepare urea without requiring a kidney or an animal, man or dog."

In spite of Wöhler's success the 'vital force' theory lingered on in chemistry until the middle of the century. It disappeared gradually as, one by one, other 'organic' compounds were prepared in the laboratory, and chemists began to realize that there was no fundamental difference between these and 'inorganic' compounds. The outstanding feature of organic compounds was that they all contained the element carbon, and this slowly led to a change in the conception of organic chemistry. From about 1850 onward organic chemistry became *the study of the compounds of carbon*.

The Influence of Liebig

Many organic compounds well-known to-day were discovered in the first half of the nineteenth century. Benzene was first obtained in 1825 by Faraday, who isolated it from the substances produced by heating fish-oil. Trichloromethane (chloroform) was discovered in 1831, and ethyne (acetylene) in 1836. Trichloromethane was first prepared by the German Justus von Liebig (1803–73), who was professor of chemistry at the University of Giessen. Liebig was the chief founder of modern organic chemistry. Not only did he discover new substances, but he greatly improved the method of analysing organic compounds. He also investigated the chemical changes which food undergoes in the body, and he was the first to divide foods into the two classes of energy-providers (sugars and fats) and body-builders (proteins).

The most important contribution of Liebig, however, to the progress of organic chemistry was to revolutionize the method of teaching the subject. At that time not a single university or school possessed a laboratory for the teaching of chemistry, and students spent most of their time in listening to lectures and reading text-books. Liebig set up a laboratory for his pupils and taught them to carry out experiments, mostly of a research character. In this way he equipped them with a technique for tackling original problems of their own. Liebig's novel method of teaching promoted great enthusiasm for chemistry, and students flocked to Giessen from all parts of Germany. On the foundations thus laid, Germany built up a pre-eminence in organic chemistry which lasted until the First World War.

Another country which benefited from Liebig's influence was Britain. In 1840 Queen Victoria married Prince Albert, a German prince, who was keenly interested in science and who was familiar with Liebig's work. The new Prince-Consort played a prominent part in the founding of the Royal College of Chemistry in London in 1845. When the appointment of the professor of chemistry arose no Englishman with suitable qualifications could be found, and Prince Albert sought the help of Liebig in filling the post. Liebig responded by sending from Germany one of his most talented graduates, a young man called Hofmann. Hofmann himself was a brilliant chemist and teacher, and the systematic study of organic chemistry in this country may be said to date from his appointment.

The Theory of Valency

The theory of valency was advanced by Frankland in 1852 to explain the different combining powers of atoms of different elements. This theory enabled chemists to deduce the manner in which the atoms were arranged in a molecule. The theory was applied to organic compounds with great success by the Scottish chemist Couper and the German chemist Kekulé. These men showed how the formulæ of quite large molecules could be explained by the tetravalency of the carbon atom and the joining together of carbon atoms in chains. Couper was the first to make use of lines to represent valency bonds and to show the arrangement of atoms in a molecule, as we do nowadays, by means of a *graphic formula*. For example, in ethanol (ethyl alcohol), C_2H_6O, the carbon, oxygen, and hydrogen atoms, which have valencies of four, two, and one respectively, are joined together as shown.

$$
\begin{array}{ccc}
 & H & H \\
 & | & | \\
H- & C- & C-O-H \\
 & | & | \\
 & H & H \\
\end{array}
$$

ethanol

This method of indicating the structure of a compound proved extremely valuable. It helped chemists to understand the properties of compounds and the reactions by which the compounds were formed. Instead of groping in the dark, research workers were able to direct their experiments to producing new molecules of definite patterns. Their success can be judged from the number of new organic compounds discovered between 1850 and 1900. In the middle of the century less than a thousand organic compounds in all were known. By the end of the century the total had increased to over 100 000.

Manufacture of Organic Chemicals

The lives of ordinary people were little affected by discoveries in organic chemistry before 1850. A few organic products such as beer, wine, and vinegar had been manufactured from ancient times, but the chemical processes involved were not understood. From the middle of the century onward organic chemistry played an increasing part in enriching the material resources of mankind. The chief reason for this was the finding of the chemical treasures contained in coal-tar, a by-product in the coal-gas industry. Coal-gas was first used for lighting in the early years of the century, and its use extended rapidly. By the 1850s the large amounts of coal-tar being produced were becoming a nuisance. A little was used for preserving timber, but most manufacturers got rid of it by burning it. The black evil-smelling liquid had attracted the interest of Hofmann in the days when he was working under Liebig, and he had succeeded in extracting from it several compounds which were later to prove very important. One of these was phenylamine (aniline).

In 1856 one of Hofmann's students at the Royal College of Chemistry was the eighteen-year-old William Perkin. Hofmann suggested to Perkin that he should try to prepare quinine, the anti-malarial drug, from the chemicals in coal-tar. Perkin took up the problem enthusiastically in his laboratory at home, but all his attempts to make quinine ended in failure. In the course of one of his experiments, however, he obtained a brilliant mauve, or lavender, dye. This proved to be far superior in lustre and fastness to any of the lavender dyes then in use. With the help of his father the youthful Perkin started a factory for the manufacture of the dye (called 'mauveine'), and when Queen Victoria appeared in public wearing a beautiful dress dyed with the new compound its commercial success was assured. The discovery of mauveine stimulated chemists into trying to make other dyes from coal-tar, and in the next twenty or thirty years some dozens of valuable new dyes were prepared from this source.

Dyes were not the only prizes, however, won from coal-tar. From this unpromising material came disinfectants, perfumes, flavouring agents, explosives, solvents, photographic chemicals, and many other useful substances. Once organic chemistry got into its stride it advanced rapidly on all sides, and soon the new discoveries were bringing about important social changes. Thus, the discovery of celluloid (by Parkes in 1865) made it possible to prepare photographic film, and this in turn paved the way for the introduction of the cinema.

In the present century manufacture of organic chemicals has expanded at an ever-increasing rate. The last fifty years have seen not only a great extension in the range of dyes, drugs, explosives, antiseptics, etc., but also the discovery of entirely new kinds of materials such as synthetic fibres, synthetic rubber, plastics, and detergents. Many of these are described in subsequent chapters.

Nowadays the chief raw material of the organic chemical industry is no longer coal, but petroleum. The change was brought about by the great increase in popularity of the motor-car in the 1920s and 1930s. As more and more petroleum was processed to obtain petrol, manufacturers were left with ever larger amounts of the lighter and heavier constituents of the oil, and for these there was little demand. However, chemists soon discovered not only how to convert the unwanted compounds into new products, but also how to use them for making existing chemicals more cheaply. As a result a flourishing 'petrochemicals' industry sprang up in the U.S.A. before the Second World War. After the War the new industry spread to Britain, and has since grown so rapidly that about 75 per cent of British organic chemicals are now made from petroleum.

After petroleum, coal is the most important source of organic chemicals. Subsidiary raw materials include wood, cotton, animal oils and fats, vegetable oils, and sugar.

FEATURES OF ORGANIC COMPOUNDS

Although there is no fundamental difference between organic and inorganic compounds, the division is still retained for the sake of convenience. Since organic compounds have certain distinctive features, they can be studied more easily as a separate class of compounds. It should be mentioned, however, that some compounds, such as carbonates and the oxides of carbon, which, strictly speaking, belong to the organic section of chemistry, are more conveniently grouped with inorganic substances.

Differences between Organic and Inorganic Compounds

(i) *Number*. Organic compounds are much more numerous than inorganic. It is impossible to say with any certainty how many individual compounds of the two classes have been prepared.

Estimates for organic compounds vary from three million to five million, but, whatever the number, it is probably at least ten times larger than that of inorganic compounds.

(*ii*) *Composition*. In spite of their large number, organic compounds are composed of relatively few elements. In order of frequency of occurrence these fall roughly into three groups: (*a*) carbon, hydrogen, and oxygen; (*b*) nitrogen, halogens, and sulphur; (*c*) silicon, phosphorus, and metals. Compounds containing other elements are known, but are relatively rare.

(*iii*) *Action of Heat*. Although there are many exceptions, organic substances are on the whole less stable to heat than inorganic ones. When they are heated in the absence of air they usually break up, the majority decomposing between 300°C and 500°C. Inorganic compounds usually require higher temperatures than this to decompose them.

When organic compounds are heated in air they generally burn away, often leaving a residue of carbon. This is readily understood when it is remembered that carbon and hydrogen enter largely into their composition. All the common fuels (coal, wood, petrol, methylated spirit, etc.) consist of organic compounds. Very few inorganic substances burn in air.

(*iv*) *Complexity*. Organic molecules, generally speaking, are much bigger than inorganic ones. A molecule of sulphuric(VI) acid, H_2SO_4, is a relatively large inorganic molecule, whereas a molecule of sucrose (cane sugar), $C_{12}H_{22}O_{11}$, is only a medium-size organic molecule. In the case of some of the more complex

organic compounds (e.g. proteins and plastics) molecules may consist of thousands of atoms. Thus hæmoglobin, the red colouring-matter in blood, contains about 9000 atoms in one molecule.

(*v*) *Isomerism*. In inorganic chemistry a molecular formula, such as H_2SO_4, usually represents only one chemical compound, but in organic chemistry the same molecular formula can stand for two, or even more, substances. This is due to the different ways in which atoms can join together to form molecules. Thus C_2H_6O is the molecular formula not only of ethanol, but also of methoxymethane (dimethyl ether). In the molecules of these two compounds the atoms are combined together as shown.

$$
\begin{array}{ccc}
& H & H \\
& | & | \\
H - & C - & C - O - H \\
& | & | \\
& H & H
\end{array}
$$

ethanol

$$
\begin{array}{ccc}
& H & H \\
& | & | \\
H - & C - O - C - H \\
& | & | \\
& H & H
\end{array}
$$

methoxymethane

The occurrence of two or more compounds with the same molecular formula is called isomerism, and the different compounds are referred to as *isomers*, or *isomerides*. Ammonium cyanate and urea (mentioned earlier) are another example of two isomeric compounds.

(vi) Homologous Series. Organic compounds fall into distinctive groups when considered from the point of view of structure and chemical properties. Each group is called an *homologous series*. An homologous series is defined as *a series of compounds in which successive members (a) possess a similar structure and similar chemical properties, (b) can be represented by a general chemical formula, and (c) differ in formula by a constant amount* CH_2. One example of an homologous series is that of the alcohols—methanol (CH_3OH), ethanol (C_2H_5OH), propanol (C_3H_7OH), etc. All of these alcohols are represented by the general formula $C_nH_{2n+1}OH$. The various members of an homologous series are called *homologues*. Another homologous series is the alkanes (paraffin hydrocarbons) of general formula C_nH_{2n+2}.

(vii) Chemical Bonding. The great majority of organic compounds do not conduct electricity in the liquid (or fused) state, nor, if they are soluble in water, in aqueous solution. It may be deduced from this that most organic compounds are composed, not of ions, but of molecules containing only covalent bonds. In contrast a fairly high proportion of inorganic compounds have an ionic structure (e.g. $Na^+ Cl^-$). Ionic organic compounds are chiefly salts of organic acids and bases.

In most organic molecules the covalent bonds are single, and consist of a pair of electrons shared between two atoms. Some organic molecules, however, contain double or triple covalent bonds formed by the sharing of two pairs or three pairs of elec-

trons respectively. Examples of single, double, and triple covalent bonds are found in the molecules of the hydrocarbons methane, ethene (ethylene), and ethyne (acetylene). The valency bonds in these molecules are represented below, firstly in the traditional manner and secondly by means of the shared electrons.

$$
\begin{array}{cc}
\text{H} & \text{H} \\
| & .. \\
\text{H}-\text{C}-\text{H} & \text{H}:\text{C}:\text{H} \\
| & .. \\
\text{H} & \text{H}
\end{array}
$$

methane

$$
\begin{array}{ccc}
\text{H}\ \text{H} & \text{H}\ \text{H} & \\
\text{C}=\text{C} & \text{C}::\text{C} & \text{H}-\text{C}\equiv\text{C}-\text{H} \\
\text{H}\ \text{H} & \text{H}\ \text{H} & \text{H}:\text{C}:::\text{C}:\text{H}
\end{array}
$$

ethene   ethyne

The nature of covalent bonds and their influence on the properties of organic compounds are discussed in Chapter 4 and subsequent chapters.

Naming of Organic Compounds

Chemical nomenclature has undergone numerous changes in the course of time. Nowadays the tendency is to make chemical names increasingly systematic, so that they reflect more accurately the molecular structure of the compounds. Modern 'recommended' names are based on the 'substitutive' system of the International Union of Pure and Applied Chemistry (I.U.P.A.C.).

In this system most compounds are regarded as formed by replacing one or more hydrogen atoms in a molecule of a parent compound by other atoms or groups, and they are named accordingly. The parent compounds are usually hydrocarbons like methane (CH_4) or benzene (C_6H_6). Thus, instead of the conventional name chloroform for the compound $CHCl_3$, the name trichloromethane is recommended, the latter showing that three chlorine atoms have replaced three hydrogen atoms from the methane molecule. Other 'recommended' names are methoxymethane instead of dimethyl ether (CH_3—O—CH_3) and methylbenzene in place of toluene ($C_6H_5.CH_3$).

It must be emphasized that the older conventional names are not wrong, but it is expected that in time they will be discarded by chemists in favour of the 'recommended' names. A number of the older names have been retained for convenience and rank equally with newer ones. When in this book one name is followed by another in brackets the first is the 'recommended' name and the second is the conventional name. If two names are given as alternatives (e.g. acetic acid, or ethanoic acid) either is acceptable.

Radicals and Functional Groups

A *radical* is a group of atoms which behaves as a unit and retains its identity through a series of reactions. For example, starting with ethanol, we can prepare firstly iodoethane (ethyl iodide) and then ethyl cyanide. In these changes the ethyl radical (C_2H_5—) persists unchanged.

$$C_2H_5OH \rightarrow C_2H_5I \rightarrow C_2H_5CN$$

The ethyl radical and other radicals with the general formula C_nH_{2n+1} are called *alkyl* radicals. The first four alkyl radicals have 'trivial' names, but subsequent ones are named according to the number of carbon atoms present.

Methyl:	CH_3—	Butyl:	C_4H_9—
Ethyl:	C_2H_5—	Pentyl:	C_5H_{11}—
Propyl:	C_3H_7—	Hexyl:	C_6H_{13}—

etc.

The members of an homologous series usually consist of a radical attached to an atom or group which is common to the series. Thus alcohols of general formula $C_nH_{2n+1}OH$ contain an alkyl radical and a hydroxyl group. The latter is called the *functional group* of the series. The functional group is responsible for the members showing a general resemblance to each other in chemical behaviour. For example, one general reaction of alcohols associated with the —OH group is that they liberate hydrogen when treated with sodium (compare the reaction of water, H—O—H, with this metal).

The properties of organic compounds, however, are determined by all the groups present in their molecules. Thus the reactions of alcohols are not solely those brought about by the hydroxyl group. Some are due to the alkyl radical. Also, the same functional group sometimes occurs in the members of two different homologous series, and its behaviour may differ in the two cases. For example, the —OH group in phenol (C_6H_5OH) does not always react in the same way as the —OH group in ethanol (C_2H_5OH). This is because the C_6H_5— radical and the C_2H_5— radical affect the —OH group differently. As will be seen later, these differences can be explained by means of the electronic structures of the two molecules.

Architecture of Organic Molecules

The special features of organic compounds are due to the remarkable ability of carbon atoms to join together in the form of 'open' chains and 'closed' chains, or rings. A few other elements (e.g. silicon) possess this property, but only to a very limited extent. Open chains of carbon atoms may be as short as one or two atoms long, or they may consist of hundreds of atoms joined together. Some organic compounds are described as *straight-chain* compounds because the carbon atoms form one continuous chain. In fact, the chains are not straight, but follow a zigzag pattern. For convenience in writing formulæ, however, they are usually represented as being straight. In other organic molecules *branched* chains of carbon atoms are present. Ignoring valency considerations, the different kinds of chains can be depicted as shown below.

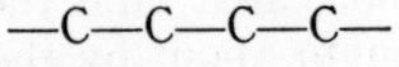

—C—C—C—C—   —C—C—C—
straight chain 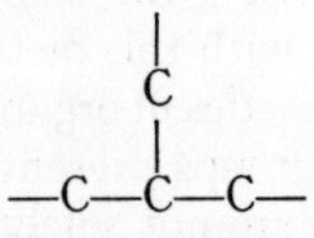   branched chain

Compounds containing only straight chains or branched chains of carbon atoms are called *aliphatic* compounds. These constitute one of the two main classes of organic compounds, and are studied in Part II of this book. Examples of straight-chain and branched-chain compounds are the hydrocarbons propane and 2,2-dimethylpropane (so-called because two methyl radicals

have replaced two hydrogen atoms attached to the second carbon atom of the propane chain.

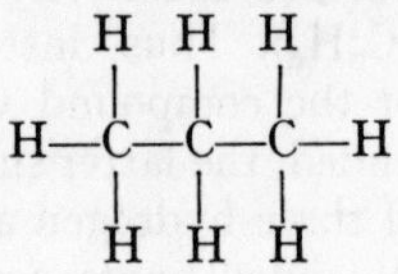

propane

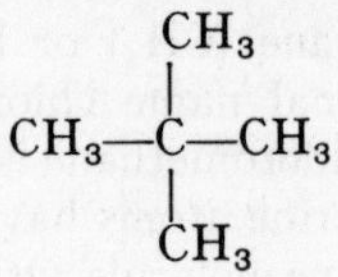

2,2-dimethylpropane

Compounds containing closed chains, or rings, of carbon atoms are described as *cyclic* compounds. The closed chains may have as few as three, or as many as twenty carbon atoms. The two simplest compounds of this type are cyclopropane and cyclobutane.

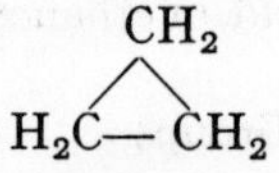

cyclopropane

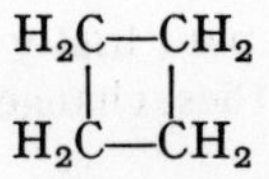

cyclobutane

With the exception of one particular group, closed chain compounds resemble the corresponding open chain compounds.

They are described as *alicyclic*. Some of them are found in petroleum, and a few (notably cyclohexane) have become important as solvents or as intermediates in the manufacture of other compounds. Otherwise they merit little attention.

The particular group of closed-chain compounds which are not alicyclic consists of benzene and related compounds. The benzene molecule (C_6H_6) is similar to that of cyclohexane (C_6H_{12}) in containing a ring of six carbon atoms, but the structures of the two rings are different, as shown below.

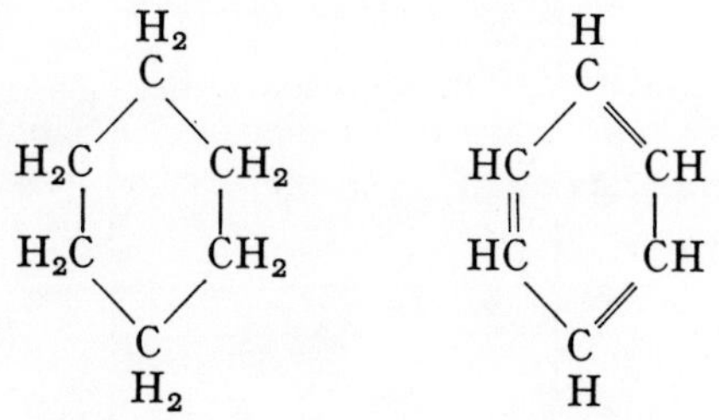

cyclohexane   benzene

Benzene and the compounds related to it are called *aromatic* compounds, and are described in Part III. Their properties differ in many ways from those of aliphatic and alicyclic compounds.

The almost unlimited capacity of carbon atoms to join together in open or closed chains distinguishes carbon from all other elements. The reasons for this unique property is discussed in Chapter 5. Bearing in mind that some, at least, of the hydrogen atoms in the hydrocarbons mentioned above can be replaced by other atoms (or groups of atoms like —OH), it is obvious that a very large number of compounds can be derived

from these hydrocarbons alone. Again, since the same atoms can join together in different ways, they can give rise to substances with quite different properties. This accounts for the existence of isomeric compounds such as ethanol and methoxymethane (dimethyl ether).

EXERCISE 1

(*Note*. The 'quiz' questions given here and at the ends of other chapters are intended to help students to test themselves on their grasp of information and understanding of basic principles. Longer questions of a numerical character will be found in Miscellaneous Problems at the end of the book.)

1. From a consideration of valencies state which of the following chemical formulæ are incorrect:

(*a*) CH_3—CH—OH;   (*b*) CH_3—O—CH_2—CH_3;
(*c*) $CH_2 = CH_2 - Cl$; (*d*) HO—CH_2—CH_2 Br

2. A solid compound X gave the following experimental results:

(*a*) It had a low melting point;
(*b*) It did not conduct electricity when melted;
(*c*) It burned when heated in air;
(*d*) When heated in the absence of air it decomposed and gave off carbon dioxide and water.

Which (if any) of these results proved that X was an organic compound?

3. Which (if any) of the following compounds are isomeric?

(a) $H—\overset{\overset{\displaystyle O}{\|}}{C}—O—CH_3$; (b) $HO—CH_2—CH_2—OH$;

(c) $CH_3—\overset{\overset{\displaystyle O}{\|}}{C}—OH$

4. In the homologous series of hydrocarbons CH_4, C_2H_6, C_3H_8, C_4H_n, what is the value of n?

5. The hydrocarbon C_3H_6 contains a double bond in its molecule. Make up a formula to show how the atoms are joined together.

6. Which of the following statements are true?

(a) An inorganic compound is a compound which does not contain carbon;

(b) A double covalent bond is formed when two atoms share two electrons;

(c) Successive members of an homologous series differ in formula by a constant amount CH_3;

(d) Ethanol (C_2H_5OH) and phenol (C_6H_5OH) contain the same functional group.

7. Each of the following carbon structures can be described by two of the terms: straight-chain, branched-chain, closed-chain, aliphatic, alicyclic. Which two terms apply in each case?

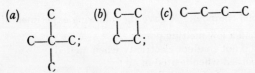

Note. Answer to Exercises will be found on page 360 and following.

2 Purification of Organic Compounds

Whether an organic compound occurs naturally, or whether it is artificially produced, it usually has to be purified before it can be used. Many organic reactions take place only slowly, and do not proceed to completion. The products may therefore be mixed with varying amounts of the reactants as well as with impurities derived from side-reactions, which often occur at the same time as the main reaction. For these reasons purification of a product may take longer than its preparation in the first 'crude' state. In the descriptions of preparations given later it may be stated that a substance is purified by 'recrystallization from ethanol' or by 'fractional distillation.' It is of the utmost importance that the student should understand how these, and other, purification processes are carried out.

The chief methods used in the purification of organic compounds are the following:

1. recrystallization;
2. distillation (including fractional distillation, distillation under reduced pressure, and distillation in steam);
3. sublimation;
4. adsorption;
5. chemical purification (including the use of drying agents).

The precise manner in which these operations are carried out depends on the scale of preparation which has been used. Nowadays 'small-scale' working is favoured in the laboratory because it is more economical in materials, time, and effort. In small-scale preparations the amounts of substances used are of the order 1 g–10 g, and specially designed apparatus with ground-glass joints is employed. Industrial methods of purification are similar in principle to those of the laboratory, although the apparatus used is often different in form.

1. RECRYSTALLIZATION

Most organic solids are purified by recrystallization from a suitable solvent. *A suitable solvent is one in which the compound has a small solubility at ordinary temperatures but a relatively large solubility at higher temperatures.* Thus, if a concentrated solution of the impure compound is made in the hot solvent the compound is precipitated when the solution is cooled. Any impurity should be either insoluble in the solvent (so that it can be removed by filtration) or readily soluble (so that it stays in solution). In practice the impurity usually dissolves in the hot solvent, but, being relatively small in amount, it remains dissolved when the solution is cooled. Recrystallization consists of three main stages as now described.

(*i*) *Finding a Suitable Solvent.* Solvents commonly employed in recrystallization include water, ethanol, propanone (acetone), ethanoic, or acetic, acid, trichloromethane (chloroform), petrol, and benzene. Sometimes a mixture of two liquids (e.g. ethanol and water) forms a better recrystallizing medium than either liquid separately. Ethoxyethane (diethyl ether) and carbon disulphide, although excellent solvents for many compounds, are generally avoided owing to the flammability of their vapours.

A preliminary determination of the most suitable liquid to use is first carried out on a small scale, using the method of 'trial and error'. About 0·5 g of the compound is put into a dry test-tube, 1–2 cm³ of the liquid on trial are added, and the tube is shaken.

If the compound dissolves in the cold liquid the latter is un-suitable, because the compound would not be precipitated when a hot solution is cooled. If the compound fails to dissolve in the cold the test-tube is carefully heated over a *very small* flame until the liquid begins to boil. Should the compound still not dissolve completely further portions of 1–2 cm³ of liquid are added until the test-tube is half-full and heating is continued. If the solid still remains undissolved another solvent is tried. If, however, the compound dissolves in the hot liquid the solution is cooled to room temperature, when crystals of the substance should be deposited. In this case the liquid is suitable to use as a solvent for recrystallization of the main portion of the compound.

(*ii*) *Recrystallizing the Main Portion*. The next step is to dis-solve the main portion of the cystals in the boiling solvent. If the solvent is water the dissolving can be carried out in a small conical flask placed on a tripod and gauze. If the solvent is an organic liquid which gives off a flammable or poisonous vapour, a 50 cm³ pear-shaped flask is used. This is attached to an upright, or reflux, condenser (Fig. 2-1). The vapour of the solvent is thus condensed and the solvent returned to the flask. The shape of the flask is designed to give the liquid a smaller surface area, so that it vaporizes less rapidly and boils more steadily. As a further aid to steady boiling a little powdered pumice is often added to the flask (broken porcelain should *not* be used as it scratches the inside of the flask and leads to cracking). 'Bumping' of the liquid is caused by superheating owing to the absence of nuclei for the formation of bubbles of vapour. Pumice supplies these nuclei in the form of minute air bubbles present in the pores.

Before assembly of the apparatus the crystals are placed in the flask together with 1–2 cm³ of solvent (much less than the quan-tity required to dissolve the whole of the solid). The flask is heated

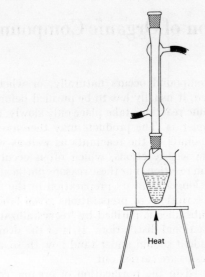

Fig. 2–1. Use of reflux condenser and water-bath

by means of a water-bath, which is placed on a tripod and gauze above a Bunsen flame. When the liquid begins to boil the flame is turned down and further portions of 1–2 cm³ of solvent are added down the condenser until the solid has completely dissolved. When this stage is reached it is advisable to add one more portion of solvent, so that the solution is not completely saturated.

The flask is now detached from the condenser, and the hot solution is filtered to remove any suspended impurities. If the amount of solution is less than about 10 cm³ it can be filtered into a dry beaker through a previously warmed funnel (preferably short-stemmed) with a filter-paper folded in the normal manner.

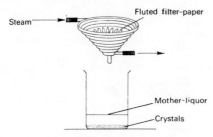

Fig. 2–2. Method of filtering with steam-funnel

If the amount of solution, however, is more than 10 cm³, the liquid would take so long to filter that it would cool and deposit crystals in the filter-paper. The solution is therefore kept hot during filtering by placing the glass funnel in a heated steam-funnel. This is made from a coil of lead tubing, through which steam is passed from a boiler (Fig. 2-2). The flame under the boiler should be extinguished before the hot solution is poured into the funnel. To ensure rapid filtering the filter-paper is 'fluted,' so that the complete surface area of the paper is used. The filtered liquid is collected in a dry beaker, and crystals of the solid are deposited as the liquid cools. The remaining saturated solution, containing the impurities, is called the *mother-liquor*.

(*iii*) *Separation and Drying of the Crystals*. It is desirable to separate the crystals from the mother-liquor as quickly as possible because impurities may be deposited on the crystals if much evaporation of the mother-liquor occurs. The crystals are therefore filtered by suction, using one of the devices shown.
A Buchner funnel (Fig. 2-3) is made of porcelain and has a base, perforated with small holes, over which a filter-paper is placed. The stem of the funnel fits into a rubber stopper in the neck of a

strong conical flask. A side-arm on the flask is connected by thick pressure-tubing to a water-pump fastened to the tap.

A device used for filtering small amounts of material by suction is the Wilstätter 'nail.' This is made by softening one end of a thin glass rod about 7 cm long in a Bunsen flame and pressing the softened end on a charcoal block, so that it forms a circle about 8 mm in diameter. The 'nail' is introduced into the stem of an ordinary glass funnel, as shown in Fig. 2-4. A disc of filter-paper to cover the top of the 'nail' is cut by a cork-borer of suitable size.

When the crystals have been filtered at the pump they are washed once or twice by adding cold solvent to the funnel, and the tap is left running for two or three minutes to drain away the solvent from the crystals. The filter-paper is now removed from

Fig. 2–3. Filtration with a Buchner funnel   Fig. 2–4. Filtration with a Wilstätter 'nail'

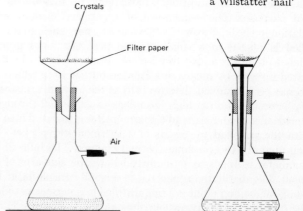

the funnel, and the crystals are scraped on to a wad of several thicknesses of drying-paper. A similar wad is placed over the crystals, which are then pressed between the two layers. The final drying of the crystals is carried out in a desiccator (if water has been the solvent) or in a slightly warm oven (in other cases).

A suitable material for an exercise in recrystallization is 1,3-dinitrobenzene contaminated with about 5 per cent of naphthalene. Ethanol is used as the recrystallizing medium. This exercise is conveniently combined with the melting point test (given shortly) for purity.

2. DISTILLATION

Distillation is chiefly used in the purification of liquids, although it is occasionally applied to solids of low boiling point. The basic apparatus consists of a distillation flask (or 'still'), a condenser, and a receiver, but additional parts may be introduced for special purposes. The usual form of apparatus for small-scale distillation in the laboratory is shown in Fig. 2-5. The liquid to be distilled is placed in a small pear-shaped flask, and a little powdered pumice is added to promote steady boiling. The flask is heated directly by means of a Bunsen burner from which the barrel has been removed. The gas is lit at the jet and adjusted to give a flame about 2 cm high. The flask is connected through a 'stillhead' to a condenser, and a thermometer is inserted into the neck of the stillhead by means of a thermometer pocket containing mercury to give good thermal contact. The bulb of the thermometer is opposite, or slightly below, the side-arm of the stillhead. The distillate is collected in a small conical flask. If a liquid of low boiling point is being distilled it is necessary to lead off the distillate and any vapour into a cooled receiver. This is

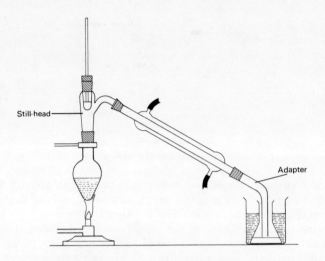

Fig. 2-5. Apparatus used in small-scale distillation (with use of adapter and cooled receiver)

done by attaching an *adapter* to the end of the condenser tube. The cooling bath is a beaker containing cold (preferably iced) water.

An impure liquid can be purified by simple distillation only when the impurities themselves are non-volatile. This is seldom the case in organic chemistry. Usually the required liquid is mixed with other liquids, and has to be purified by 'fractional distillation' as now described.

Fractional Distillation

This process is used to separate two or more *completely miscible* liquids, that is, liquids which form one homogeneous solution when mixed together. Examples of completely miscible liquids are methanol (CH$_3$OH) and water, and benzene and methylbenzene (toluene). Fractional distillation cannot be applied to *immiscible* liquids, that is, liquids which form separate layers when added to each other (e.g. benzene and water). Fractional distillation depends on the difference in boiling points of the liquids to be separated, and, providing the difference is large enough (normally not less than 5°C) the liquids can usually be separated completely. Exceptions occur in the case of certain pairs of liquids which give rise on distillation to 'constant-boiling,' or 'azeotropic' mixtures. One example is ethanol (C$_2$H$_5$OH) and water. This case is discussed in Chapter 8.

Methanol (b.p. 65°C) and water (b.p. 100°C) may be taken as a typical pair of liquids to illustrate the general principles involved in fractional distillation. When a mixture of methanol and water is distilled changes occur progressively in (a) the com-

Fig. 2-6. Temperature-composition diagram for methanol (methyl alcohol) and water

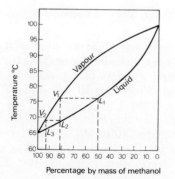

Percentage by mass of methanol

position of the liquid mixture, (b) the composition of the vapour evolved, and (c) the boiling point of the mixture. These changes can be followed with the help of the temperature-composition graphs given in Fig. 2-6.

Let us suppose that the mixture being distilled contains 50 per cent by mass of both methanol and water. The mixture is represented by L_1 in Fig. 2-6. Boiling commences when the vapour-pressure of the liquid becomes equal to the atmospheric pressure (assumed to be standard). This occurs at a temperature (t_1) somewhere between the boiling points of the pure liquids and, as will be seen from Fig. 2-6, temperature t_1 is about 77°C. The vapour evolved has *not* the same composition as the liquid, but *always contains a higher proportion of the more volatile constituent than the liquid in the flask at the same temperature*. The composition of the vapour given off at temperature t_1 is represented by V_1.

If vapour V_1 is condensed it gives a liquid L_2, containing about 82 per cent of methanol. When this liquid is put into another distilling-flask and distilled it begins to boil at a temperature (t_2) of about 69°C, and the vapour given off has the composition V_2, corresponding to a still higher proportion of methanol. Condensation of this vapour yields a liquid L_3, containing about 93 per cent of methanol. It is clear that by repeating the process a sufficient number of times pure methanol can ultimately be obtained.

The liquids represented by L_1, L_2, L_3, etc., have progressively lower boiling points, and finally, when the liquid is pure methanol, the boiling point is 65°C. The upper curve in Fig. 2-6 shows the composition of the vapour which is in equilibrium with the boiling mixture at the various temperatures.

Since the proportion of the more volatile constituent (methanol) decreases, the boiling point of the remaining liquid rises as the distillation proceeds. The lower curve in Fig. 2-6 shows the variation in the boiling point of the liquid in the flask as distillation progresses. It will be seen that *the composition of the liquid in the flask alters in the ascending direction of the boiling point— composition curve.* The final drops of liquid consist of practically pure water and the boiling point is then approximately 100°C.

The number of distillations required to obtain two pure liquids from a mixture of equal proportions of the two depends on the difference in the boiling points of the liquids. If the liquids are methanol and water it is possible to obtain the separate constituents in three distillations by means of the following procedure. During the first distillation the receiver is changed for each rise of about 5°C in the reading of the thermometer. The first and last fractions (which represent the bulk of the distillate) are separately redistilled, and this time the receiver is changed for each 2°C rise in temperature. Again the first and last fractions respectively are redistilled. The first few drops of distillate collected from the *first* fraction are practically pure methanol, while the final few drops from the *last* fraction consist of practically pure water. This method of separation is slow and yields only very small amounts of the pure constituents. In practice fractional distillation is performed more efficiently with the help of a 'fractionating column,' the use of which is described in the next section.

Fractionating Columns

We have seen that several redistillations may be required to separate two liquids by ordinary fractional distillation. A fractionating column is a device whereby the redistillations are

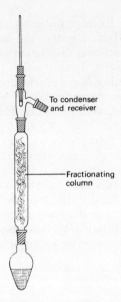

Fig. 2-7. Use of a fractionating column

combined in one operation. Fractionating columns have various forms. One form used in small-scale working is shown in Fig. 2-7. This column simply consists of a plain glass tube filled with short lengths of glass rod or small coils of metal foil (aluminium or stainless steel).

The fractionating column is inserted into the neck of the flask containing the liquid mixture. When the mixture is distilled the bulk of the more volatile liquid is obtained as a distillate over a small range of temperature extending just above the boiling

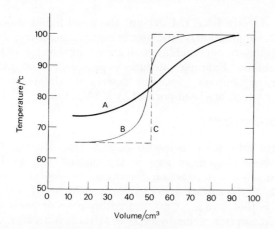

Fig. 2-8. Distillation curves for a mixture of methanol and water

point of the liquid (see Fig. 2-8). After the major portion of the more volatile liquid has distilled the temperature rises rapidly and the less volatile liquid distils—again over a small range of temperature. A single redistillation of each of the two main fractions is then sufficient to yield the two pure liquids. It is even possible to dispense with redistillation altogether under suitable conditions, namely, when there is a sufficient difference in the boiling points of the liquids, an adequate length of column, and carefully controlled distillation.

The principle of different types of fractionating column is the same. When the column is in use the temperature decreases progressively from the bottom to the top and the ascending vapour is partially condensed. Condensation of the vapour is assisted by the large surface area of the 'filling' and by the liquid already condensed. *The vapour of the less volatile constituent of the mixture condenses more readily than that of the more volatile constituent,* and, therefore, the higher the vapour ascends into the column the richer it becomes in the more volatile constituent. For the column to work at maximum efficiency it is necessary for an equilibrium to be established between the ascending vapour and the descending liquid at the different levels; the 'filling' of the column is designed to promote the setting up of these equilibria.

To obtain the best results a very large proportion of the vapour given off must be condensed and returned to the flask. This means that the heating must be regulated so that the distillate is collected at a slow rate (not exceeding one drop per five seconds). Also, the column should be maintained under approximately adiabatic conditions, so that no heat is lost to the outside air. For this reason the column is often 'lagged' by wrapping it in a jacket made of asbestos or some other poorly conducting material.

The curves given in Fig. 2-8 show the results obtained when a mixture of 50 cm³ of methanol and 50 cm³ of water is distilled (*i*) without a fractionating column (curve A), and (*ii*) with a lagged fractionating column (curve B). These curves are constructed by plotting the temperature of the vapour passing into the condenser against the volume of distillate collected. The broken line (C) shows the ideal form of curve B, and is theoretically obtainable with a fractionating column operating under perfect conditions.

Fractional Distillation under Reduced Pressure

A liquid can be made to distil below its normal boiling point by reducing the pressure above the liquid. This process is called 'distillation under reduced pressure,' 'distillation *in vacuo*,' or 'vacuum distillation' (these terms do not imply a complete vacuum). The method is often used to separate and purify liquids which tend to decompose at temperatures approaching their normal boiling points. A special form of distillation apparatus is employed in which the liquid is sealed off from the atmosphere, and a suction pump attached to the apparatus is used to lower the pressure to the required value.

Distillation under reduced pressure has a number of important applications in industry. Simple vacuum distillation is used to remove water from orange-juice, milk, and beef-extracts and thus obtain dry powders, which can be reconstituted by adding water. A modern development of the process is *high vacuum distillation*, or *molecular distillation*, in which pressures of the order of a millionth of an atmosphere are used. At these low pressures liquids distil at temperatures far below their normal boiling points (water can be made to boil at 0°C). It is thus possible to separate and purify organic compounds, such as vitamins, which are extremely heat-sensitive.

Distillation in Steam

Liquids which are immiscible with water (or mix only slightly with water) can often be distilled below their normal boiling points with the help of steam. Steam distillation is used in the preparation of phenylamine (aniline) and is described in Chapter 23.

3. SUBLIMATION

Sublimation is really a special case of distillation, since it consists of vaporizing a substance by heat and then condensing the vapour by cooling. Cooling, however, results in the formation of a solid deposit directly from the vapour, the usual liquid stage being omitted. Sublimation is little used in organic chemistry, although a few common compounds are well known for the ease with which they sublime. Examples include naphthalene, and camphor. 'Flowers of camphor' (compare 'flowers of sulphur') consist of camphor which has been purified by sublimation.

4. ADSORPTION

The term 'adsorption' is used to describe the adhering of a substance (solid, liquid, or gas) to the surface of a solid. Since adsorption is a surface action, the amount of substance adsorbed depends on the surface area of the adsorbing agent. The phenomenon is therefore most marked with porous or finely divided solids. Adsorption is due to forces of attraction between the surface of the adsorbent and molecules of the substance adsorbed.

A common adsorbing agent is carbon (animal charcoal), which is used both in the laboratory and in industry to get rid of unwanted colour from compounds. In many preparations (particularly those of aromatic compounds) the crude product is coloured yellow or brown by traces of tarry impurities. The colouring matter can be removed by boiling a solution of the compound with animal charcoal. The latter is also used in the manufacture of white sugar. Another industrial adsorbent is kieselguhr, a fine white earth composed of silicon(IV) oxide (SiO_2).

An adsorption technique widely used for the separation and purification of both organic and inorganic compounds is *chromatography*. This has various forms, which are described in textbooks of physical chemistry. The use of 'paper partition' chromatography for separating amino-acids derived from proteins is described in Chapter 16.

5. CHEMICAL PURIFICATION

Chemical methods of purification are used in organic chemistry in two ways. Sometimes impurities are removed from a compound by treating it with chemicals which combine with the impurities, but not with the compound. Thus, acidic impurities are usually removed from a product by treating it with aqueous sodium(I) hydroxide or sodium(I) carbonate. Alternatively, the compound itself may be made to combine with some substance with which the impurities do not react, the compound being liberated subsequently by suitable treatment. This is illustrated by the laboratory method (described in Chapter 11) of purifying acids like ethanoic, or acetic, acid. The impure acid is first treated with copper(II) carbonate, which converts the acid into its copper(II) salt. The latter is crystallized and then distilled with concentrated sulphuric(VI) acid, when the organic acid distils over.

Some of the methods used for removing small amounts of water from organic compounds are purely physical (e.g. evaporation of moisture from a solid in a warm oven). As a rule, however, dehydrating agents are employed, and these form hydrates or other compounds with the water. Care must be taken to select a drying agent which does not react with the compound to be dried. Common drying agents include anhydrous salts (e.g. calcium(II) chloride and sodium(I) sulphate (VI)), potassium(I) hydroxide, calcium(II) oxide (quicklime), metallic sodium, and metallic calcium. Less frequently used are silicon(IV) oxide (silica) gel, magnesium(II) chlorate(VII) (magnesium perchlorate), and phosphorus(V) oxide (phosphorus pentoxide). Examples of the application of drying agents are given in later chapters.

TESTS FOR PURITY OF ORGANIC COMPOUNDS

The purity of an organic solid is usually tested by observing its melting point. If the compound is a liquid the boiling point is determined. A pure compound melts and boils sharply at *definite* temperatures, whereas an impure compound (with certain exceptions mentioned later) melts and boils over a *range* of temperature.

Determination of Melting Point

The substance being tested is first dried by leaving it in a desiccator for twenty-four hours. A little of the powdered substance is then introduced into a 'melting point tube,' a glass capillary tube about three inches long and sealed at one end. The tube is attached to a thermometer by means of a rubber band so that the lower end of the tube is close to the thermometer bulb.

Two forms of apparatus for investigating melting points are shown in Fig. 2-9 *a* and *b*. The first is the more accurate form. In this form the thermometer and capillary tube are warmed in a small round-bottomed flask which has a long neck so that most of the thermometer stem is enclosed. The flask contains liquid paraffin or glycerine, and warming is carried out by moving a small, non-luminous flame over the glass where it is in contact with the liquid. The rubber band should be above the surface of the liquid; if it is immersed it will dissolve and discolour the liquid.

The rate of heating must be adjusted, so that as the melting point of the compound is approached the temperature does not rise more than two degrees per minute. (In the case of an unknown compound a preliminary experiment is carried out, in which the

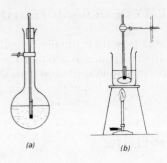

(a)   (b)

Fig. 2-9. Forms of melting-point apparatus

heating is performed more quickly; this shows the approximate melting point of the compound.) To help in observing the change of state and in reading the thermometer a small hand-lens is used. The temperature is noted when the solid begins to melt and again when the last trace of solid disappears. Even if the compound is pure there is usually a small difference between these two temperatures. This is because the thermometer continues to take up heat while the substance is melting. Providing the difference does not amount to more than one degree, however, it may be assumed that the compound is pure. If the compound is impure there will be a range, or difference between the two readings, of several degrees.

In the simpler form of apparatus, shown in Fig. 2-9 *b*, the heating liquid is contained in a tall beaker and is kept stirred by means of a stirrer. Owing to the relatively large portion of thermometer stem not enclosed the temperature readings are invariably two or three degrees below the true values. This apparatus is satisfactory, however, for finding whether the compound melts sharply

or whether it melts over a range of temperature.

It should be noted that melting points are always observed with a *rising* temperature. Theoretically the melting point of a solid is the same as the freezing point, or solidifying point, of the liquefied compound. In practice 'supercooling' of the liquid often occurs, and the substance remains in the liquid state, although the temperature has fallen below the melting point.

Explanation of Range of Melting

If a little of a solid compound A is added as impurity to a solid compound B the melting point of B is lowered (the melting point of an impure substance, which melts over a range of temperature, is taken to be the temperature at which the last trace of solid disappears). Conversely, the addition of a little B to A lowers the melting point of A. The effects on melting point of mixing increasing amounts of

Fig. 2-10. Melting point curves for a mixture of two solids

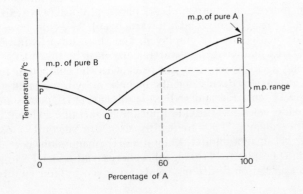

28

A with B, and of B with A, are shown by the curves PQ and RQ in Fig. 2-10. These curves meet at Q, and a mixture of A and B with the composition represented by Q melts at a definite temperature. Such a mixture is called a *eutectic mixture,* and the temperature at which it melts the *eutectic point,* or *eutectic temperature.*

If, however, a mixture containing, say, 60 per cent of A and 40 per cent of B is heated, partial melting occurs as soon as the eutectic temperature is reached The liquid formed has the composition of the eutectic mixture. Melting then continues until the temperature has risen to the value on the curve RQ corresponding to the original composition. The interval between these two temperatures is the range of melting point.

Since a eutectic mixture of two compounds melts sharply, we cannot be certain that a given substance is pure because it melts at a definite temperature (it is *probably* pure if this occurs). To make sure that the substance is pure after it has been shown to melt sharply it should be recrystallized from a suitable solvent. If, after recrystallization, it still melts sharply at the same temperature its purity has been proved. Recrystallization of a eutectic mixture alters the proportion of the compounds present, and the mixture subsequently gives a melting point range.

Method of Mixed Melting Points

This is a quick way of identifying an unknown organic solid (e.g. when the label has been lost from its container). The melting point of the solid is first determined approximately. Let us suppose this is 130°C. Now there is usually more than one organic compound which melts at, or near, a given temperature. In this case there are benzamide (m.p. 130°C), urea (m.p. 132°C), and benzene-1,2-dicarboxylic anhydride (phthalic anhydride) (m.p. 132°C). A little of the unknown compound is mixed separately with each of these compounds, and the melting points of the different combinations are determined. If the unknown is one of the three compounds a sharp melting point

is obtained only when it is mixed with this particular compound. Any other combination shows a range of melting.

Determination of Boiling Point

The boiling point of a liquid is usually determined by distilling it in a distillation apparatus (see Fig. 2-5). If the liquid is pure the distillation temperature remains constant at the boiling point of the liquid. In practice the boiling point found may differ from that given in text-books for the following reasons:

(*i*) the thermometer may not be accurate (it should be checked before use);

(*ii*) part of the thermometer stem is exposed to the air, resulting in a slightly lower reading; and

(*iii*) the atmospheric pressure may differ from the standard value ($101\ 325\ N\ m^{-2}$).

As mentioned earlier, some pairs of liquids form constant-boiling mixtures when they are in a certain proportion; that is, the mixture boils at a constant temperature. Thus one cannot be sure that a given liquid is pure because it boils at a definite temperature. The composition of a constant-boiling mixture varies, however, with the pressure at which distillation takes place. As a further test, therefore, some of the liquid is distilled at a reduced pressure. If it is a constant-boiling mixture the distillate obtained has a different composition from the original liquid. Hence, if the distillate is now redistilled at atmospheric pressure, it will be found to boil over a range of temperature.

1. An organic solid has the following solubilities (in mol kg^{-1}) at 17°C and 70°C:

	At 17°C	At 70°C
In water	0·12	0·16
In ethanol	0·37	2·84
In carbon disulphide	0·51	5·65
In benzene	1·04	1·86

Taking into account any other relevant factors, which solvent would you select for recrystallizing an impure specimen of the compound?

2. In recrystallizing an organic compound the hot solution is filtered rapidly (a) to reduce evaporation of the solvent, (b) to cool the solution more quickly, (c) to avoid depositing crystals in the filter, (d) to obtain bigger crystals. Which is correct?

3. In distilling an organic liquid powdered pumice is added because (a) it lowers the boiling point of the liquid, (b) it raises the boiling point of the liquid, (c) it adsorbs impurities from the liquid, (d) it gives off small air bubbles. Which is correct?

4. Which (if any) of the following are true? Fractional distillation can be used to separate two liquids only when (a) they are immiscible, (b) they are completely miscible, (c) their boiling points are close together, (d) they do not form a constant-boiling (azeotropic) mixture.

5. A mixture containing 90 per cent by mass of methanol (b.p. 65°C) and 10 per cent by mass of water (b.p. 100°C) is distilled until only a few drops of liquid remain in the distilling-flask. Which of the following alternatives apply?

(a) The mixture begins to boil at/above 65°C;

(b) The vapour given off at first contains more/less than 90 per cent of methanol;

(c) The vapour given off near the end contains more/less than 90 per cent of methanol;

(d) The boiling point of the liquid being distilled increases/decreases,

(e) The last few drops of liquid are almost pure methanol/water.

6. Which of the following are correct? When a fractionating column is used in fractional distillation of a mixture of two liquids (a) the temperature increases up the column, (b) most of the vapour condenses in the column, (c) the vapour of the more volatile liquid condenses more readily, (d) the proportion of the more volatile constituent increases up the column.

7. In an experiment an organic compound was found to melt at 79°–79·5°C, whereas the 'official' melting point was 81°C. State which of the following might account for the discrepancy: (a) the compound was impure; (b) the heating liquid was not stirred sufficiently; (c) the thermometer was faulty; (d) part of the thermometer stem was exposed to the air.

3. Composition and Structure of Organic Molecules

To understand the behaviour of an organic compound it is necessary to investigate the nature of the individual molecules. Primarily the investigation is designed to answer the following questions:

(*i*) What elements are present in the compound?

(*ii*) How many atoms of each element are contained in one molecule of the compound?

(*iii*) Which atoms are joined to which within the molecule?

(*iv*) How are the atoms in the molecule arranged in space?

The answers to the above questions enable the chemist to construct a series of chemical formulæ for the compound, each formula summarizing in a concise manner increasing amounts of information about the molecules.

The empirical formula *of a compound is the simplest formula which shows the ratio of the atoms of the different elements in the compound.* For example, the empirical formula of hydrogen peroxide is HO, while that of benzene is CH.

The molecular formula *of a compound is the formula which shows the actual number of atoms of the different elements in one molecule of the compound.* Thus the molecular formula of hydrogen peroxide is H_2O_2, and that of benzene is C_6H_6. The molecular formula is a whole-number multiple of the empirical formula.

The graphic formula *of a compound is the formula which shows the arrangement of the atoms in the molecule by means of valency*

bonds. For example, the graphic formula of ethanol is

ethanol

It must be emphasized that the graphic formula is not a picture of the molecule in space. Graphic formulæ are written in two dimensions, whereas atoms are usually distributed in three dimensions. Graphic formulæ are too cumbersome for ordinary use, and therefore the arrangement of the atoms in a molecule is usually shown by means of the simpler 'structural formula.'

The structural (or constitutional) formula *of a compound is an abbreviated form of the graphic formula, and shows the arrangement of the atoms in the molecule by means of groups of atoms.*

In a structural formula a valency bond between two carbon atoms is often represented by a dot, although sometimes a line is used for this purpose. Thus, the structural formula of ethanol can be written either as $CH_3.CH_2OH$ or as $CH_3—CH_2OH$. One condition attached to the use of structural formulæ is that the groups of atoms represented must have only one possible constitution when due regard is paid to the valencies of the elements. Another way of writing the structural formula of ethanol is C_2H_5OH because the $C_2H_5—$ group can have only one possible structure if the carbon and hydrogen atoms are to have valencies of four and one respectively. The group $C_3H_7—$, however, cannot be used in such a formula because there are two possible arrangements of the carbon and hydrogen atoms in this group, as shown.

$$(i) \quad CH_3—CH_2—CH_2—$$

$$(ii) \quad \begin{matrix} CH_3 \\ \diagdown \\ CH— \\ \diagup \\ CH_3 \end{matrix}$$

The geometrical form of a molecule, or the arrangement of its atoms in space, is sometimes represented by a space formula (examples are given in Chapter 17). The best method of showing the geometrical form of a molecule, however, is to construct a *space model*. Molecular models have varying degrees of complexity according to the amount of information which they are intended to convey. One of the simplest forms is the 'ball-and-spring' model, an illustration of which is given in Fig. 3-4. In this the atoms are represented by plastic spheres of different size and colour, and the valency bonds by springs. Ball-and-spring models show which atoms are joined to which, and indicate (roughly) the arrangement of the atoms in space. They are misleading, however, in implying that the atoms are separated by relatively large distances. As explained shortly, we know from electron density 'maps' of molecules that the atoms are in close contact.

While a knowledge of the spatial distribution of atoms in molecules helps us to understand many of the properties of compounds, it is insufficient to explain the 'finer' details of their chemical behaviour. To do this it is necessary to establish the manner in which the electrons are arranged in the valency bonds between the atoms. Although, as stated earlier, the bonds are usually covalent, they may differ in the way in which the electrons are shared between the atoms. This aspect of molecular structure is dealt with in the next chapter.

The first step in finding the molecular structure of a compound is to determine its empirical formula. This involves firstly *qualitative* analysis, which shows the elements present, and secondly *quantitative* analysis, which gives their proportion by mass.

QUALITATIVE ANALYSIS

The elements composing organic compounds are few in number, and, as regards frequency of occurrence, they fall into three groups: (A) carbon, hydrogen, and oxygen; (B) nitrogen, halogens, and sulphur; and (C) silicon, phosphorus, and metals. In practice compounds seldom contain more than two or three other elements in addition to carbon. The tests used to detect the presence of the more common elements (groups A and B) are now described.

(A) Tests for Carbon and Hydrogen

It is seldom necessary to test for the presence of these elements unless there is some uncertainty as to whether the given substance is an organic or inorganic compound. An organic compound can usually be recognized by heating it on porcelain, when it burns away, leaving no residue except, perhaps, a black one of carbon. An inorganic compound, when treated similarly, usually does not burn and generally leaves a residue. There are exceptions, however, in both cases. If there is any doubt as to the nature of the compound it is heated with excess of black copper(II) oxide, which oxidizes any carbon and hydrogen in the compound to carbon dioxide and water. These are easily identified.

Experiment

Both the organic substance to be tested and the copper(II) oxide should be dried by leaving them in a desiccator for 24 hours. Mix 0·5 g of glucose with about ten times as much copper(II) oxide. Place the mixture in a dry test-tube and clamp the tube so that it

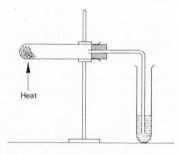

Fig. 3-1. Oxidation of glucose by copper(II) oxide

slopes slightly downward (Fig. 3-1). Warm the mixture gently, and pass any gases evolved through calcium(II) hydroxide solution (lime-water). In a few moments the latter turns milky, showing that carbon dioxide is being given off. At the same time drops of a colourless liquid condense in the cooler part of the tube. Test these with white anhydrous copper(II) sulphate(VI), which turns blue, proving that the liquid is water.

Tests for Oxygen

There is no general test for the presence of oxygen in organic substances. It is often possible to infer, however, that a compound contains oxygen from its chemical reactions. Thus oxygen must be present if the compound (after careful drying) gives off moisture when heated, or if it can be shown to possess a hydroxyl (—OH) group or some similar oxygen-containing group.

(B) Tests for Nitrogen, Sulphur, and Halogens. 1. Preliminary Tests (*i*) *Nitrogen.* A quick test for nitrogen is to heat the compound with sodium(I) hydroxide solution. With this reagent many nitrogen-containing compounds evolve ammonia, which can be recognized by its smell and the turning blue of damp red litmus paper. The test is given only by certain classes of nitrogen compounds (e.g. acid amides, nitriles, and ammonium salts of organic acids). Hence a negative result does not prove the absence of nitrogen.

(*ii*) *Halogens* (*Beilstein's Test*). This test is used to indicate the *probable* presence or absence of a halogen in an organic compound. It does not show which halogen is present.

Experiment

Heat the end of a length of copper wire in the oxidizing zone of a Bunsen flame until the wire ceases to impart a green colour to the flame. Now dip the hot end of the wire into a small amount of a halogen-containing organic compound (e.g. chloroacetic acid), and introduce the wire again into the flame. The latter at once turns green.

33

The Beilstein test depends on the formation of a volatile copper halide by interaction of the organic compound and copper(II) oxide on the wire. Unfortunately, the test is not reliable. If it is applied to certain nitrogen-containing compounds volatile copper(I) cyanide may be produced, and this also colours the flame green.

2. The Lassaigne Test. The general tests for nitrogen, sulphur, and halogens depend on the fact that when these elements are present in a compound they can be converted into easily recognizable ions by heating the compound with sodium. Nitrogen is left as cyanide ions (CN^-), sulphur as sulphide ions (S^{2-}), and halogens as halide ions. In the Lassaigne test for nitrogen the cyanide ions are identified by producing from them a precipitate of Prussian blue. Sulphide ions and halide ions are recognized by the methods now described.

Experiment

Put a pellet of sodium about one-quarter the size of a pea into an ignition tube. Wrap a little paper round the top of the tube, and hold the tube by gripping the paper in a pair of tongs. Warm the tube over a medium flame until the sodium melts. Now hold the tube over a sand-tray and add about 0·5 g of the organic compound in three portions. (If the compound is a liquid add three small drops from a dropping-tube at intervals.) Continue heating the tube over a medium flame for one minute, and then heat strongly for two minutes. While the end of the tube is red-hot plunge it into about 20 cm³ of distilled water in an evaporating-dish, holding a clock-glass over the dish with the other hand to guard against an explosion. The end of the tube breaks, and any cyanide ions formed are left in solution. Filter the solution and use portions of the filtrate for the tests given below.

(*i*) *Test for Nitrogen.* Test one portion of the filtrate with red litmus paper to make sure that it is alkaline (if the paper does not turn blue make the liquid alkaline by adding a little sodium(I) hydroxide solution). Add to the liquid an equal volume of iron(II) sulphate(VI) solution, which gives a dark green precipitate of iron(II) hydroxide. Warm the tube gently for a few moments. The cyanide ions and iron(II) hydroxide interact to give hexacyanoferrate(II) (ferrocyanide) ions in solution.

$$6CN^- + Fe(OH)_2 \rightarrow Fe(CN)_6^{4-} + 2OH^-$$

Cool the tube under the tap, add two drops of iron(III) chloride solution, and finally acidify the contents with concentrated hydrochloric acid. The appearance of a precipitate of Prussian blue shows that the organic compound contains nitrogen. The acidifying of the liquid is necessary because Prussian blue is not precipitated in alkaline solution.

(*ii*) *Test for Sulphur.* Add one or two drops of the filtrate to a freshly prepared solution of sodium(I) pentacyanonitrosylferrate(II) (sodium nitroprusside), $Na_2Fe(CN)_5(NO)$. An intense violet colour shows that the original compound contains sulphur.

(*iii*) *Tests for Halogens.* There is a slight difference in the procedure according to whether nitrogen and sulphur have been found present in the compound or not. In the *absence* of nitrogen and sulphur acidify the third portion of the filtrate from the sodium fusion with dilute nitric(V) acid and add to it silver(I) nitrate(V) solution. If the compound contains chlorine, bromine, or iodine a precipitate of silver(I) halide is now formed. This can be identified by its colour and solubility in dilute aqueous ammonia or as follows:

White, readily soluble in aqueous ammonia . . . . . . chloride
Pale yellow, soluble in aqueous ammonia with difficulty . . . bromide
Yellow, insoluble in aqueous ammonia iodide

If either nitrogen or sulphur is *present* in the compound, boil the third portion of the filtrate with excess of dilute nitric(V) acid for five minutes. This is to destroy cyanide ions (which evolve hydrogen cyanide) or sulphide ions (which give off hydrogen sulphide). Both cyanide ions and sulphide ions react with silver(I) nitrate(V) solution, the former giving a white precipitate of silver(I) cyanide (soluble in aqueous ammonia) and the latter a black precipitate of silver(I) sulphide. After boiling the liquid, cool it and test with silver(I) nitrate(V) solution as described previously.

Note. The Lassaigne test should not be performed with trichloromethane (chloroform) or tetrachloromethane (carbon tetrachloride). These compounds are liable to explode if they are heated with sodium.

Organic compounds containing metals (Group C) are chiefly salts of organic acids. When the salts are strongly heated in a crucible the organic matter is decomposed and oxidized, and the metal is usually left as the oxide or carbonate. For example, copper(II) acetate is converted into copper(II) oxide and sodium(I) acetate into sodium(I) carbonate. Silver salts of organic acids leave a residue of metallic silver. In every case the residue is investigated and the metal identified by the ordinary methods of inorganic analysis.

QUANTITATIVE ANALYSIS

Estimation of Carbon and Hydrogen

These two elements are estimated together. A simplified form of the apparatus used is shown in Fig. 3-2. A known mass of the compound is contained in a platinum boat in a combustion tube. (If the compound is a liquid it is weighed in a small bulb, which has been drawn out into a long narrow neck and sealed. The end

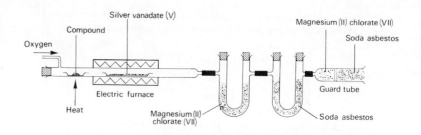

Fig. 3-2. Simplified form of the apparatus used in estimation of carbon and hydrogen

of the neck is broken off just before the bulb is placed in the boat.) The compound is oxidized to carbon dioxide and water partly by oxygen gas and partly by an oxidizing agent, silver(I) vanadate(V), $Ag_3 VO_4$. The latter is used because it also catalyses the oxidation by oxygen and it absorbs any halogens or sulphur dioxide produced.

The compound is vaporized by gentle heat from a Bunsen burner, while the silver(I) vanadate(V) is strongly heated in an electric furnace. The water and carbon dioxide formed are absorbed by magnesium(II) chlorate(VII) (magnesium perchlorate), $Mg(ClO_4)_2$, and soda-asbestos respectively. The latter consists of sodium(I) hydroxide deposited on asbestos. The absorbents are contained in weighed U-tubes, and the increase in mass of these gives the masses of the water and carbon dioxide produced. A guard tube, which also contains magnesium(II) chlorate(VII) and soda-asbestos, prevents backward diffusion of atmospheric moisture and carbon dioxide into the U-tubes.

35

The masses of hydrogen and carbon in the sample are calculated from the masses of water and carbon dioxide as now shown.

$$\underbrace{\underset{2+16}{H_2O}}_{18} \qquad \underbrace{\underset{12+32}{CO_2}}_{44}$$

Mass of hydrogen $= \frac{2}{18}(= \frac{1}{9})$ of the mass of water.
Mass of carbon $\quad= \frac{12}{44}(= \frac{3}{11})$ of the mass of carbon dioxide.

Estimation of Oxygen

Although a direct method is now available for finding the mass of oxygen in a given amount of an organic compound, this element is usually estimated 'by difference.' The combined masses of the other elements present are subtracted from the mass of the compound, or the combined percentages of the other elements are subtracted from 100.

Estimation of Nitrogen

Two methods are in use for the estimation of nitrogen. One of these, the *Dumas method*, is applicable to any nitrogen-containing compound, but requires considerable time for preparation of the apparatus. The other, *Kjeldahl's method*, can be carried out more rapidly, but its use is restricted to certain classes of compounds.

(*i*) *The Dumas Method.* In this method a known mass of the compound is heated with a large excess of copper(II) oxide. Carbon and hydrogen are oxidized to carbon dioxide and water, and nitrogen is liberated partly as nitrogen gas and partly as oxides of nitrogen. The nitrogen oxides are reduced to nitrogen by passing the gases over red-hot copper. The nitrogen is collected over a concentrated solution of potassium(I) hydroxide (which absorbs the carbon dioxide) and its volume is measured at laboratory temperature and pressure. The volume at standard temperature and pressure is found, and the mass of the nitrogen is then calculated.

(*ii*) *Kjeldahl's Method.* This is used for estimation of nitrogen in foods, fertilizers, and drugs. It depends on the fact that when a known mass of the substance is refluxed with concentrated sulphuric(VI) acid for several hours any nitrogen present is converted into ammonia, which forms ammonium sulphate(VI) with excess of the acid. The ammonia is then liberated from the ammonium sulphate(VI) by distillation with sodium(I) hydroxide solution.

$$NH_4^+ + OH^- \rightarrow NH_3 + H_2O$$

The liberated ammonia is passed into a measured volume (excess) of standard acid. The remaining acid is estimated by titration. From the amount of acid neutralized by the ammonia the amount of ammonia, and hence the amount of nitrogen can be calculated.

Example

The following results were obtained in a Kjeldahl experiment:

Mass of organic compound $\quad= 0.319$ g
Volume of 0.5 M H_2SO_4 used for ammonia neutralization $= 20$ cm^3
Volume of M NaOH used to neutralize excess acid $\quad= 16.2$ cm^3

Calculate the percentage of nitrogen in the compound.

Volume of 0·5 M H₂SO₄ neutralized by ammonia = 3·8 cm³

But 3·8 cm³ of 0·5 M H₂SO₄ = 3·8 cm³ of M ammonia solution

$$= \frac{3\cdot8 \times 17}{1000} \text{ g of ammonia (NH}_3\text{)}$$

$$= \frac{3\cdot8 \times 14}{1000} \text{ g of nitrogen}$$

$$= 0\cdot053\,2 \text{ g of nitrogen}$$

Percentage of nitrogen $= \dfrac{0\cdot053\,2}{0\cdot319} \times \dfrac{100}{1}$

$$= 16\cdot7 \text{ per cent}$$

Estimation of Halogens

Several methods are available for estimating halogens in organic compounds. The one selected in a particular case depends on the nature of the halogen and the type of compound. A typical method often used in the estimation of chlorine, bromine, or iodine is to fuse a known mass of the compound with sodium(I) peroxide (Na_2O_2), a strong oxidizing agent. This is done in a small 'bomb,' which consists of a strong nickel crucible fitted with a screw-lid. The latter is screwed down tightly before the mixture is ignited by heating the crucible with a Bunsen flame.

Carbon and hydrogen are oxidized to carbon dioxide and water, and the halogen is converted to sodium(I) halide. In the case of bromine or iodine a little bromate(V) or iodate(V) is also formed. The residue is dissolved in water, and any bromate(V)

or iodate(V) is reduced to bromide or iodide by addition of a solution of sodium(I) hydrogensulphate(IV) (sodium bisulphite), $NaHSO_3$. The liquid is then acidified with nitric(V) acid, and excess of silver(I) nitrate(V) solution is added. The precipitate of silver(I) halide is filtered, washed, dried, and weighed. The mass of halogen is calculated from the mass of silver(I) halide with the aid of relative atomic masses.

AgCl	AgBr	AgI
$\underbrace{108 + 35\cdot5}$	$\underbrace{108 + 80}$	$\underbrace{108 + 127}$
143·5	188	235

It will be seen that

143·5 g of silver(I) chloride contain 35·5 g of chlorine;
188 g of silver(I) bromide contain 80 g of bromine; and
235 g of silver(I) iodide contain 127 g of iodine.

Estimation of Sulphur

This element is estimated by heating a known mass of the compound with an oxidizing agent (e.g. fuming nitric(V) acid), so that the sulphur is oxidized to sulphuric(VI) acid. The remaining liquid is then treated with excess of barium(II) chloride solution. A precipitate of barium(II) sulphate(VI) is formed, and this is filtered, washed, dried, and weighed. From the mass of barium(II) sulphate(VI) obtained the mass of sulphur is calculated with the aid of relative atomic masses.

DETERMINATION OF EMPIRICAL FORMULA

When an organic compound has been analysed qualitatively and quantitatively its empirical formula can be found by dividing the masses of the different elements present in a known mass of the compound by the relative atomic masses of the elements. Sometimes it is more convenient to divide the percentages of the elements present by their relative atomic masses. The smallest of the quotients is then divided into the other quotients to obtain the simplest whole number ratio of the atoms. The following example illustrates the method of calculation:

Example

0·283 5 g of an organic compound containing carbon, hydrogen, oxygen, and chlorine gave on oxidation 0·263 7 g of carbon dioxide and 0·080 9 g of water. In a halogen estimation 0·189 1 g of the same compound produced 0·286 8 g of silver(I) chloride. Calculate (i) the percentages of the elements present in the compound, (ii) the empirical formula of the compound. ($H = 1$, $C = 12$, $O = 16$, $Cl = 35·5$ $Ag = 108$.)

Mass of carbon $\quad = \frac{3}{11}$ of mass of carbon dioxide

$\qquad\qquad\qquad = \frac{3}{11}$ of 0·263 7 g = 0·071 9 g

Percentage of carbon $\quad = \dfrac{0·071\ 9}{0·283\ 5} \times \dfrac{100}{1} = \textbf{25·36 per cent}$

Mass of hydrogen $\quad = \frac{1}{9}$ of mass of water

$\qquad\qquad\qquad = \frac{1}{9}$ of 0·080 9 g = 0·009 0 g

Percentage of hydrogen $\quad = \dfrac{0·009\ 0}{0·283\ 5} \times \dfrac{100}{1} = \textbf{3·17 per cent}$

Mass of chlorine $\quad = \dfrac{35·5}{143·5}$ of mass of silver(I) chloride

$\qquad\qquad\qquad = \dfrac{35·5}{143·5}$ of 0·286 8 g = 0·071 0 g

Percentage of chlorine $\quad = \dfrac{0·071\ 0}{0·189\ 1} \times \dfrac{100}{1} = \textbf{37·55 per cent}$

Total percentages of C, H, and Cl $\quad = 25·36 + 3·17 + 37·55 = 66·08$

∴ percentage of oxygen $\quad = 100 - 66·08 = \textbf{33·92 per cent}$

$$
\begin{array}{ccccc}
& C & : & H & : & Cl & : & O \\
\text{Ratio of atoms} & = \dfrac{25·36}{12} & & \dfrac{3·17}{1} & & \dfrac{37·55}{35·5} & & \dfrac{33·92}{16} \\
& = 2·11 & : & 3·17 & : & 1·06 & : & 2·12 \\
& = 2 & : & 3 & : & 1 & : & 2
\end{array}
$$

approximately

∴ the empirical formula is $\textbf{C}_2\textbf{H}_3\textbf{ClO}_2$

DETERMINATION OF MOLECULAR FORMULA

The relation between the molecular formula of a compound and its empirical formula is given by the equation:

$$\text{Molecular formula} = \text{empirical formula} \times n$$

where n is usually a simple whole number. The precise multiple of the empirical formula to be taken as the molecular formula can be found by determining the relative molecular mass of the compound as outlined below. In the case of very large molecules such as those of polythene, $(CH_2)_n$, n may have a value of many thousands and it may not be possible to find the exact value.

Determination of Relative Molecular Masses

Methods of finding relative molecular masses (formerly called 'molecular weights') are described in text-books of physical chemistry. The chief ones are those now given.

(*i*) *From the Relative Density of the Vapour*. The relative density of the vapour of a liquid or volatile solid can be measured by means of the apparatus of Dumas or Victor Meyer. The relative molecular mass is then obtained by multiplying the relative density of the vapour by two. The method cannot be used for compounds like sucrose (cane sugar) which decompose when heated.

The use of the relative densities of vapours in fixing molecular formulæ is illustrated by the following example. Suppose the empirical formula of a certain compound is CH_2O and that by the method of Dumas or Victor Meyer the relative density of the vapour of the compound has been found to be 29·8. Then

The relative molecular mass = relative density of vapour × 2
$$= 29·8 × 2 = 59·6.$$

The molecular formula $= (CH_2O)_n$.
The relative molecular mass derived from this formula

$$= 12n + 2n + 16n$$
$$= 30n$$

$\therefore \quad 30n = 59·6$

and $\qquad n = 2$ (to the nearest whole number).

$\therefore \qquad$ the molecular formula is $C_2H_4O_2$.

(*ii*) *From Freezing Point Depression or Boiling Point Elevation*. These methods, which are similar in principle, are used to measure relative molecular masses when the relative vapour density method cannot be applied. Their use is limited by the fact that comparatively few organic compounds dissolve in water, and association into molecular aggregates often occurs in organic solvents. For example, ethanoic, or acetic acid dissolved in benzene consists of double molecules, $(C_2H_4O_2)_2$. When the relative molecular mass of a compound has been determined by the freezing point, or boiling point, method the molecular formula is found from the empirical formula as illustrated previously.

(*iii*) *By Mass Spectrometer*. This instrument is widely used in industrial laboratories for determining relative molecular masses. When a molecular compound is vaporized at a low pressure and the vapour is bombarded with electrons one or more electrons become detached from the molecules, and positively charged 'molecular ions' are produced. Thus methanol (CH_4O) may give rise to ions CH_4O^+ and CH_4O^{2+}. Since the masses of the detached electrons can be ignored, the mass of the CH_4O molecule can be found by measuring the masses of the ions. The relative molecular mass is the ratio of the average molecular mass to one-twelfth of the mass of an atom of carbon-12 (^{12}C).

(*iv*) *Special Methods*. Special methods are used for fixing the relative molecular masses of carboxylic acids (Chapter 11), proteins (Chapter 16), and plastics (Chapter 26).

DETERMINATION OF GRAPHIC OR STRUCTURAL FORMULA

When the molecular formula has been established the next stage in characterizing an organic compound is to find which atoms are joined to which within the molecule. It is then possible to write the graphic formula or its abbreviated counterpart, the structural formula. The evidence used for this may be chemical or physical, but in either case the problem resolves itself into finding (*i*) what functional groups are present, and (*ii*) the nature of the carbon 'skeleton,' that is, the framework of carbon atoms to which the functional groups are attached. For simple molecules this information can usually be obtained from chemical evidence combined with valency considerations, which limit the number of possible structures. For more complex molecules physical evidence is usually required as well.

Chemical Evidence

In general chemical evidence of molecular structure is derived from the reactions of the compound, the methods by which it can be prepared, and the nature of the products formed when the compound is broken down into simpler substances. The last of these is particularly useful in the case of complex molecules such as those of proteins (Chapter 16). Ethanol is a relatively simple compound which illustrates the use of chemical evidence in finding molecular structure.

Structure of Ethanol. The empirical formula of ethanol obtained by qualitative and quantitative analysis is C_2H_6O. This is also the molecular formula, as shown by determination of the relative density of the vapour by Victor Meyer's method. From valency considerations the molecule C_2H_6O can have only two possible structures as shown.

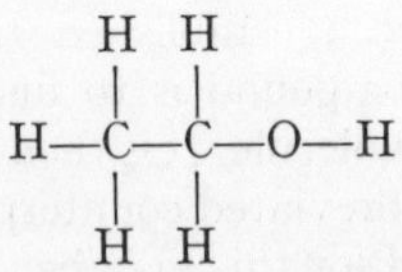

(A) 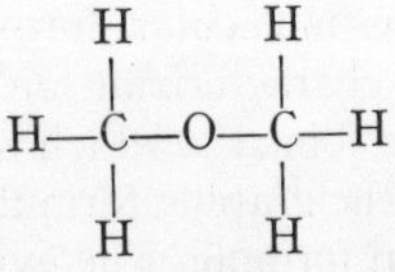   (B)

That structure A is the correct one for ethanol is indicated by the following evidence:

(*i*) It can be deduced that the molecule contains a hydroxyl (—OH) group because ethanol liberates hydrogen when treated with metallic sodium. This is a common test for the presence of the —OH group. Other tests are the reactions with phosphorus pentachloride (Chapter 8) and acetyl, or ethanoyl, chloride (Chapter 12). If a known mass of pure ethanol is treated with excess of sodium and the hydrogen evolved is measured, it is found that only one-sixth of the hydrogen of the alcohol has been displaced. This confirms that one of the six hydrogen atoms present in the molecule is combined differently from the rest.

(*ii*) The reactions of ethanol closely resemble those of methanol, and the molecules of both can therefore be expected to have a similar type of structure. From valency considerations, however, a molecule of methanol can have only one structure, which is represented by the formula CH_3—OH. Hence the structure of the ethanol molecule is represented by the analogous formula $CH_3 . CH_2$—OH.

(*iii*) Ethanol can be obtained from ethane by substituting a hydrogen atom of the latter by a bromine atom and boiling the resulting bromoethane with sodium(I) hydroxide solution.

$$C_2H_6 \xrightarrow{Br_2} C_2H_5Br \xrightarrow{NaOH} C_2H_5OH$$

Now the two carbon atoms must be joined together in the ethane molecule, and, from the reactions given, it is reasonable to suppose that they are still joined together in the molecule of ethanol. This evidence supports structure A, but not structure B.

It should be mentioned that a compound having the structure B is known. This compound is methoxymethane (dimethyl ether), which is isomeric with ethanol. Its reactions are quite different from those of the latter (see Chapter 8).

Physical Evidence

Physical methods of investigating molecular structure can be applied to both simple and complex compounds. They are particularly valuable, however, in the latter case because the information derived by chemical methods is often insufficient to settle the structure conclusively. Physical evidence of molecular structure can be obtained by several techniques, the most widely used of which is infared analysis.

Infrared Analysis

A valency bond between two atoms may be likened to a spiral spring connecting two small masses. The spring and the masses can vibrate in different ways. In one mode of vibration, 'stretching,' the masses alternately approach, and recede from, each other, while in another, 'rocking,' they move from side to side. Similar kinds of vibration occur in the valency bonds of organic molecules. Absorption of infrared radiation causes the atoms to vibrate more vigorously, but since absorption of energy can take place only in definite amounts (called 'quanta') a given valency bond responds only to certain wavelengths of radiation. Different quanta of energy are absorbed according to the atoms joined by the bond, the nature of the bond itself (whether single, double, or triple), and the type of vibration (e.g. stretching or rocking).

In an infrared spectrometer infrared radiation with a suitable range of wavelengths is passed through a thin film of the compound in the form of solid, liquid, or a solution. The absorption spectrum, which is recorded automatically on paper by a moving pen, takes the form of a graph, in which the percentage of radiation transmitted is plotted against the wavelength, λ (lambda), of the radiation. Pronounced dips in the graph show wavelengths at which strong absorption takes place.

Infrared absorption spectra show the presence in molecules of particular types of bond. Thus, stretching vibration for a C—H bond in a CH_3— group occurs at a characteristic wavelength of $3\cdot4$ μm (I micrometre, μm, $= 10^{-6}$ metre), while for the same bond in a —CH_2— group the wavelength is slightly higher ($3\cdot5$ μm). For the C=O bond in esters like methyl octadecanoate (methyl stearate) the characteristic wavelength for stretching vibration is $5\cdot8$ μm. A long chain of carbon atoms vibrates as a whole, giving an absorption band at about $13\cdot8$ μm. All these absorption bands can be seen in the infrared absorption spectrum given for methyl octadecanoate in Fig. 3-3. Methyl octadecanoate has the structure shown below. Other absorption

$$CH_3\!-\!(CH_2)_{16}\!-\!\overset{\displaystyle O}{\overset{\displaystyle \|}{C}}\!-\!O\!-\!CH_3$$

methyl octadecanoate

bands in the spectrum are caused by other atomic groups (such as C—O—C) and other modes of vibration.

The infrared absorption spectrum of an organic compound is so characteristic that it can be used as a 'finger-print' to identify that compound. Industrial laboratories maintain files of infrared spectra of compounds in which they are interested. An unknown compound can then be quickly recognized by comparing its spectrum with those in the files.

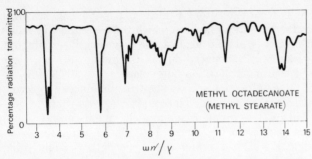

Fig. 3-3. Infrared absorption spectrum of methyl octadecanoate
(methyl stearate)

Amongst other physical methods which provide information on molecular structure are *X-ray diffraction* and *electron diffraction*. These are used primarily to find the spatial distribution of atoms in molecules and are considered in the next section.

SPACE STRUCTURE OF MOLECULES

The manner in which the atoms are arranged in a molecule determines the shape of the molecule. Molecular shapes are important because they influence the properties, and hence the uses, of compounds. The study of the space structure of molecules forms the special branch of chemistry known as *stereochemistry*.

The Tetrahedral Carbon Atom

Long before modern methods of investigating molecular geometry were developed, chemists had come to believe that molecules in general have three-dimensional structures (this does not exclude the possibility that some may be flat or linear). They

reached this conclusion because it was the only way to explain the existence of isomeric forms of certain compounds (see Chapter 17). In 1874 van't Hoff and Le Bel put forward the theory of the tetrahedral distribution of the four valencies of the carbon atom. According to this theory a carbon atom joined to four other atoms can be regarded as situated at the centre of a regular tetrahedron, and has its four valencies directed outwards towards the corners of the tetrahedron. Thus, in the methane molecule (CH_4) the carbon atom and the four hydrogen atoms are arranged as shown in Fig. 3-4. In this arrangement the angles between the valency bonds are all 109° 28′.

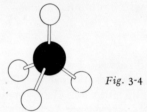

Fig. 3-4

Confirmatory evidence of the spatial distribution of the four carbon valencies is provided by dichloromethane (CH_2Cl_2). If the five atoms of this molecule were all in the same plane dichloromethane would be expected to exist in two forms, in which the atoms would be arranged as shown below. Although these two

```
      Cl                  Cl
      |                   |
 H—C—Cl             H—C—H
      |                   |
      H                   Cl
     (i)                 (ii)
```

hypothetical isomers might have similar chemical properties, they would certainly show differences in physical properties (e.g. boiling point), by which they could be distinguished.

Actually, only one form of dichloromethane is known, and only one form is possible if the atoms are distributed in space in the same way as in methane. Modern experimental methods have confirmed the tetrahedral distribution of the four valencies of the carbon atom in the molecules of both methane and dichloromethane.

X-ray Diffraction and Electron Diffraction.

X-ray diffraction is used to investigate the space structure of organic molecules in crystalline solids. It can be carried out with a single large crystal, but is more commonly applied to a mass of small crystals (the 'powder method'). When X-rays of fixed wavelength are directed on to the crystal mass at a suitable angle they are partly transmitted and party reflected. If the reflected rays are allowed to fall on a photographic plate or film, the latter, when developed, shows a series of light and dark rings of different width and intensity. This diffraction pattern, which is due to the orderly arrangement of the molecules in the crystal planes, is characteristic for a given compound and, like an infrared spectrum, can be used to identify the compound.

An X-ray diffraction pattern can also be used to find the electron density in different parts of a molecule. Diffraction of X-rays is caused by the electrons round the atomic nuclei, and the effect is greatest where the electron density is highest, that is, in the immediate neighbourhood of the nuclei. Hence by measuring the intensities of the scattered rays the electron densities can be calculated and used to construct an electron density

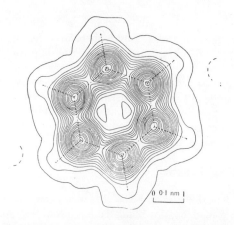

Fig. 3-5. Electron density map of the benzene molecule. (From "The Crystal Structure of Benzene at $-3°C$," by E. G. Cox, D. W. J. Cruickshank, and J. A. S. Smith, 1958, *Proc. Roy. Soc.* A, **247**, p. 7.)

'map,' in which 'contour' lines join points of equal electron density.

Figure 3-5 shows an electron density map of the flat benzene molecule (C_6H_6) obtained by diffraction of X-rays by a frozen benzene crystal. In this diagram the positions of the six carbon atoms in the benzene ring are shown by electron density maxima. Hydrogen atoms are too small to scatter X-rays appreciably, so that the positions of the six hydrogen atoms are not indicated by similar maxima. They are shown, however, by curved projections in the contour lines extending outwards from the carbon atoms. The C—C and C—H bonds are represented by broken lines.

Electron density maps of the kind shown in Fig. 3-5 reveal

how the atoms are joined together in a molecule. They also yield precise information about bond lengths and the angles at which the bonds meet. From this information the shape of the molecule can be deduced.

Diffraction patterns, similar to those given by X-rays with crystals, can be obtained by passing a beam of electrons through a gas (or vapour of a liquid) at low pressure. These *electron diffraction* patterns can likewise be used to measure bond lengths and bond angles in molecules.

Covalent Bond Lengths

By 'bond length' is meant the distance between the nuclei of the atoms joined by the bond. It has been customary to give bond lengths in ångstrom units, one ångstrom (Å) being equal to 10^{-10} metre. In the Système Internationale (SI), which has now been adopted in science, bond lengths are expressed in sub-multiples of the metre. The recommended sub-multiples are those involving powers which are multiples of three (that is, 10^{-3}, 10^{-6}, etc.). Thus in the new system an ångstrom unit $= 10^{-10}$ m $= 0.1$ nm, where nm stands for nanometre (10^{-9} m). For the present bond lengths will be expressed both in nanometre units and ångstrom units.

It is found that, subject to one condition, the length of a covalent bond between two specified atoms is approximately the same in different kinds of molecules. The condition is that the bond is not affected by 'mesomerism' (this is explained in Chapter 4). The length of the C—H bond may vary between 0.106 nm (1.06Å) in ethyne (acetylene), C_2H_2, and 0.110 nm (1.10 Å) in ethane, C_2H_6. Average values for the lengths of some other single bonds involving the carbon atom are shown below.

C—O	C—Cl	C—Br
0.143 nm	0.176 nm	0.194 nm
(1.43 Å)	(1.76 Å)	(1.94 Å)

The length of a covalent bond between two given atoms depends on whether it is a single, double, or triple bond. The length is less for a double bond than a single bond, and less again for a triple bond than a double bond. Thus, the carbon—carbon bonds in ethane (H_3C—CH_3), ethene (H_2C=CH_2), and ethyne (HC≡CH) have the following lengths:

C—C	C=C	C≡C
0.154 nm	0.133 nm	0.120 nm
(1.54 Å)	(1.33 Å)	(1.20 Å)

Thus four shared electrons hold the atoms more closely together than two, and six more closely than four.

EXERCISE 3

(*Note.* Calculations on percentage composition, empirical formulæ, and molecular formulæ are given in Miscellaneous Problems at the end of the book.)

Relative atomic masses: H = 1, C = 12, O = 16, Cl = 35.5.

1. Supply the missing words in the following definition:
'The molecular formula of a compound is the formula which shows the actual number of atoms of the different elements in ...... the compound.'

2. Which elements are shown to be present in a compound by the following experimental results?

(a) When X was heated with aqueous sodium(I) hydroxide it gave off a gas which turned red litmus blue;

(b) When X was heated in the absence of air it yielded a colourless liquid which turned white anhydrous copper(II) sulphate(VI) blue;

(c) X was fused with sodium, and the residue was dissolved in water. After filtration a portion of the filtrate gave a violet colour with a solution of sodium(I) pentacyanonitrosylferrate(II) (sodium nitroprusside).

(d) Another portion of the filtrate obtained in (c) was boiled for some time with dilute nitric(V) acid and then cooled. When silver(I) nitrate(V) solution was added it gave a yellow precipitate, which failed to dissolve in ammonia solution.

3. What would be the masses of carbon dioxide and water formed by complete oxidation of 4·6 g of ethanol (C_2H_6O)?

4. The empirical formula of a compound is C_2HCl. What is its molecular formula if its relative molecular mass is 181·5?

5. The empirical formula if a compound is CH_2. What is its molecular formula if the relative density of its vapour is 42?

6. An organic liquid has a molecular formula C_3H_8O. It does not liberate hydrogen when treated with sodium. What is its structural formula?

7. Which of the methods infrared spectroscopy, mass spectrometry, X-ray diffraction, electron diffraction are used in establishing (a) relative molecular masses, (b) the nature of the bonds present in molecules, (c) bond lengths in gaseous molecules, (d) bond lengths in molecules of solids?

8. The lengths of carbon—carbon bonds in aliphatic molecules are usually about 0·120 nm (1·20 Å), 0·133 nm (1·33 Å), or 0·154 nm (1·54 Å). Which of these lengths would you expect to find in the carbon—carbon bonds (in order) of the following molecules: (a) CH_3—$C{\equiv}CH$, (b) CH_3—$CH{=}CH_2$?

45

4   Electronic Theory and Organic Compounds

Nature of the Covalent Bond

Nowadays an electron is regarded, not as travelling round the nucleus in a fixed path, but as occupying a certain region in space round the nucleus. This region is called the *orbital* of the electron. Electron orbitals in a free atom are described as *atomic orbitals*. Their shapes, which can be calculated with the aid of spectral data, vary with the energy level and sub-level occupied by the electron in the atom. In the case of the solitary electron of the hydrogen atom the atomic orbital is a sphere surrounding the nucleus (Fig. 4-1).

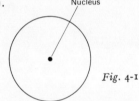

Nucleus

Fig. 4-1

Modern theory explains the formation of a single normal covalent bond between two atoms by overlapping of the atomic orbital of an electron from one atom with the atomic orbital of an electron from the second atom. This is illustrated for the H—H bond in Fig. 4-2.

Fig. 4-2

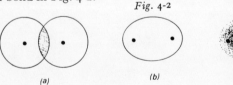

(a)   (b)   (c)

When two atomic orbitals overlap they coalesce and produce a *molecular orbital*. The molecular orbital of the two electrons forming the H—H bond is represented (in two dimensions only) in Fig. 4-2. This is the space in which the electrons are located. When two electrons occupy the same orbital they spin about their axes in opposite directions (clockwise or anticlockwise), and are said to be 'paired.' It is a fundamental principle that an orbital (atomic or molecular) cannot be occupied by more than two electrons at the same time.

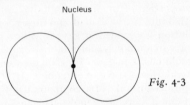

Nucleus

Fig. 4-3

Calculations show that the probability of one or other of the electrons in the molecular orbital being located at any instant at a certain point is not equal for all points in the orbital. The variations in probability can be represented by means of dots, as shown in Fig. 4-2. This is the picture (in two dimensions) which would be obtained if the dots indicated the positions of the electrons averaged over a suitable interval of time. This probability distribution, or *charge cloud*, shows the electron density in different parts of the molecule. Electron density is greatest in the central region between the two nuclei. The concentration of negative charge between the positively charged nuclei is responsible for holding the atoms together against the mutual repulsion of the nuclei. *The attraction of the two nuclei for the concentration of negative charge constitutes the covalent bond.*

Tetravalency of Carbon

If an atom has more than one atomic orbital containing a solitary ('unpaired') electron it will be able to form more than one covalent bond; that is, it will be 'polyvalent.' The three most common polyvalent elements in organic compounds are carbon, nitrogen and oxygen, which are neighbours in the Periodic Table. In the atoms of these elements, the electrons in the outer valency shell occur in two energy sub-levels, which are denoted by the letters s and p. Electrons in the s sub-level have a somewhat lower energy than those in the p sub-level. The shapes of the s and p atomic orbitals also differ. The latter are not spherical, but have the shape of a figure 8 in three dimensions, with the nucleus in the middle (Fig. 4-3). Only one atomic orbital is available for s electrons, while there are three possible orbitals for p electrons. The three orbitals have the same shape, but differ in their orientation in space, their directions being mutually at right angles to each other. They are distinguished by labelling them p_x, p_y and p_z. In the carbon, nitrogen and oxygen atoms the electrons in the outer valency shell are distributed as follows:

	s	p_x	p_y	p_z
C	2	1	1	
N	2	1	1	1
O	2	2	1	1

When atoms combine together they usually do so in such a way that their outer valency shells acquire a stable rare-gas pattern of electrons (two in the case of helium and eight in other cases). Thus the oxygen atom completes its 'octet' by using its two unpaired electrons to form covalent bonds with two hydrogen atoms (H_2O). All four orbitals in the outer shell are then filled. To achieve the same result the nitrogen atom must combine with three hydrogen atoms (NH_3).

$$
\text{H:}\overset{..}{\underset{..}{\text{O}}}\text{:} \qquad \text{H:}\overset{\text{H}}{\underset{..}{\overset{..}{\text{N}}}}\text{:}
$$
$$
\text{H} \qquad\qquad \text{H}
$$

water   ammonia

Since the carbon atom resembles the oxygen atom in having only two unpaired electrons, it might be expected to be similarly divalent. If it formed only two covalent bonds, however, its outer shell would contain only six electrons, that is, three filled orbitals. In practice the normal valency of carbon is four. This comes about as follows: When the carbon atom combines with other atoms it absorbs energy and passes into an 'excited' state. The two paired electrons in the s orbital of the outer shell become unpaired, and one is 'promoted' to the vacant p orbital. This gives the shell an electron arrangement of 1 (s), 1 (p_x), 1 (p_y) and 1 (p_z). A process called *hybridization* of atomic orbitals now occurs. The charge clouds of the four electrons interact and produce four new orbitals, each containing one electron. Since the new orbitals are obtained by combination of one s orbital and three p orbitals, they are described as 'sp^3 hybrid' orbitals. These are the orbitals used when the 'excited' carbon atom unites with four hydrogen atoms to give the CH_4 molecule.

Fig. 4-4 shows (in two dimensions) the shape calculated for one of the sp^3 hybrid orbitals of the carbon atom. It also shows how the orbital overlaps with the orbital of a hydrogen electron

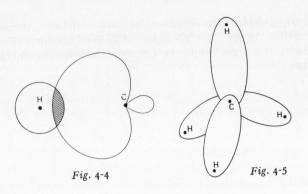

Fig. 4-4   *Fig. 4-5*

to form a C—H bond. Fig. 4-5 represents the tetrahedral distribution of the molecular orbitals of the four C—H bonds in the methane molecule. Hybridization of atomic orbitals occurs with other elements besides carbon (e.g. with boron and silicon).

Bond Angles of Carbon

When a polyvalent atom forms only single covalent bonds the chief factor governing the angles between the bonds is the number of filled orbitals in the outer shell of the atom. It is a general rule that *the orbitals of the electron pairs tend to become as widely separated as possible*, that is, the bonds tend to form the maximum possible angles with each other. This is because the electrical field of one electron pair repels the electrical field of another electron pair. For four filled orbitals the maximum angle is the tetrahedral one of $109° 28'$, which is the value found experimentally for the bond angles in methane and tetrachloromethane (carbon tetrachloride), CCl_4.

The angles for the four single bonds of carbon differ from $109° 28'$ if the attached atoms or groups are appreciably different in size. Bulky atoms or groups interfere with each other and tend to become more widely separated. Thus, in dichloromethane (CH_2Cl_2) the chlorine atoms are much larger than the hydrogen atoms. As a result, the Cl—C—Cl angle is increased to $112°$, while the other bond angles are reduced. In such cases the tetrahedral arrangement of the bond angles is maintained, but the angles are not those of a regular tetrahedron.

dichloromethane

In some compounds carbon atoms are joined to only three, or two other atoms, and the molecules contain a double, or triple covalent bond. Examples are methanal (formaldehyde), $H_2C=O$, and hydrogen cyanide, $H—C≡N$. Electron diffraction experiments show that in these molecules the valency bonds of the carbon atoms are not arranged tetrahedrally. This is because (as explained later) hybridization of the atomic orbitals of the carbon atoms takes place in a different way from that in methane. The general principles given earlier for bond angles can still be applied if the multiple bonds are thought of as equivalent to single bonds. Thus, if the carbon atom in methanal is regarded as having only three bonds, the result is a maximum bond angle (for three equal angles) of $120°$. This is the angle found experimentally. For bond angles of $120°$ the carbon atom and the three attached atoms must be in the same plane; that is, the molecule is flat.

$$
\begin{array}{c}
\text{H} \\
\diagdown \\
120° \quad \text{C}{=}\text{O} \\
\diagup \\
\text{H}
\end{array}
$$

methanal

In a similar manner it can be deduced that the hydrogen cyanide molecule is linear, with bond angles of 180°. This is found to be the case in practice.

$$
\overset{180°}{\text{H}{-}\text{C}{\equiv}\text{N}}
$$

hydrogen cyanide

IONIC CHARACTER IN COVALENT BONDS

The Inductive Effect

The two electrons forming the covalent bond in a simple diatomic molecule such as H—H or Cl—Cl are shared equally between the two atomic nuclei. Usually this is not the case when two different atoms are joined by a covalent bond. Thus, in the hydrogen chloride molecule (H—Cl), the chlorine atom is more electronegative, or has a greater attraction for electrons, than the hydrogen atom. Hence the bond electrons spend more of their time in the neighbourhood of the chlorine nucleus than in that of the hydrogen nucleus. This is often expressed by saying that there is an 'electron displacement' towards the chlorine nucleus owing to the *inductive effect* of the chlorine atom.

As a result of the electron displacement the chlorine atom acquires a partial negative charge, that is, a charge smaller than the full charge of an electron. The partial charge is represented by the sign $\delta-$ (delta minus). The hydrogen atom acquires a corresponding partial positive charge $\delta+$. Electron displacement in the bond is shown either (*i*) by placing an arrowhead at the end of the bond line to indicate the direction of the displacement, or (*ii*) by labelling each atom with its partial charge.

$$
\text{H} \rightarrow \text{Cl} \qquad \overset{\delta+}{\text{H}}{-}\overset{\delta-}{\text{Cl}}
$$
$$
(i) \qquad\qquad (ii)
$$

A molecule like HCl which is electrically neutral as a whole, but which has partial charges of opposite sign on different parts, is said to contain an *electric dipole*. The bonds in which electron displacements are present are described as *polarized* bonds. Such bonds are intermediate in character between 'pure' covalent bonds and ionic bonds. Thus, the structure of the hydrogen chloride molecule can be regarded as a combination of two 'contributing' structures. One of these is the conventional structure (with equal sharing of the bond electrons), while the other is an ionic structure. The actual structure of the molecule may be represented by writing the formulæ of the two contributing structures and connecting these by arrows pointed at both ends:

$$
\text{H}{-}\text{Cl} \leftrightarrow \text{H}^+ \text{Cl}^-
$$

A molecule which has an electronic structure intermediate between two (or more) contributing structures is said to be a *resonance hybrid*.

The actual structure of a resonance hybrid depends on the relative energies of the contributing structures. The higher the energy of a particular structure the smaller is its contribution to the actual structure. In the case of hydrogen chloride the ionic structure makes a contribution of about 17 per cent to the structure of the resonance hybrid. Thus the covalent bond in the HCl molecule is 17 per cent ionic.

Experimental Evidence of Bond Polarization

(i) *Electric Dipole Moments*

The existence of electric dipoles in molecules can usually be shown by their effect on the capacitance of a condenser. If the capacitance is first measured with air and then with hydrogen chloride as the dielectric (the medium between the plates), the capacitance is greater in the second case. This is because molecules with electric dipoles tend to set their axes in the direction of the electric field, and thus reduced the strength of the field (Fig. 4-6).

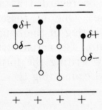

Fig. 4-6. Orientation of electric dipoles in an electric field

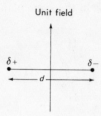

Fig. 4-7. Electric dipole moment $= \delta+ \times d$

To maintain the axis of an electric dipole at right angles to the direction of the electric field (Fig. 4-7) a certain couple would be needed. The magnitude of this couple when the field has unit intensity is called the *electric dipole moment* (p). It is measured by the product of one of the charges, $\delta+$ or $\delta-$, and the distance between the atomic nuclei. Until recently electric dipole moments were expressed in debye units (named after Debye, an early worker in this field). In the modern International System (SI) the units for electric charge and distance are the coulomb (C) and the metre (m). 1 debye unit $= 3 \cdot 336 \times 10^{-30}$ C m. From the values of electric dipole moments we can calculate the extents to which ionic structures contribute to the actual structures of resonance hybrids.

In certain cases polarization of covalent bonds is not revealed by measurement of electric dipole moments. Thus tetrachloromethane (CCl_4) has zero dipole moment, although there are strong electron displacements in the C—Cl bonds towards the chlorine atoms. The zero dipole moment is the result of the chlorine atoms being symmetrically distributed about the carbon atom, so that the effects of the four dipoles cancel each other.

Similarly, carbon dioxide, which has linear molecules ($O{=}C{=}O$), has zero dipole moment.

(ii) Infrared Absorption Spectra

These are obtained only in the case of molecules which have polarized bonds. If a molecule does not contain an electric dipole it is not affected by the oscillating electric field associated with the radiation. Thus, chlorine does not yield an infrared spectrum. It may be noted that both tetrachloromethane and carbon dioxide give infrared spectra, showing that their bonds are polarized.

(iii) Bond Energies

The energy evolved (per mole) when a covalent bond is formed between two free atoms in the gaseous state is called the *bond energy*, or, more accurately, the *bond energy of formation*. We can find its value for different bonds from either spectroscopic or thermochemical measurements. In SI bond energies (E) are expressed in kilojoule per mole. Some values are shown in the following table.

Bond energies
$E/\text{kJ mol}^{-1}$

H—H	430
Cl—Cl	238
H—Cl	430
C—C	335
C—Cl	327

It may be assumed that a bond A—A or B—B between two similar atoms is a 'pure' covalent bond (that is, one without any ionic character). Then, if two different atoms A and B were to form a pure covalent bond A—B, one could expect the energy evolved in forming the bond to be the average of the energies for the A—A bond and the B—B bond. In practice the energy of the A—B bond is usually larger. Thus, the energy of formation of the H—Cl bond is greater (by 96 kJ mol^{-1}) than the average energies of the H—H and Cl—Cl bonds. A similar difference is found in the case of the C—Cl bond. The giving out of additional energy means that a stronger bond has been formed than if the bond electrons were equally shared. The difference between the actual bond energy and that calculated for a pure covalent bond is called the *resonance energy*.

Relaying of the Inductive Effect

In hydrocarbons like methane (CH_4) and ethane (C_2H_6) there is little, if any, electron displacement in the C—H bonds because the carbon and hydrogen atoms have about the same attraction for the bond electrons. Electron displacements may be produced in the C—H bonds, however, by relaying, or transmission, of the inductive effect of an electronegative atom. In chloromethane (CH_3Cl) or chloroethane (C_2H_5Cl) there is an electron displacement in the C—Cl bond towards the chlorine atom. The effect is also felt in neighbouring bonds, and similar, although smaller, displacements occur in the C—H (and C—C) bonds, as shown.

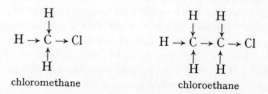

chloromethane   chloroethane

The inductive effect of an electronegative atom can thus be relayed along a chain of carbon atoms. The effect, however, diminishes rapidly with increasing distance. Hence in the series of alkyl halides given below the electron displacements in the C—H and C—C bonds become progressively smaller with greater distance from the chlorine atom. Beyond the second carbon atom they are almost negligible. This is shown by the values of the electric dipole moments (p) becoming practically constant.

$$CH_3 \rightarrow Cl \quad CH_3 \rightarrow CH_2 \rightarrow Cl \quad CH_3 \rightarrow CH_2 \rightarrow CH_2 \rightarrow Cl$$
$$6 \cdot 20 \qquad\qquad 6 \cdot 71 \qquad\qquad\qquad 6 \cdot 81 \qquad p/10^{-30}\,C\,m$$

Relaying of inductive effects influences the properties of many compounds. This is one of the chief ways in which the behaviour of an atom or group in a molecule may be modified by the presence of another atom or group.

$+I$ and $-I$ Effects

In representing an inductive effect it is often convenient to treat a group of atoms as a unit. Thus the formula $CH_3 \rightarrow Cl$ indicates an electron displacement from the methyl group (CH_3—) as a whole towards the chlorine atom. Owing to relaying of the inductive effect of the chlorine atom to the C—H bonds the CH_3— group releases electrons more readily than the hydrogen atom in $H \rightarrow Cl$. This means that the CH_3— group is 'electron-repelling' *in comparison with a hydrogen atom*, and that there is a bigger electron displacement towards the chlorine atom in $CH_3 \rightarrow Cl$ than in $H \rightarrow Cl$. The inductive effect of a group like CH_3— which is electron-repelling in comparison with a hydrogen atom is described as a $+I$ effect.

The hydroxyl group (—OH), on the other hand, has a stronger attraction than a hydrogen atom for bond electrons. This is due to the presence of the strongly electronegative oxygen atom. If the group is substituted for a hydrogen atom it decreases the electron density on an adjacent atom. The inductive effect of an atom or group which is electron-attracting in comparison with a hydrogen atom is said to be a $-I$ effect. The relative inductive effects of some common atoms and groups are shown below.

$+I$ *Groups* (electron repelling compared with H)
$$C_2H_5— > CH_3— > H—$$

$-I$ *Atoms and Groups* (electron attracting compared with H)

$$—NO_2 > —CN > —F > —Cl > —Br > —I > —OH > —NH_2$$
$$> —H$$

From the $+I$ or $-I$ effects of the various atoms or groups the different extents to which they increase or decrease the electron density in other parts of a molecule can be assessed. This often helps to explain variations in the properties of similar types of compounds. For example, the fact that ethanoic, or acetic acid ($CH_3.COOH$) is a stronger acid than propanoic acid ($C_2H_5.COOH$) is due to the CH_3— group having a smaller $+I$ effect than the C_2H_5— group. This is explained in Chapter 11.

Mesomerism.

This is the name given to an entirely different manner in which ionic character can arise in covalent bonds. It may be explained by reference to the structure of the carboxyl group (—COOH).

The conventional structure of the group is shown in formula I. In this structure the carbon atom and the two oxygen atoms possess complete octets of electrons, and the bond electrons are supposed to be equally shared.

(I)   (II)

There is a second way in which the electrons in the valency shells of the three atoms can be arranged to give complete octets. The second structure, II, is derived from structure I by transferring a pair of electrons from the double bond to the solitary oxygen atom, and *at the same time* another pair of electrons from the second oxygen atom to the C—O bond. These changes confer a negative charge on the first oxygen and a positive charge on the second one. Also, the double bond is converted into a single bond, and the single bond into a double bond.

There is evidence that the actual structure of the carboxyl group corresponds to neither I nor II, but is intermediate between them. The molecules containing the group are again *resonance hybrids*. In mesomerism two (or more) bonds are simultaneously

affected. At least one of the bonds must be a double (or triple) bond, and as a rule one of the atoms has an unshared pair of electrons which can be transferred to a bond. For conciseness the electron transfers are often represented by curved arrows, as shown in III.

(III)

As in the case of hydrogen chloride discussed earlier, the extent to which the actual structure of the carboxyl group is determined by each of the contributing structures depends on the relative energies of the latter. The energy of structure II with separate positive and negative charges is relatively high. Hence this structure makes a much smaller contribution than the conventional one to the structure of the resonance hybrid.

Mesomerism in the *carboxylate ion* ($RCOO^-$) differs from that in carboxylic acids in that it does not create ionic character in the bonds, but merely brings about a redistribution of ionic character already present. It is found experimentally that the distances between the carbon atom and the two oxygen atoms in the ion are the same. This means that the oxygen atoms cannot be joined to the carbon atom by a double bond and a single bond as represented in the conventional formula, but must be in a similar state of combination. Furthermore, the carbon–oxygen distance (0·127 nm or 1·27 Å) is intermediate between the distances for a single bond and a double bond (Chapter 3). This agrees with the ion being a resonance hybrid of the two structures now given.

$$\ddot{O}: \qquad\qquad \overset{-}{\underset{\cdot\cdot}{\ddot{O}}}: $$
$$R\!-\!C \qquad\longleftrightarrow\qquad R\!-\!C$$
$$\underset{\cdot\cdot}{\overset{-}{O}}: \qquad\qquad \underset{\cdot\cdot}{O}:$$

Since the two contributing structures are identical they make equal contributions to the actual structure. Each oxygen atom thus has a charge of $-\frac{1}{2}$.

The occurrence of ionic character in the bonds of a molecule owing to mesomerism does not exclude the possibility that ionic character, due to the inductive effect, may also be present. The permanent ionic character of the bonds depends on both effects if both are operating. In some cases the two effects reinforce each other, while in others they act in opposition.

INTERMOLECULAR ATTRACTION AND PHYSICAL PROPERTIES

Why Molecules Attract Each Other

Owing to their thermal energy molecules are in a state of constant movement, which tends to separate them as widely as possible. Therefore, the fact that they can come together and form a liquid or a solid shows that they must have an attraction for each other. Molecules containing permanent electric dipoles attract each other because of their oppositely charged parts. This is called *dipole–dipole* attraction. Exceptions occur in the case of symmetrical molecules (e.g. CO_2 and CCl_4), in which the effects of two or more dipoles cancel each other.

$$\overset{\delta+}{H}\!\!-\!\!\overset{\delta-}{Cl} \quad \overset{\delta+}{H}\!\!-\!\!\overset{\delta-}{Cl}$$
dipole—dipole
attraction

Even in the absence of permanent dipoles molecules still attract each other, and tend to cohere. This is shown by the fact that even simple diatomic gases like hydrogen and chlorine can be liquefied and solidified. In these cases intermolecular attraction is brought about by *temporary* electric dipoles in the molecules.

The manner in which temporary dipoles are thought to arise may be illustrated by means of the hydrogen molecule. The two electrons forming the H—H bond travel round both atomic nuclei, but at any instant it is unlikely that both will be situated exactly midway between the nuclei. Usually they will be more under the influence of one nucleus than the other. In this case one half of the molecule will carry a partial negative charge, while the other half will have a partial positive charge; that is, the molecule will possess a temporary dipole. Adjacent molecules will similarly contain temporary dipoles.

It might be expected that in a large assembly of molecules the temporary dipoles would be orientated in every possible direction, so that attraction and repulsion between them would cancel. This is not true for the following reason. Adjacent molecules containing dipoles polarize each other by induction in such a way that charges of opposite sign are increased, while those of similar sign are decreased. (Two bar magnets act on each other in the same way.) Thus, on balance, the net attractive force is bigger than the net repulsive force.

Attraction due to temporary dipoles occurs only when the molecules are close together. In the case of small molecules the force is extremely weak, but it increases as the molecules become larger and contain more electrons.

Attraction due to temporary dipoles operates between all molecules, whether these are similar or different. Superimposed on this in many cases is the attraction resulting from permanent dipoles. These forces of attraction are known collectively as the *van der Waals forces* (after the Dutchman van der Waals, who was the first to recognize their importance in connection with the failure of gases to obey the gas laws strictly).

Hydrogen Bonding

Normally dipole–dipole attraction is only a very weak force. It is considerably stronger, however, in the case of molecules which contain hydroxyl (—OH) groups. After fluorine oxygen is the most strongly electronegative of all the elements, so that the electron displacement O ← H is unusually large. In addition, the small size of the hydrogen atom enables a 'lone' pair of electrons from another oxygen atom to approach closely to the hydrogen nucleus and set up an electrostatic attraction. The combination of dipole–dipole attraction and electrostatic attraction results in a weak kind of bond, called a *hydrogen bond*, between the oxygen atom of one hydroxyl group and the hydrogen atom of an adjacent hydroxyl group. Hydrogen bonds (represented by broken lines) are formed between molecules of water and also between molecules of methanol, as now shown.

water - - - H—O - - - H—O - - -   - - - H—O - - - H—O - - - methanol

(above the left pair: H, H; above the right pair: CH_3, CH_3)

It must be emphasized that only in ice and solid methanol are permanent hydrogen bonds formed between the molecules. The energy of a hydrogen bond between —OH groups is only about a twentieth of that of a covalent bond, so that the bond is easily broken. In liquid water and liquid methanol the thermal movement of the molecules is sufficient to prevent any lasting linkage, and the hydrogen bonds are constantly being broken and re-formed between different molecules.

Experimental evidence of hydrogen bonding is provided by infrared spectra. When hydrogen is bonded to —OH the stretching frequency of the O—H bond shows a characteristic decrease, which results in shifting of the position of the absorption band.

Physical Properties of Organic Compounds

Hardness of Crystals. Crystals composed of ions (e.g. common salt crystals) are very hard, and have a high resistance to penetration. This is due to the strength of the electrovalent bonding between the ions. Crystals of many organic compounds (e.g. naphthalene) are soft and easily penetrated. This is because the crystals are composed of molecules which are held together only by a weak van der Waals force of attraction. The crystals of some organic compounds such as sucrose (cane-sugar) are harder than usual because, like ice, they contain hydrogen bonds. The crystals fracture easily, however, owing to the weakness of the bonds.

Melting Point and Boiling Point. Melting points are a measure of the energy required to break down crystal structures, while boiling points are a measure of the energy needed to vaporize the molecules. These energies are relatively small for molecular compounds since they have to overcome only a weak van der Waals force of attraction between the molecules. Hence most organic compounds have low melting points and boiling points, the majority melting below 150°C and boiling below 300°C.

The van der Waals force increases with the size of the molecules, that is, with their relative molecular mass. Therefore melting points and boiling points also tend to increase with relative molecular mass. Compounds of low relative molecular mass are gases or liquids at ordinary temperatures, whereas those of high relative molecular mass are solids. Within an homologous series there is usually a progressive rise in the melting points and boiling points of the members as the relative molecular mass increases.

Hydrogen bonding between molecules leads to an increase in melting point and boiling point because additional energy has to be supplied to break these bonds. This is the chief reason why methanol (CH_3OH) is a liquid, while ethane (C_2H_6), which has about the same relative molecular mass, is a gas.

The shape of the molecules is also a factor in some cases. Molecular shape chiefly affects melting points, isomers with a more symmetrical structure tending to have higher melting points than those with a less symmetrical structure. This is because the more symmetrical molecules fit together more closely in a crystal, giving a higher van der Waals force of attraction. This is illustrated in Chapter 5 by the melting points of the isomeric pentanes (C_5H_{12}).

Solubility. A general rule for organic compounds is that "Like dissolves like." Thus compounds containing —OH groups are usually soluble in water, but insoluble in liquid hydrocarbons (like benzene), while hydrocarbons are almost insoluble in water, but dissolve readily in other (liquid) hydrocarbons. If no other factors intervened, two liquids would always mix together owing to the thermal movement of their molecules. Thus benzene

(C_6H_6) and methylbenzene ($C_6H_5.CH_3$) mix in all proportions because there is nothing to prevent them from doing so. When benzene is added to water, however, two separate layers are formed, the water molecules clinging together as a result of hydrogen bonding. Only the most energetic benzene molecules succeed in penetrating the water surface, and hence the solubility of benzene in water (or vice-versa) is very small.

Providing their relative molecular masses are not too high, organic compounds containing hydroxyl groups can dissolve freely in water because their molecules form hydrogen bonds with the water molecules. This is illustrated below in the case of methanol.

$$
\begin{array}{ccc}
H & & CH_3 \\
| & & | \\
---H\!-\!O---H\!-\!O--- &
\end{array}
$$

As has been seen, however, the van der Waals force of attraction between organic molecules increases with their size. Hence, in the homologous series of alcohols a stage is reached when the tendency of the molecules to cohere exceeds their tendency to form hydrogen bonds with water molecules. Thus, the higher members of the series are insoluble in water. The possibilities of hydrogen-bonding are increased, however, if the organic molecules contain more than one —OH group. In this case compounds of quite high relative molecular mass may be soluble in water. For example, the dissolving of sucrose (cane-sugar) in water depends on the eight —OH groups present in the sugar molecule ($C_{12}H_{22}O_{11}$).

Groups like the hydroxyl group which promote solubility in water are described as *hydrophilic* (water-loving). Other groups of this type are —COOH, —NH$_2$ and —SO$_3$H, all of which form hydrogen bonds with water molecules. Some groups are *hydrophobic* (water-hating), and tend to decrease the solubility of a compound in water. These are usually hydrocarbon groups such as CH$_3$—, C$_2$H$_5$—, and C$_6$H$_5$—.

CLASSIFICATION OF REACTIONS AND REACTION MECHANISMS

Classification of Reactions. Organic reactions can be classified in different ways. One method is to divide them simply into exothermic and endothermic reactions according to whether heat is evolved or absorbed in the changes. The classification given below is based on the various kinds of overall change which the reacting molecules can undergo.

(*i*) *Decomposition.* This is the breaking up of a compound into simpler substances. Thus, under the influence of heat sucrose decomposes into water, carbon dioxide, and other products. A less drastic form of decomposition in which a simple molecule breaks away from a more complex molecule is often described as *elimination*. For example, when ethanol vapour is passed over heated aluminium(III) oxide (a catalyst), water is eliminated and ethene (ethylene) is formed.

$$C_2H_5OH \rightarrow C_2H_4 + H_2O$$
ethene

(*ii*) *Addition.* Addition consists of direct combination of two substances to give a single new substance. Ethene and chlorine undergo addition to yield dichloroethane (ethylene dichloride).

$$C_2H_4 + Cl_2 \rightarrow C_2H_4Cl_2$$
dichloroethane

Sometimes the molecules which combine additively are of the same kind. The addition is then described as a *polymerization*. Thus under suitable conditions ethene (ethylene) molecules join together to form poly(ethene), the well known plastic commonly called polyethylene.

$$n\, C_2H_4 \rightarrow (C_2H_4)_n$$
poly(ethene)

(*iii*) *Condensation.* This term is used to describe the combination of molecules when it is attended by the elimination of water or some other simple substance. An example is the combination of ethanoic, or acetic, acid with ethanol to give ethyl ethanoate, or acetate, and water.

$$CH_3.COOH + C_2H_5OH \rightarrow CH_3.COOC_2H_5 + H_2O$$
ethanoic   ethanol   ethyl ethanoate
acid

(*iv*) *Substitution.* Substitution is replacement of one atom or group in a molecule by another atom or group. Thus, chlorine brings about substitution of hydrogen in methane to give chloromethane (methyl chloride) and hydrogen chloride.

$$CH_4 + Cl_2 \rightarrow CH_3Cl + HCl$$
chloromethane

Substitution reactions are often described more specifically by terms which indicate the atom or group introduced. For example, the above reaction is known as the 'chlorination' of methane.

(*v*) *Rearrangement*. As the name indicates, this consists of rearrangement of the atoms in a molecule to produce a molecule of an isomeric compound. An illustration is the conversion of ammonium cyanate to urea on heating (Chapter 1).

$$NH_4CNO \rightarrow CO(NH_2)_2$$
$$\text{urea}$$

In the petrochemicals industry (Chapter 5) rearrangement is usually called *isomerization*. It is used for changing straight-chain hydrocarbons into branched-chain isomers.

(*vi*) *Acid–Base Reactions*. An acid is defined as *any substance which can give up protons (H^+) to a base*, and a base as *any substance which can combine with protons*. Thus, hydrogen chloride behaves as an acid in aqueous solution because its molecules lose hydrogen atoms (in the form of protons) to water molecules. Here water acts as a base.

$$\overset{\delta+}{H}-\overset{\delta-}{\underset{\underset{\overset{|}{\underset{\delta+}{H}}}{}}{O}}: + \overset{\delta+}{H}-\overset{\delta-}{Cl}: \rightleftharpoons \overset{+}{H}-\underset{\underset{H}{|}}{O}-H + :\overset{..}{\underset{..}{Cl}}:^-$$

oxonium ion

The ability of any molecule or ion to act as a base depends on one of its atoms having an unshared pair of electrons which can be 'donated' to form a co-ordinate covalent, or dative, bond with a proton.

In the reaction between water and hydrogen chloride the 'donor' atom is the oxygen atom, which is left with a positive charge. The reason for the positive charge is as follows. The two electrons forming the new bond may be regarded as belonging to the oxygen atom for half of their time and to the hydrogen atom for the remaining half. The effective charge of each electron is thus $-\frac{1}{2}$ in each atom. Since the oxygen atom provides both electrons, the net result is a reduction of one negative charge, or an increase of one positive charge, on the oxygen atom. The chlorine atom is converted into Cl^- because it retains both electrons of the H—Cl bond.

The organic acids most frequently encountered are the carboxylic acids (e.g. ethanoic, or acetic, acid), which contain the carboxyl group (—COOH). In aqueous solution some of the carboxyl groups give up their hydrogen atoms (as protons) to water molecules, leaving carboxylate ions. Other well known

$$CH_3.COOH + H_2O \rightleftharpoons H_3O^+ + CH_3.COO^-$$

organic acids are the sulphonic acids (Chapter 22), the phenols (Chapter 24) and hydrocyanic acid (HCN).

Most organic bases owe their basic character to the presence of a nitrogen atom in their molecules. A common basic group is the *amino* group (—NH$_2$). Thus, when methylamine (CH_3NH_2) is dissolved in dilute hydrochloric acid the 'lone-pair' electrons of the nitrogen atom are used· to form a co-ordinate covalent bond with a proton from the acid. The remaining solution contains ions of the salt methylammonium chloride. As will be seen

$$CH_3\overset{..}{N}H_2 + HCl \rightarrow [CH_3\overset{+}{N}H_3]Cl^-$$
$$\text{methylammonium}$$
$$\text{chloride}$$

later, the basic properties of some organic compounds (e.g.

alcohols and ethers) are due to lone-pair electrons of oxygen atoms. These compounds are usually very weak bases.

(*vii*) *Redox Reactions*. The term 'redox' is an abbreviation of 'reduction–oxidation.' Oxidation usually consists of addition of oxygen (or some other electronegative element), removal of hydrogen, or both simultaneously. Reduction is the converse. According to electronic theory a redox reaction consists of transfer of electrons from the reducing agent to the oxidizing agent; that is, a reducing agent is an electron donor, while an oxidizing agent is an electron acceptor.

Redox reactions in organic chemistry are more easily understood by using the concept of *oxidation numbers*. Atoms of free elements are given an oxidation number of zero. When two different atoms are joined by a covalent bond, the bond electrons are regarded as belonging to the more electronegative atom and the bond as ionic. The 'oxidation number' is then simply the number of electrons given up or received by each atom. The less electronegative atom is in a positive oxidation state, and the more electronegative atom in a negative oxidation state. Thus in H—Cl (regarded as H^+Cl^-) hydrogen has an oxidation number of I, and chlorine of —I. Combined oxygen always has an oxidation number of —II (except in peroxides, where it is —I).

Many elements, including carbon, can have more that one oxidation number according to how they are combined. When a carbon atom is attached to four hydrogen atoms (in CH_4), it is assumed that the hydrogen atoms give up electrons to the carbon atom. Hence in CH_4 carbon has an oxidation number of —IV, while that of hydrogen is again I. In CO_2 and CCl_4 the carbon atoms are joined to more electronegative atoms, and have an oxidation number of IV. If two carbon atoms are linked together as in CH_3—CH_3 (ethane) and CH_2═CH_2 (ethene) the electrons

of the carbon–carbon bonds are assumed to be equally shared, and do not contribute to the oxidation numbers of the carbon atoms. Thus, the oxidation numbers of the carbon atoms in ethane and ethene are —III and —II respectively.

If there are two or more carbon atoms in the same molecule their oxidation numbers are not necessarily the same. When the structural formula of the compound is known the oxidation number for any particular carbon atom is calculated by using one or other of the following rules:

(*i*) The sum of the oxidation numbers of all the atoms in the molecule is zero (for an ion the sum is equal to the charge on the ion);

(*ii*) If the carbon atom is attached to another carbon atom, the sum of the oxidation numbers for the group containing either carbon atom is zero.

Carbon can have oxidation numbers from —IV to IV, including zero. The complete range of values is represented by the carbon atoms which are specified in the formulæ shown.

$$CH_4 \quad CH_3.CH_3 \quad CH_3OH \quad CH_3.CH_2OH$$
$$\text{—IV} \quad \text{—III} \qquad \text{—II} \qquad\qquad \text{—I}$$

$$CH_2O$$
$$O$$
$$CH_3.CHO \quad HCOOH \quad CH_3.COOH \quad CO_2$$
$$\text{I} \qquad\quad \text{II} \qquad\qquad \text{III} \qquad\quad \text{IV}$$

Oxidation of an organic compound involves an increase in the oxidation number of one or more carbon atoms, and reduction a decrease. Thus, in the oxidation of methanol (CH_3OH), firstly to methanal (formaldehyde), CH_2O, and then to methanoic, or formic, acid, HCOOH, the oxidation number of the carbon atom increases from —II to o, and then from o to II.

Oxidation numbers play an important part in the systematic naming of compounds. Examples include most binary compounds (e.g. copper(I) oxide), co-ordination compounds (e.g. potassium (I) hexacyanoferrate(II), $K_4Fe(CN)_6$), and oxoacids and their anions. In the last two, where oxygen atoms are attached to a central atom, the oxidation number of the atom is specified in naming the acid or anion. Thus H_2SO_4 is the formula of sulphuric (VI) acid, the central sulphur atom having an oxidation number of VI. The systematic name for H_2SO_3 is not sulphurous acid, but sulphuric(IV) acid. The corresponding anions are designated as the sulphate(VI) ion and the sulphate(IV) ion. In a similar way HNO_3 and HNO_2 are called nitric(V) acid and nitric(III) acid, while the anions are the nitrate(V) ion and the nitrate(III) ion. It should be noted that oxidation numbers are not included in the names of organic acids (including carbonic acid), nor in those of their anions.

Reaction Mechanisms

Organic reactions usually involve the breaking of covalent bonds and the making of new ones. The reactions, however, are seldom as simple as they appear from the conventional equations. A few take place in a single step, but the majority have a number of intermediate stages, in which unstable particles (molecules or ions) exist for a brief period. Mixing of the reactants may be compared with putting a coin into a weighing-machine, from which a stamped card is obtained. Between the insertion of the coin and the delivery of the card a series of events takes place inside the machine, each event depending on the preceding one and each determining the final result. In the same way the inter-mediate stages in a reaction govern the products which appear.

The detailed path by which the products of a reaction are obtained is called the *reaction mechanism*. A knowledge of reaction mechanisms is important because it enables a closer control to be maintained over the end-products. Thus the same reactants may give rise to different products because two reaction paths are followed simultaneously. If these paths are known, it may be possible to suppress the mechanism which produces the unwanted compound. Experimental methods of investigating reaction mechanisms cover a wide range of techniques, some of which are outlined later.

Classification of Reaction Mechanisms. When a single covalent bond between two atoms is broken the electrons forming the bond can be disposed of in two, and only two, ways, as now described.

(i) *Heterolytic Fission.* In this method both electrons are left with one or other of the atoms, and if both atoms are initially neutral oppositely charged ions are formed.

$$A:B \rightarrow A^+ + :B^-$$

Or,

$$A:B \rightarrow A:^- + B^+$$

Heterolytic bond fission is characteristic of reactions which occur by an *ionic mechanism*. A simple illustration is seen in the rupture of the H—Cl bond when hydrogen chloride is dissolved in water, both electrons of the bond remaining with the chlorine atom.

$$H_2O + H:Cl \rightarrow H_3O^+ + :Cl^-$$

Ionic mechanisms occur under conditions which aid the formation of ions. Thus, the presence of ionic character in a bond favours heterolytic fission of the bond. Ionization is also facilitated by solvents (like water) which have high dielectric constants.

(ii) *Homolytic Fission.* In this case one electron remains with each atom, and free atoms (or free radicals) are produced. These are electrically neutral.

$$A:B \rightarrow A\cdot + \cdot B$$

Homolytic bond fission is characteristic of reactions which take place by a *free radical mechanism.* This type of mechanism is illustrated shortly.

Classification of Reagents in Ionic Mechanisms. In the reaction between hydrogen chloride and water the HCl molecule attacks the H_2O molecule at the oxygen atom because the latter has a partial negative charge, or high electron density. Reagents like hydrogen chloride which seek a point of high electron density for their attack are called *electrophilic* reagents, or *electrophiles.* They are electron acceptors. Common electrophilic reagents include the following:

Electrophilic reagents: H^+ (or H_3O^+) and other positive ions, water, acids, halogens, oxidizing agents.

Equally, it may be said that in the reaction between water and hydrogen chloride the water molecule attacks the hydrogen chloride molecule at the hydrogen atom, which has a partial positive charge, or low electron density. Reagents like water which react at centres of low electron density are described as *nucleophilic* reagents, or *nucleophiles.* These are electron donors. Molecules or ions containing an atom with an unshared pair of electrons commonly act as nucleophilic reagents. The following are examples:

Nucleophilic reagents: OH^- and other negative ions, water, alcohols, ammonia, reducing agents.

It should be noted that the term 'nucleophilic reagent' has a wider meaning than 'base.' The basic reactions of a nucleophilic reagent are those in which its lone pair of electrons is shared with a proton. The lone pair, however, may also be shared with some other atom or ion.

Water can behave either as an electrophile or as a nucleophile. When it brings about ionization of an acid it shows nucleophilic character. In its reaction with ammonia or a similar base it acts as an electrophile.

$$H_3N: + H—O—H \rightleftharpoons NH_4^+ + OH^-$$

Ionic Mechanisms

(*i*) *A One-step Ionic Mechanism.* An organic reaction may take place in one, two or several steps. A simple one-step reaction is the hydrolysis of iodomethane (methyl iodide) by aqueous sodium(I) hydroxide that is, by OH^- ions. ('Hydrolysis' is a substitution reaction in which an atom or group is replaced by the —OH group). The reaction, which is reversible, is represented by the equation.

$$\underset{\text{iodomethane}}{CH_3I} + OH^- \rightleftharpoons \underset{\text{methanol}}{CH_3OH} + I^-$$

Evidence that the reaction has a one-step mechanism is obtained by measuring the rate of the reaction under controlled conditions. As iodomethane is almost insoluble in water the experiment is carried out in a mixture of water and ethanol. A solution containing known concentrations of the two reactants is kept in a thermostat.

The course of the reaction is followed by withdrawing portions of the reaction mixture at regular time-intervals and titrating them with standard acid. This gives the concentrations of alkali remaining at different times. Also, since one OH^- ion is known to react with one molecule of iodomethane, the amounts of the latter which have reacted and the concentrations left at the different time-intervals can be calculated. If the various concentrations are expressed as molar concentrations, it is found that the rate of the reaction at any instant is proportional to the concentrations of *both* reactants.

$$\text{Rate of reaction} \, \alpha \, [CH_3I] \, [OH^-]$$

This is characteristic of a reaction in which the products are obtained by direct interaction of the two reactants and therefore by a one-step mechanism.

If the mechanism is examined more closely it will be noticed that the iodine atom appears in the products as a negative ion. Thus heterolytic fission of the C—I bond must occur. This type of fission is favoured by the iodine atom having a partial negative charge due partly to its own attraction for the bond electrons and partly to the $+I$ (electron-repelling) effect of the CH_3— group. The partial positive charge, or low electron density, on the carbon atom makes it vulnerable to attack by the nucleophilic OH^- ion. A new covalent bond is formed with the carbon atom by means of a lone pair of electrons donated by the oxygen atom. A clearer picture of what happens may be obtained if the equation for the reaction is re-written as follows:

$$\overset{-}{HO:} + \overset{\delta+}{C}H_3 \overset{\delta-}{-I} \rightleftharpoons [HO \text{---} CH_3 \text{---} I] \rightleftharpoons HO\text{---}CH_3 + I^-$$
$$\text{transition state}$$

Fission of the C—I bond and formation of the new C—OH bond are not consecutive processes, but take place simultaneously. If the hydroxyl ion is imagined to approach the molecule from the opposite side to the iodine atom (this is the most likely direction because there is least repulsion by the electron fields in the molecule), the formation of the C—OH bond begins as soon as the electric field of the ion affects that of the electrons of the C—I bond. These electrons are repelled towards the iodine atom, the length of the C—I bond increases, and the bond becomes weaker. With a still closer approach of the hydroxyl ion the iodine atom recedes further, and a *transition state* is reached, in which the carbon atom is joined to both the hydroxyl group and the iodine atom by weak bonds of abnormal length.

If the reacting particles collide with just the right amount of energy to form the transition state, there is an equal chance of the latter decomposing into either reactants or products. If, however, the combined energies exceed this value (the 'activation energy') the iodine atom with its two bond electrons is repelled to such a distance that it becomes a free negative ion, leaving the hydroxyl group joined to the carbon atom by a new covalent bond.

The transition state can also be reached from the opposite direction, that is, from methanol and iodide ions. In this case a larger activation energy is required to form the transition state, and hence the backward reaction occurs less readily.

(ii) A Two-step Ionic Mechanism. The compound 2-chloro-2-methylpropane (tertiary butyl chloride), $(CH_3)_3CCl$, resembles iodomethane in being hydrolysed to the corresponding alcohol by alkalies in aqueous–ethanolic solution. The mechanism of the hydrolysis, however, is quite different. This is shown by the fact

that the rate of the reaction depends only on the concentration

$$\text{Rate of reaction} \propto [(CH_3)_3CCl]$$

of the organic compound. From the point of view of the rate of reaction it is immaterial whether alkali is added or not. Therefore the hydrolysis can be represented as brought about simply by the solvent water.

$$(CH_3)_3CCl + H_2O \rightleftharpoons (CH_3)_3COH + HCl$$

Further experiments indicate that the reaction takes place in two steps as shown.

$$(i) \qquad (CH_3)_3\overset{\delta+}{C} - \overset{\delta-}{Cl} \underset{\xrightarrow{\text{slow}}}{\rightleftharpoons} (CH_3)_3C^+ + Cl^-$$

$$(ii) \qquad (CH_3)_3C^+ + H_2O \underset{\xrightarrow{\text{fast}}}{\rightleftharpoons} (CH_3)_3COH + H^+$$

The first step consists of ionization of the halogen compound by the solvent. This takes place relatively slowly. The resulting cation, containing a positively charged carbon atom, is called a *carbonium* ion, and is extremely reactive.

The second step consists of attack by the carbonium ions on the water molecules, and is very rapid. When a reaction has two or more steps, which occur at different speeds, the rate of the over-all reaction is governed by that of the slowest step. In this case the slower step is ionization of the organic compound. Hence the rate of hydrolysis is proportional to the concentration of the latter, and is independent of the concentration of the alkali. Hydroxyl ions, however, displace the equilibrium point to the right by combining with the hydrogen ions produced.

A Free-Radical Mechanism. An illustration of the second type of mechanism is found in the substitution of hydrogen in methane to give chloromethane and hydrogen chloride.

$$CH_4 + Cl_2 \rightarrow CH_3Cl + HCl$$

At ordinary temperatures this reaction does not take place unless sunlight is present. This is because the reaction depends on formation of some free chlorine atoms from chlorine molecules. Absorption of light energy by a molecule causes homolytic fission of the Cl—Cl bond.

$$(\text{I}) \qquad\qquad Cl:Cl \rightarrow 2Cl\cdot$$

Free atoms are neutral particles containing an odd electron, which makes them extremely reactive. When a chlorine atom collides with a methane molecule it captures a hydrogen atom, including one of the two electrons of the C—H bond. This leaves a free methyl radical, which is also very reactive. The radical can remove a chlorine atom from a chlorine molecule to give a free chlorine atom again.

$$(2) \qquad\qquad Cl\cdot + CH_4 \rightarrow HCl + CH_3\cdot$$

$$(3) \qquad\qquad CH_3\cdot + Cl_2 \rightarrow CH_3Cl + Cl\cdot$$

Steps (2) and (3) can now be repeated over and over again, so that dissociation of a single molecule of chlorine into free atoms may result in thousands, or even millions, of chloromethane molecules. A self-sustaining reaction of this kind, which occurs in a series of repeating steps is called a *chain reaction*. This chain is eventually broken by two of the free atoms or free radicals colliding and forming a stable molecule. This can happen in three ways, as shown.

(4) $$Cl\cdot + \cdot Cl \rightarrow Cl:Cl$$
(5) $$CH_3\cdot + \cdot Cl \rightarrow CH_3:Cl$$
(6) $$CH_3\cdot + \cdot CH_3 \rightarrow CH_3:CH_3$$

Chain reactions have three well-defined steps—*initiation*, *propagation* and *termination*. In the reaction between methane and chlorine initiation is represented by equation (1), propagation by equations (2) and (3), and termination by equations (4), (5) and (6).

Experimental evidence supporting the mechanism described is obtained in several ways. For example, if chlorine is exposed to ultraviolet radiation the absorption spectrum of the gas shows the presence of atomic chlorine, but this rapidly disappears when methane is mixed with the gas. Again, methane and chlorine react in complete darkness if they are mixed with a trace of tetramethyl-lead(IV) (lead tetramethyl) and heated to 140°C. The bonds between the lead atoms and the carbon atoms in this compound are only weak, and the compound decomposes on heating to give lead and free methyl radicals which initiate the chain reaction.

$$Pb(CH_3)_4 \rightarrow \cdot \overset{\cdot}{\underset{\cdot}{Pb}}\cdot + 4CH_3\cdot$$

At about 250°C chlorine itself begins to dissociate into free atoms. Hence above this temperature neither sunlight nor tetramethyl-lead(IV) are required for the substitution to proceed.

The conditions favouring free-radical mechanisms are those leading to the formation of free radicals. We can therefore expect to find this type of mechanism in reactions which require high temperatures, the presence of sunlight, or assistance from compounds known to yield free radicals. Most of the reactions which occur in flames have free-radical mechanisms.

EXERCISE 4

1. Give the approximate values (109°, 120° or 180°) of the bond angles marked a, b, c, etc., in the following molecules:

(i)
$$\begin{array}{c} CH_3 \\ |a \\ C\equiv CH \end{array}$$

(ii)
$$\begin{array}{c} CH_3 \\ |b \\ H_2C-Cl \end{array}$$

(iii)
$$\begin{array}{c} H \\ |c \\ H-C=O \end{array}$$

(iv)
$$\begin{array}{c} H\ \ H \\ d|\ \ e| \\ CH_3-C-C=O \\ | \\ H \end{array}$$

2. Which of the following statements are true of a resonance hybrid?

(a) It consists of two or more kinds of molecules (or ions) in equilibrium.

(b) It always has the conventional structure as one of its contributing structures.

(c) It has a lower energy than any of its contributing structures.

(d) If it has two contributing structures, the one with the higher energy makes the larger contribution to the actual structure.

3. Which of the following molecules would you expect to contain electric dipoles? (a) CH_3-O-H; (b) CH_3-CH_3; (c) $S=C=S$; (d) CH_2Cl_2; (e) CBr_4.

4. Which of the molecules in Question 3 would you expect to possess electric dipole moments?

5. Which of the following statements are correct?

(a) A chlorine atom in a molecule has a $+I$ effect.

(b) A hydrogen bond is a bond between two hydrogen atoms.

(c) Two ethane molecules (C_2H_6) attract each other more strongly than two methane molecules (CH_4).

(d) Ethanol, C_2H_5OH, would be expected to have a higher boiling point than its isomer methoxymethane (dimethyl ether), CH_3—O—CH_3.

(e) Naphthalene ($C_{10}H_8$) would be expected to dissolve readily in benzene (C_6H_6).

6. Classify each of the reactions represented below as elimination, substitution, condensation, rearrangement, or acid—base reactions.

(a) $CH_3I + NH_3 \rightarrow CH_3NH_2 + HI$.

(b) $CH_2{=}CH—OH \rightarrow CH_3—CHO$.

(c) $HCOOH + CH_3OH \rightarrow HCOOCH_3 + H_2O$.

(d) $HCOOH + NH_3 \rightarrow HCOO^- + NH_4^+$.

(e) $C_2H_5OH \rightarrow C_2H_4O + H_2$.

7. By considering the oxidation numbers of the carbon atoms state whether the first compound in the following conversions is oxidized or reduced, or whether neither oxidation nor reduction occurs:

(a) $CH_4 \rightarrow CH_3Cl$; (b) $CH_2{=}CH_2 \rightarrow CH_3—CH_3$;

(c) $CH_3OH \rightarrow CH_3Cl$; (d) $H_2C{=}O \rightarrow CH_3OH$;

(e) $H_2C{=}O \rightarrow H—C{=}O$
$\qquad\qquad\qquad\quad\ \ |$
$\qquad\qquad\qquad\quad OH$

8. Give the terms used to describe the following:

(a) Covalent bond fission in which both bond electrons remain with one of the atoms.

(b) A reagent which attacks a molecule at an atom with a low electron density.

(c) The minimum energy which two colliding particles must have in order to form a transition state.

(d) An ion like $(CH_3)_3 C^+$, which contains a positively charged carbon atom.

(e) The three steps characteristic of a chain reaction.

part two

ALIPHATIC COMPOUNDS

5 Alkanes

A hydrocarbon *is a compound composed of hydrogen and carbon only*. Open-chain hydrocarbons fall into three different homologous series, which, until fairly recently, were distinguished by the names *paraffins*, *olefins*, and *acetylenes*. A more systematic method of naming the three series has now been adopted. Paraffins are called *alkanes*, olefins *alkenes*, and acetylenes *alkynes*. The general formulæ of the three classes are shown below.

Alkanes C_nH_{2n+2} Alkenes C_nH_{2n} Alkynes C_nH_{2n-2}

This chapter is concerned only with the alkanes. Table 5-1 shows the first ten members of the series. The first four have common, or trivial, names, but subsequent members are named according to the number of carbon atoms in the molecule. From butane onward the alkanes exist in isomeric forms, in some of which the carbon chains are branched. Isomerism of alkanes and the manner in which they are named systematically are described later in this chapter. The names listed in Table 5-1 now apply only to the straight-chain compounds; e.g.

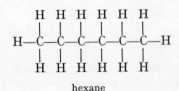

hexane

As was seen in Chapter 1, successive members of an homologous series differ in molecular formula by a constant amount, CH_2. Any member can be regarded as derived theoretically from the preceding one by introducing a CH_3— group into the molecule in place of a hydrogen atom. Thus the first four alkanes are methane—CH_4, ethane—$CH_3.CH_3$, propane—$CH_3.CH_2.CH_3$, and butane —$CH_3.CH_2.CH_2.CH_3$.

Table 5-1. STRAIGHT-CHAIN ALKANES
(General formula: $C_n H_{2n+2}$)

Name	Molecular formula	Melting point/°C	Boiling point/°C	Normal physical state
Methane	CH_4	−183	−161	Gas
Ethane	C_2H_6	−172	−88	Gas
Propane	C_3H_8	−187	−44	Gas
Butane	C_4H_{10}	−135	−0.5	Gas
Pentane	C_5H_{12}	−130	36	Liquid
Hexane	C_6H_{14}	−95	69	Liquid
Heptane	C_7H_{16}	−91	98	Liquid
Octane	C_8H_{18}	−57	126	Liquid
Nonane	C_9H_{20}	−54	151	Liquid
Decane	$C_{10}H_{22}$	−30	174	Liquid

It will be seen from Table 5-1 that there is a general rise in melting point and boiling point with increase in relative molecular mass (some of the melting points are exceptional). As noted in Chapter 4, this characteristic is explained by an increase in the van der Waals force of attraction between the molecules as the latter become larger. The first four members of the series are gases. Liquids first appear at pentane, and from hexadecane ($C_{16}H_{34}$) onward the members are solids at ordinary temperature. Paraffin wax is a mixture of some of these solids. The highest straight-chain alkane which has been obtained is heptacontane ($C_{70}H_{142}$).

Alkanes are almost insoluble in water, but dissolve readily in ether, benzene, and petrol. The liquid members have densities (at 15°C) between 0.62 and 0.80 g cm^{-3}, and hence they float on water. They are all flammable compounds.

Occurrence of Alkanes. The lower alkanes are found naturally in association with coal and petroleum. Methane is the chief constituent of 'fire damp,' which has often caused explosions in coal-mines. It is also formed by the action of bacteria on the cellulose in vegetable matter. Thus, methane is present in 'marsh gas,' which bubbles up from the bottom of stagnant ponds. It is also obtained in the biochemical treatment of sewage by special strains of bacteria.

Petroleum is a complex mixture of hydrocarbons containing up to 500 carbon atoms in the molecule. The hydrocarbons are mostly alkanes, but varying amounts of aromatic hydrocarbons are also present. Methane is the main constituent of 'natural gas,' which may occur either by itself or along with petroleum. In 1959 an abundant supply of natural gas was discovered at a depth of one to two miles below the bed of the North Sea. The gas is now piped to various parts of Britain for use in the home or in chemical manufacture. The composition of natural gas varies with its source, a typical composition (expressed in percentages by volume) being as follows:

Methane 76, ethane 11, propane 8, butane 2, other gases 3.

Laboratory Preparation of Alkanes. Methane can be obtained by heating fused (anhydrous) sodium(I) ethanoate, or acetate, with soda-lime in a hard-glass tube. The gas is collected over water. Soda-lime (made by slaking calcium(II) oxide with aqueous sodium(I) hydroxide) is preferred to solid sodium(I) hydroxide because it attacks glass less readily. The reaction is described as a *decarboxylation* because the carboxyl group of the sodium(I) ethanoate is replaced by a hydrogen atom.

$$CH_3.COONa + NaOH \rightarrow CH_4 + Na_2CO_3$$

Methane prepared by the above method contains hydrogen and ethene as impurities. The method is unsuitable for making higher alkanes because the sodium salts of higher acids yield mixtures of different alkanes. The following are 'general' methods for the preparation of alkanes.

(i) *Reduction of an Alkyl Halide.* Alkyl halides are represented by the general formula RX, R standing for the alkyl radical and X for a halogen atom. In the laboratory alkyl iodides are usually preferred to the other halides because they are the most reactive, the order of reactivity being.

$$iodides > bromides > chlorides > fluorides$$

The usual reducing agent is a metal couple such as Al/Hg or Na/Hg, used in conjunction with an alcohol (methanol or ethanol). The latter supplies the hydrogen required for the reduction of the alkyl iodide. The reaction is complex, but in the case of iodomethane (methyl iodide) it can be represented simply as follows:

$$CH_3I + 2H \rightarrow CH_4 + HI$$

In general,

$$RX + 2H \rightarrow RH + HX$$

In these equations '2H' represents the reducing agent as a whole. It does not mean that the reduction is brought about by free hydrogen atoms.

Experiment

First prepare an aluminium-mercury couple in the flask which is to be used for the preparation of the methane. Cut about 1 g of aluminium foil into small pieces, and put the pieces into a 100 cm^3 distilling-flask. Clean the surface of the foil from metal oxide by covering the pieces with dilute sodium(I) hydroxide solution.

Fig. 5-1. Small-scale preparation of methane

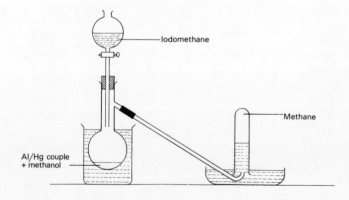

After two minutes pour away the alkali, wash the metal with water, and then immerse it for two minutes in a dilute (2 per cent) solution of mercury(II) chloride. Mercury is displaced from solution, and an amalgam is formed on the surface of the aluminium. Pour away the remaining mercury(II) chloride solution, and wash the couple immediately, firstly with water and then with methanol. Without delay cover the couple with fresh methanol (aluminium(III) oxide is rapidly formed if the couple is exposed to the air).

Fit the flask with a dropping-funnel (Fig. 5-1) and into this put 2-3 cm^3 of iodomethane. Cool the flask by placing it in a beaker containing cold water (heat is evolved in the reaction). Arrange to collect the gas in test-tubes over water. Add the iodomethane gradually to the flask, so that methane is given off steadily. The gas obtained contains about one per cent of hydrogen.

(ii) Electrolysis of an Aqueous Solution of the Sodium (or Potassium) Salt of a Carboxylic Acid (Kolbe's Method). This method is little used, but it has considerable theoretical interest

Fig. 5-2. Electrolysis of sodium (I) ethanoate solution.

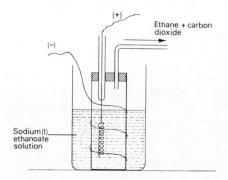

because of the free-radical mechanism involved. It is illustrated by the preparation of ethane from sodium(I) ethanoate, or acetate.

The electrolytic cell (Fig. 5-2) consists of a small porous pot placed in a wider vessel, both containing a concentrated aqueous solution of sodium(I) ethanoate. It is necessary to have a high current density at the anode, and to obtain this an anode consisting of a spiral of platinum wire is used. The cathode is a platinum wire wound round the pot. The electrodes are connected to a 12 volt source of direct current, and a mixture of ethane and carbon dioxide is given off at the anode. The carbon dioxide is removed by passing the mixture through a U-tube containing soda-lime, and the ethane is collected over water.

The reactions which occur in the above experiment are explained as follows. In an aqueous solution of sodium(I) ethanoate the ionic dissociations shown take place.

$$CH_3.COONa \rightleftharpoons CH_3.COO^- + Na^+$$
$$H_2O \rightleftharpoons H^+ + OH^-$$

At the cathode hydrogen ions are discharged in preference to sodium ions.

$$2H^+ + 2e \rightarrow H_2$$

At the anode ethanoate ions are preferentially discharged, and oxidized to free ethanoate radicals. These are unstable and decompose to carbon dioxide and free methyl radicals. The latter then combine in pairs, giving ethane.

$$CH_3.COO^- - e \rightarrow CH_3.COO\cdot$$
$$CH_3.COO\cdot \rightarrow CH_3\cdot + CO_2$$
$$2CH_3\cdot \rightarrow CH_3-CH_3$$

69

Evidence for the free radical mechanism is the fact that if a solution containing the sodium salts of two different acids is electrolysed *three* alkanes are obtained. Thus a mixture of sodium(I) ethanoate ($CH_3.COONa$) and sodium(I) propanoate ($C_2H_5.COONa$) yields not only ethane (C_2H_6) and butane C_4H_{10}), but also propane (C_3H_8). Presumably the latter is formed by combination of free methyl radicals and free ethyl radicals.

(*iii*) *Reaction of Water with a Grignard Reagent.* This method is described in Chapter 9.

(*iv*) *Reduction of an Aldehyde or Ketone by Clemmensen's Method.* See Chapter 10.

Reactions of Alkanes

Alkanes were originally called paraffins because of their inertness towards the usual laboratory reagents (Latin *parum* = little, *affinis* = affinity). Thus, alkanes are not affected by concentrated acids or alkalies, nor by the common oxidizing reagents or reducing agents. Their lack of reactivity can be understood from the electronic structures of the molecules. As mentioned earlier, carbon and hydrogen atoms have roughly the same attraction for electrons, and therefore in a C—H bond of an alkane there is no appreciable displacement of the bond electrons towards either atom. This means that the molecule has neither a point of high electron density which can be attacked by an electrophilic reagent, nor one of low electron density which can attract a nucleophilic reagent.

By using special conditions (e.g. high temperatures and catalysts) alkanes can be made to undergo a number of useful reactions. Most of these involve free-radical mechanisms.

1. Thermal Decomposition (Pyrolysis). Alkanes decompose when they are strongly heated, the higher members splitting up more readily than the lower ones. This is the basis of the 'cracking' process (described later), whereby petroleum oils are broken down into simpler, and more useful, hydrocarbons. The lower alkanes eliminate hydrogen and form alkenes or, in the case of methane an alkyne. The reactions are used in the manufacture of ethene (ethylene) and ethyne (acetylene).

$$C_2H_6 \xrightarrow{850°C} C_2H_4 + H_2$$
$$\text{ethane} \qquad \text{ethene}$$

$$2CH_4 \xrightarrow{1500°C} C_2H_2 + 3H_2$$
$$\text{methane} \qquad \text{ethyne}$$

2. Oxidation. Alkanes burn readily in air, yielding carbon dioxide and water; e.g.

$$CH_4 + 2O_2 \rightarrow CO_2 + 2H_2O$$

If insufficient oxygen is present, incomplete combustion results in the formation of carbon monoxide. This is important in the combustion of petrol in internal combustion engines (the exhaust gases contain about 15 per cent of carbon monoxide).

Under suitable conditions oxidation of alkanes may be even more restricted. Thus, when methane (from natural gas) is burned in a very limited supply of air extremely fine particles of carbon ('carbon black') are deposited. The carbon is used for strengthening the rubber in tyres and for making black paint, printer's ink, and shoe polish.

3. Methane-Steam Reaction. Methane and steam react together when a mixture of the two is passed over a nickel catalyst at 800°C. Several reactions occur, but the chief one yields *synthesis gas*, a mixture of carbon monoxide and hydrogen used in the synthesis of organic compounds.

$$CH_4 + H_2O \rightarrow CO + 3H_2$$

4. Substitution. In alkanes the four valencies of each carbon atom are taken up separately by other atoms, and no extra atoms can be added to the molecule. Alkanes are therefore described as *saturated* hydrocarbons. A characteristic reaction of saturated hydrocarbons consists of substitution, in which one or more hydrogen atoms in the molecule are replaced by other atoms or groups.

(*i*) *Halogenation*. This consists of substitution of hydrogen atoms by halogen atoms. Alkanes explode violently with *fluorine* even in complete darkness. If the fluorine is first diluted with nitrogen, however, substitution proceeds smoothly in the presence of cobalt(II) fluoride as a catalyst. With methane tetra-fluoromethane is formed.

$$CH_4 + 4F_2 \rightarrow 4HF + CF_4$$

The manner in which methane reacts with *chlorine* depends on the conditions. A mixture of methane and chlorine explodes when exposed to direct sunlight or when a lighted taper is applied to the mixture. Carbon is deposited and hydrogen chloride is formed.

$$CH_4 + 2Cl_2 \rightarrow C + 4HCl$$

Chlorine does not react with methane in the dark unless heated above 250°C. In diffused sunlight substitution takes place slowly by the free-radical mechanism described in Chapter 4. There are four stages of substitution, as shown below. Normally all four substitution compounds are formed, whatever proportions of methane and chlorine are mixed together.

$$CH_4 + Cl_2 \rightarrow HCl + CH_3Cl$$
chloromethane

$$CH_3Cl + Cl_3 \rightarrow HCl + CH_2Cl_2$$
dichloromethane

$$CH_2Cl_2 + Cl_2 \rightarrow HCl + CHCl_3$$
trichloromethane

$$CHCl_3 + Cl_2 \rightarrow HCl + CCl_4$$
tetrachloromethane

In the modern system the halogen compounds are named as substitution derivatives of the alkanes. In this system the name of the substituting atom is placed *before* that of the hydrocarbon. Older names for the four derivatives are: CH_3Cl—methyl chloride; CH_2Cl_2—methylene dichloride; $CHCl_3$—chloroform; CCl_4—carbon tetrachloride.

Bromine brings about substitution in methane in the same way as chlorine, but reacts more slowly. The influence of light on substitution in alkanes by bromine can be demonstrated as now described.

Experiment

Half-fill a dry test-tube with a liquid alkane (e.g. hexane), and add a drop of liquid bromine. Insert a stopper in the tube, and leave it standing in daylight. Prepare a similar tube, and leave this in a dark cupboard. In half-an-hour the colour of the bromine in the first tube will have disappeared, while that in the second tube will be unchanged. When the first tube is opened steamy fumes of hydrogen bromide appear.

Iodine does not react appreciably with alkanes. This is because hydrogen iodide is a strong reducing agent and reduces the alkyl iodide back to the alkane. Actually the reaction is a balanced one with the equilibrium point far to the left.

$$CH_4 + I_2 \rightleftharpoons CH_3I + HI$$

From the above it is seen that the order of reactivity of the halogens with alkanes is as follows:

$$F_2 > Cl_2 > Br_2 > I_2$$

Unlike methane, ethane and higher alkanes give rise to isomeric halogen derivatives when more than one hydrogen atom is substituted. Thus there are two dichloroethanes, one called 1,2-dichloroethane and the other 1,1-dichloroethane according to whether the two chlorine atoms become attached to different carbon atoms or the same one. (The naming of halogen derivatives of alkanes is explained more fully shortly).

$$CH_3.CH_3$$
$$\downarrow$$
$$CH_3.CH_2Cl$$
chloroethane
$$\downarrow$$
$$CH_2Cl.CH_2Cl \quad \text{and} \quad CH_3.CHCl_2$$
1, 2-dichloroethane   1,1-dichloroethane

(ii) Nitration. Although alkanes are not affected by nitric(V) acid at ordinary temperatures a reaction occurs when a mixture of an alkane and nitric(V) acid vapour is passed through a tube at 400°–450°C. A hydrogen atom of the alkane is substituted by the nitro group (—NO_2). This type of substitution is described as *nitration*.

$$RH + HNO_3 \rightarrow RNO_2 + H_2O$$

Methane and ethane are converted by nitration into nitromethane and nitroethane. In industry these compounds are made more conveniently by nitrating propane, which yields a mixture of nitromethane, nitroethane, and nitropropane. The three products are separated by fractional distillation and used as solvents. Nitration of alkanes is much less important than that of aromatic hydrocarbons (Chapter 22).

$$CH_3NO_2$$
nitromethane
$$C_2H_5NO_2$$
nitroethane
$$C_3H_7NO_2$$
nitropropane

Chemical Tests for Alkanes. There are no simple chemical tests for alkanes because they are inert to most reagents. Usually a process of elimination is adopted. A gaseous alkane can be distinguished from carbon monoxide and hydrogen (which also burns with a blue flame), by passing it over heated copper(II) oxide, when both water and carbon dioxide are obtained. Further tests are then carried out to prove that the gas does not belong to the alkene or alkyne series.

Two of these tests are of particular importance:

(*i*) an alkane does not immediately decolorize bromine vapour (or a solution of bromine in trichloromethane);

(*ii*) an alkane does not affect Baeyers' reagent, which consists of aqueous potassium(I) manganate(VII) (potassium permanganate) made alkaline with sodium(I) carbonate solution.

Alkenes and alkynes react with both of these reagents, as described in Chapters 6 and 7.

Shapes of 'Straight-chain' Alkane Molecules. If a 'ball-and-spring' model of the ethane molecule (CH₃—CH₃) is constructed it will be seen that one CH₃— radical can have many different orientations with respect to the other. The question arises as to which is the correct one. The answer is that there is no single orientation which applies to all the molecules, and even for a given molecule the orientation is constantly changing.

This is because different amounts of energy are associated with the different orientations and the internal energies of the molecules alter because of intermolecular collisions and changes of temperature. The orientations can vary because the CH₃— radicals are able to rotate about the C—C bond, which is therefore said to possess the property of *free rotation*.

The molecules of propane, butane, and other 'straight-chain' alkanes are represented by graphic formulæ of the type now shown.

$$H-\overset{\displaystyle H}{\underset{\displaystyle H}{C}}-\overset{\displaystyle H}{\underset{\displaystyle H}{C}}-\overset{\displaystyle H}{\underset{\displaystyle H}{C}}-\overset{\displaystyle H}{\underset{\displaystyle H}{C}}-H$$

butane

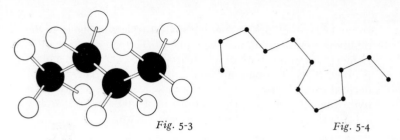

Fig. 5-3   *Fig.* 5-4

Electron diffraction (of gases) and X-ray diffraction (of solids) show that the graphic formulæ do not represent the actual distribution in space of the atoms in alkane molecules. The arrangement of the atoms in butane is indicated by a space model (Fig. 5-3), in which the angles between the C—C bonds and the C—H bonds have the value of approximately 109·5° (see Chapter 4). As in the case of ethane, 'free rotation' is possible about the C—C bonds, so that there are many possible orientations for the different groups in the molecule. Here again energy considerations determine the configuration of a particular molecule at any instant.

Even more misleading is the description 'straight-chain' when applied to normal alkanes of high relative molecular mass. In these the carbon backbone of the molecule is usually twisted in different directions. Thus, a possible configuration for the chain of twelve carbon atoms in the molecule C₁₂H₂₆ is represented (in two dimensions only) in Fig. 5-4.

Isomerism of Alkanes. The structural formulæ of methane, ethane, and propane can be written in only one way, and there are no isomers of these hydrocarbons. When, however, a molecule of butane is theoretically derived from one of propane by replacing a hydrogen atom of the latter by a CH₃— group the substitution can be carried out in two ways. A hydrogen atom from one of

73

the end CH_3— groups (it does not matter which) or one from the middle —CH_2— group may be replaced. This gives two possible isomers for butane, both of which are known. The next member, pentane (C_5H_{12}), occurs in three isomeric forms. These can be derived from the two isomers of butane again by substituting a hydrogen atom by a CH_3— group in all possible ways. The substitutions can be represented as now shown.

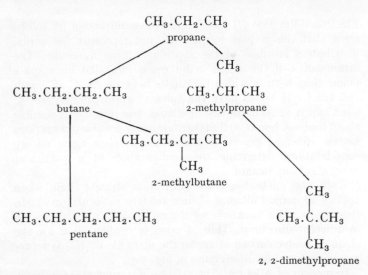

The student should convince himself that substitution of a hydrogen atom from any position other than those shown merely leads to the production of a compound identical in structure with one of those already obtained. The manner in which the modern names of the isomers are derived is explained in the next section.

In general, isomers of an alkane have similar chemical properties, but some of their physical properties differ. Thus the boiling point of butane is $-0.5°C$, while that of its isomer, 2-methylpropane, is $-12°C$. Of the isomeric pentanes 2,2-dimethylpropane has a much higher melting point ($-17°C$) than either pentane itself ($-130°C$) or 2-methylbutane ($-160°C$). This is because the first isomer has symmetrical molecules which can pack together more closely in the solid, thus giving a larger van der Waals force of attraction (see Chapter 4).

The number of isomers theoretically possible increases rapidly with the number of carbon atoms in the molecule. This is shown in Table 5-2. There is no mathematical formula connecting the number of isomers with the number of carbon atoms in the molecule, but the number of isomers can be calculated indirectly.

Table **5-2**. ISOMERS OF ALKANES

Formula	Number of Isomers
C_4H_{10}	2
C_5H_{12}	3
C_6H_{14}	5
C_7H_{16}	9
C_8H_{18}	18
C_9H_{20}	35
$C_{10}H_{22}$	75
$C_{14}H_{30}$	1858
$C_{18}H_{38}$	60 523
$C_{25}H_{52}$	36 797 588

Systematic Naming of Alkanes and Alkyl Radicals. As stated earlier, in modern systematic nomenclature the simple names butane, pentane, etc., without prefix, are used only for the straight-chain alkanes. It has been customary to describe the latter by prefixing the term 'normal' (e.g. normal, or n-, butane), while branched-chain isomers have been indicated by prefixes like 'iso,' which signified the presence of a $(CH_3)_2CH-$ group in the molecule.

Branched-chain alkanes are named as alkyl-substitution compounds of the hydrocarbon with the longest straight chain of carbon atoms. To show where substituting alkyl radicals are attached to the chain the latter is numbered. Numbering may start from either end of the chain, but the end chosen is the one which gives the lower number (or numbers) for the substituent group (or groups). This is illustrated by the following formula and name of one of the isomers of hexane:

$$CH_3-CH_2-CH_2-\overset{\displaystyle CH_3}{\underset{}{CH}}-CH_3$$
$$54321$$

2-methylpentane

Further illustrations of the system are seen in the formulæ and names of the isomers of butane and pentane (see last section).

If there is more than one kind of alkyl substituent, the radicals are included in the name in alphabetical order. This applies even when there are two alkyl radicals of the same kind. An illustration is seen in the following branched-chain isomer of nonane (C_9H_{20}).

$$CH_3-\overset{\displaystyle CH_3C_2H_5}{\underset{\displaystyle CH_3}{C}}-\overset{}{CH}-CH_2-CH_3$$

3-ethyl-2,2-dimethylpentane

The naming of alkyl radicals follows the same general rules as for alkanes except for two differences. Firstly, the ending -*ane* of the alkane is replaced by -*yl*, and, secondly, numbering of the carbon atoms in the straight chain begins at the carbon atom with the free valency bond. Thus the names 'isopropyl' and 'isobutyl' for the two radicals represented below are replaced by the systematic names shown.

$$CH_3-\overset{\displaystyle }{\underset{\displaystyle CH_3}{CH}}-$$

1-methylethyl
(isopropyl)

$$CH_3-\overset{\displaystyle }{\underset{\displaystyle CH_3}{CH}}-CH_2-$$

2-methylpropyl
(isobutyl)

Hydrides of Other Group IV Elements. Carbon is unique as an element in the number and complexity of its hydrides. The elements below carbon in Group IV of the Periodic Table are silicon, germanium, tin, and lead. All these elements have a

75

similar distribution of electrons to carbon in their outer valency shells, and all can show a valency of four as a result of sp^3 hybridization of atomic orbitals (Chapter 4). Only silicon and germanium, however, are able to form a series of hydrides analogous to alkanes, and then only to a limited extent. Thus silicon gives rise to the silanes SiH_4, Si_2H_6,........Si_6H_{14}, and germanium to the germanes GeH_4. Ge_2H_6,.........Ge_5H_{12}. Neither element, however, forms hydrides corresponding to alkenes, alkynes, or cyclic hydrocarbons. With tin and lead hydride formation is restricted to SnH_4 and PbH_4, both of which are unstable.

The failure of the other elements of Group IV to match carbon in forming hydrides is due to a number of factors, the chief of which are the following:

(*i*) The strength of the X—H bond (where X = C, Si, etc.) decreases down the Group, and therefore the bond becomes easier to break. There is a similar decrease in strength of the X—X bond. In both cases the decrease is due to the increase in size of X. As X becomes larger the bond electrons are further from the nucleus of X, and are attracted less strongly.

(*ii*) From carbon to lead the elements become less electronegative. Hence there is an increasing electron displacement in the X—H bond towards the hydrogen atom, resulting in higher partial charges on X and H. These make the hydrides more reactive.

(*iii*) Elements below carbon in Group IV are able to expand their outer valency shell to twelve electrons by co-ordination with molecules which can donate a share in a lone pair of electrons. The effect is again to enhance the chemical reactivity of the hydrides. Thus, silane (SiH_4) is rapidly hydrolysed by aqueous alkalies because OH^- ions are good electron donors.

PETROLEUM

The origin of petroleum (Latin *petra* = rock, *oleum* = oil) is obscure. One theory is that it was derived from microscopic marine organisms, of both animal and vegetable origin, which lived in the oceans in great abundance during a warmer period of the Earth's history. As the organisms died and sank to the sea-bed they became sealed in layers of mud. Petroleum was then formed as a result of bacterial action and chemical changes under the influence of heat and pressure. The chemical composition of petroleum varies with the place of origin. The chief constituents are usually straight-chain and branched-chain alkanes, but other hydrocarbons present include cycloalkanes and aromatic compounds such as benzene and methylbenzene.

Fractional Distillation of Petroleum. The crude oil is first given a preliminary heating to drive off dissolved natural gas. The degassed oil is then pumped to a fractional distillation plant. In the interests of efficiency fractional distillation is carried out differently from the method described in Chapter 2. Instead of liquids of different boiling point being distilled in turn the whole of the oil is vaporized at once, and the vapours are condensed at different temperatures in a fractionating column.

Vaporization of the oil takes place in a 'pipe-still,' which consists of a steel coil heated in an oil-furnace (Fig. 5-5a). The mixture of vapours passes to the fractionating-tower which is made of steel and is about thirty metres high. The bottom of the tower is at a higher temperature than the top, so that the heavier, and less volatile, constituents condense first, while the more volatile ones pass on up the tower. The tower is divided into sections by trays (up to sixty in number) containing openings

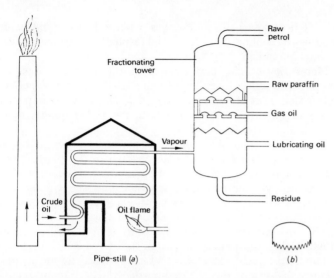

Fig. 5-5 (a) Fractional distillation of petroleum (diagrammatic);
(b) a bubble-cap

is collected. The fractions obtained in this way are not single compounds, but mixtures of hydrocarbons boiling over different ranges of temperature. These ranges overlap to some extent. The principal mixtures obtained from the crude oil, their carbon content, and boiling point ranges are shown in the table now given.

Table 5-3. CHIEF FRACTIONS OBTAINED
FROM CRUDE PETROLEUM

Fraction	Carbon content	Boiling point range/°C
Natural gas	C_1–C_5	Below 30
Raw petrol	C_5–C_{12}	30–200
Paraffin (kerosene)	C_{10}–C_{16}	150–280
Gas oil	C_{14}–C_{20}	250–375
Lubricating oil	C_{20}–C_{70}	Above 375
Residue	—	—

through which the vapour passes on its upward journey. The openings are covered by 'bubble caps,' which have serrated edges to allow the vapour to pass through (Fig. 5-5b). As the vapour passes under a bubble cap it bubbles through a layer of liquid on the floor of the tray. This liquid assists in the condensation of the less volatile portion of the vapour. Overflow pipes, known as 'downspouts', keep the liquid at a constant level in each tray, excess of liquid running down to the tray below.

By drawing off the condensed liquid at different heights, a series of products, called 'cuts,' of progressively lower boiling points,

Treatment of the Fractions. The fractions obtained from the primary distillation of petroleum are treated further either to purify them or to produce mixtures of hydrocarbons required for particular purposes. Most of the fractions have to be purified from sulphur compounds, which have an objectionable smell, darken on exposure to light, and form poisonous sulphur dioxide when burned. One method of purification is to filter the oil through fuller's earth (a fine mineral silicate). The sulphur is recovered and sold for use in sulphuric acid manufacture.

(*i*) *Gas Fraction.* This has the same kind of composition as given earlier for natural gas. The alkanes are separated and used either as fuels or in making petrochemicals. Propane and butane, which are easily liquefied by pressure are sold in liquid form (e.g. 'Propagas' and 'Calor gas').

(*ii*) *Raw Petrol Fraction.* After removal of sulphur compounds the purified liquid is fractionally distilled to yield *benzine* (not to be confused with the aromatic hydrocarbon benzene), *naphtha*, and *white spirit.* Benzine is the basis of motor spirit, but has to be blended with spirit derived from 'cracking' and other processes described shortly. Naphtha is used in one of these processes ('reforming'). White spirit is employed as 'turpentine substitute' in oil-paints in place of the more expensive turpentine.

(*iii*) *Kerosene Fraction.* This is the source of kerosene ('paraffin oil') used as an illuminating agent in lamps and as a fuel in oil-stoves, tractors, rockets, and some kinds of marine engines.

(*iv*) *Gas Oil Fraction.* Gas oil is a medium-heavy oil consisting of a complex mixture of higher hydrocarbons. Much of it is subjected to 'cracking' to break it down into simpler hydrocarbons. Gas oil is used as a fuel in oil-furnaces and Diesel engines.

(*v*) *Lubricating Oil Fraction.* The raw lubricating oil is first 'dewaxed' by treating it with liquid propane under pressure, which dissolves the wax held in solution. Paraffin wax is then obtained by evaporation of the solution (the propane being used again). Paraffin wax has a wide variety of uses. It is made into candles, mixed with oils to give petroleum jelly, and 'cracked' to yield C_{15}—C_{16} hydrocarbons. From the latter detergents of the 'Teepol' type are manufactured. The dewaxed oil is also used as a lubricant for machinery and in making 'liquid paraffin.'

(*vi*) *The Residue.* The fraction which flows from the bottom of the fractionating tower is again distilled. This yields a distil-late of heavy fuel oil, and leaves a black tarry liquid, which solidifies on cooling. The black solid (asphalt, or bitumen) is used to surface roads.

Motor Spirit. The performance of the modern motor-car, measured in miles per gallon of petrol used, is some 60 per cent better than that of its predecessor of 1930. The most important factor in this improvement has been the higher quality of the fuel.

The chief way of obtaining more power from petrol is to increase the *compression ratio* of the engine (the compression ratio is the ratio between the volume of the mixture of fuel vapour and air in the cylinder before compression and the volume after compression). If the compression ratio is too high, however, the engine begins to 'knock.' 'Knocking' is a complex phenomenon, but it consists essentially of uneven burning of the fuel, as a result of which the piston moves jerkily in the cylinder.

The 'anti-knock' value of a fuel is expressed by means of its *octane number*, a quantity based on the widely different 'knocking' properties of the straight-chain alkane heptane (C_7H_{16}) and its branched-chain isomer, 2,2-dimethyl-4-methylpentane (iso-octane), $(CH_3)_3 C.CH_2.CH(CH_3)_2$. Heptane has a very pronounced 'knocking' tendency, while in the isomer the tendency is very small. A scale of 'anti-knock' values, or octane numbers, has therefore been devised in which heptane is given the value of zero and the isomer is rated at 100. The octane number of a particular spirit is then the percentage by volume of the isomer in a mixture of the two which 'knocks' to the same extent as the spirit being tested. The octane number of a given petrol is found with the aid of a special single-cylinder engine, in which the compression ratio can be varied.

The 'straight-run' petrol used in motor-cars of forty years ago had an octane number of about 50. Present-day petrol is sold in several grades with octane numbers ranging from 91–101. Straight-run spirit consists mostly of straight-chain, or only slightly branched, alkanes with low octane numbers. To improve the octane rating the straight-run spirit is blended with a large proportion of highly branched alkanes, cycloalkanes, and aromatic hydrocarbons. These are obtained by the 'cracking' and 'reforming processes described below.

Another method of increasing the octane number of motor spirit is to add about 2 per cent of tetraethyl-lead(IV), $Pb(C_2H_5)_4$. 'Leaded' petrol, however, has the drawback that it deposits lead(II) oxide in the cylinder when burned. This can be overcome by incorporating a small amount of 1,2-dibromoethane ($CH_2Br—CH_2Br$). The latter converts the oxide to bromide, which vaporizes and passes away with the exhaust gases.

'Cracking' of Oils. This consists of vaporizing heavy oils at about 500°C in the presence of a catalyst (magnesium(II) silicate). The large hydrocarbon molecules are broken down into simpler ones, and these recombine to some extent to give branched-chain alkanes, alkenes, and aromatic hydrocarbons; e.g.

$$C_{10}H_{22} \rightarrow C_8H_{18} + C_2H_4$$
decane   octane   ethene

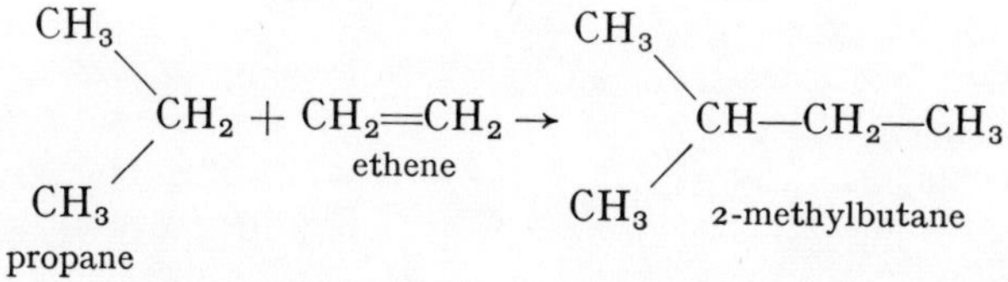

Many of the products of the 'cracking' process are suitable for blending with straight-run petrol, particularly valuable being the branched-chain alkanes with their high octane rating. The process not only provides the bulk of motor-spirit used at the present time, but also furnishes a variety of gases employed in chemical manufacture.

Catalytic 'Reforming'. This process differs from catalytic 'cracking' in using naphtha (C_6—C_{10} range) instead of heavy oil as the feedstock. Also, the chief reactions consist, not of breaking down large molecules into small ones, but of molecular rearrangements (hence the description 'reforming'). Cycloalkanes undergo dehydrogenation to aromatic hydrocarbons (*aromatization*), while straight-chain alkanes are converted either into isomeric branched-chain alkanes (*isomerization*) or aromatic hydrocarbons. In practice the naphtha vapour at a pressure of about 40 atmospheres is passed over a platinum catalyst at 500°C. Typical reactions are the following

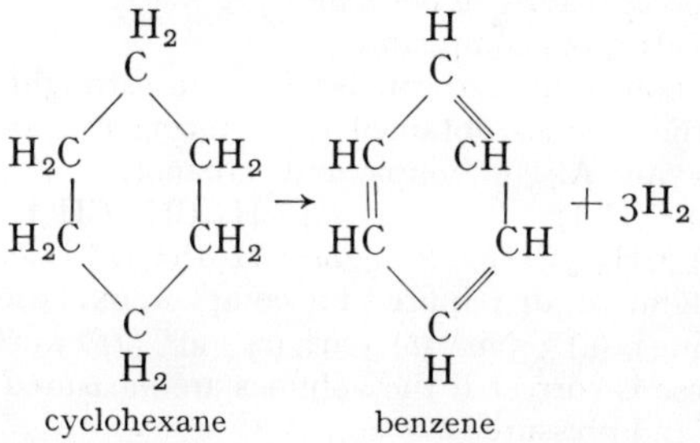

cyclohexane   benzene

$$CH_3.CH_2.CH_2.CH_2.CH_2.CH_3 \rightarrow CH_3.CH.CH_2.CH_2.CH_3$$
hexane   2-methylpentane

The above process not only supplies suitable ingredients for petrol, but is used to manufacture benzene and other aromatic hydrocarbons.

EXERCISE 5

1. Classify the following open-chain hydrocarbons as alkanes, alkenes, or alkynes: (*a*) C_3H_6; (*b*) C_5H_{12}; (*c*) C_2H_4; (*d*) C_3H_4; (*e*) C_8H_{18}.

2. Which of the following statements would you expect to be true of dodecane ($C_{12}H_{26}$)?

(*a*) It is a gas at ordinary temperatures;

(*b*) It floats on water;

(*c*) It dissolves readily in benzene;

(*d*) It is a saturated compound;

(*e*) The carbon atoms in the molecule are in a straight line in space.

3. Name the alkanes obtained by reducing the following alkyl halides with an Al/Hg couple and ethanol:

    (*a*) C_2H_5I;   (*b*) $CH_3.CH_2.CH_2I$,

    (*c*) $(CH_3)_2 CHI$;   (*d*) $CH_3(CH_2)_4 CH_2Br$;

4. The volume of air required for complete oxidation of 1 cubic metre of ethane is (*a*) 3·5 m³, (*b*) 5 m³, (*c*) 7 m³, (*d*) 14 m³, (*e*) 17·5 m³. Which of these is correct if the volumes are measured at the same temperature and pressure?

5. Which of the following statements are true of the C—H bond in alkanes?

(*a*) It is not appreciably polarized;

(*b*) It reacts with halogens by substitution;

(*c*) It reacts with halogens by an ionic mechanism;

(*d*) It reacts with bromine less readily than with chlorine.

6. Give the systematic names of the following:

    (*a*) $(CH_3)_2 CH.C_2H_5$;   (*b*) $(CH_3)_3 C.C_2H_5$;

    (*c*) $(C_2H_5)_2CH.CH(CH_3)_2$;   (*d*) $CH_3.CHCl.CHCl.C_2H_5$.

7. Give the structural formulæ of the following:

(*a*) 3-methylhexane; (*b*) 2,2-dimethylpentane; (*c*) 1,2-dichloropropane; (*d*) 1-chloro-3-methylbutane.

8. Which of the following are commonly present in petrol:

(*a*) benzene; (*b*) methanol; (*c*) tetraethyl-lead (IV); (*d*) bromethane, (*e*) dibromoethane?

9. Give the terms used to describe the following:

(*a*) A mixture of carbon monoxide and hydrogen used in manufacturing organic compounds;

(*b*) Breaking down of complex hydrocarbons into simpler ones by heat and a catalyst;

(*c*) Conversion of alkanes to aromatic hydrocarbons by heat and a platinum catalyst;

(*d*) Conversion of straight-chain alkanes to branched-chain isomers;

(*e*) Chemicals made from petroleum.

6 Alkenes

It was seen in Chapter 5 that the carbon atoms in alkanes are combined with the maximum number of hydrogen atoms when judged by the number of valency bonds available. In alkenes (olefins) the number of hydrogen atoms joined to the carbon atoms is less than the maximum number. All alkenes contain a *double bond* between two of the carbon atoms.

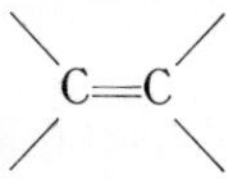

$$C=C$$

an alkene double bond

$$H_2C=CH_2$$

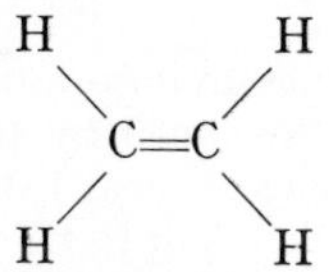

ethene

The alkenes form an homologous series of general formula C_nH_{2n}. The simplest member ($n = 1$) would be methene (methylene). This does not normally exist, although it occurs as a short-lived free radical ($\cdot CH_2 \cdot$) in some reactions, and it is present as a radical in compounds like dichloromethane (methylene dichloride), CH_2Cl_2. The systematic names of alkenes are derived by substituting the ending *-ene* for *-ane* in the names of the corresponding alkanes. Table 6-1 shows the names, formulæ and normal physical state of the lower alkenes.

Table 6-1. THE LOWER ALKENES
(General formula: C_nH_{2n})

Systematic name	Trivial name	Molecular formula	Structural formula	Normal physical state
Ethene	Ethylene	C_2H_4	$CH_2{=}CH_2$	Gas
Propene	Propylene	C_3H_6	$CH_3.CH{=}CH_2$	Gas
Butenes	Butylenes	C_4H_8	3 structural isomers	Gases

The lower alkenes are colourless flammable gases. Ethene boils at $-104°C$ at atmospheric pressure. As in other homologous series, boiling points rise with increase in relative molecular mass so that the pentenes (C_5H_{10}) and hexenes (C_6H_{12}) are liquids at ordinary temperature. Ethene has a pleasant sweet smell, which becomes slightly oily in higher members. The gaseous alkenes dissolve only slightly in water, but are moderately soluble in ethanol and ether.

Alkenes do not occur naturally because of their high reactivity. As seen in the last chapter, the lower members are produced in the 'cracking' of petroleum oils. Their manufacture from this source and their use in making petrochemicals are described shortly.

Laboratory Preparation of Alkenes. Alkenes are conveniently prepared by dehydration of the corresponding alcohols. When an alcohol is heated with a suitable catalyst one molecule of water is eliminated from each molecule of the alcohol, as illustrated below. The preparation of the alkene is carried out in two ways according to the nature of the catalyst used.

$$CH_3—CH_2OH \xrightarrow{-H_2O} CH_2{=}CH_2$$
$$\text{ethanol} \qquad\qquad \text{ethene}$$

(i) *With an Acid Catalyst*. In this method the alcohol is heated with an acid (usually concentrated sulphuric(VI) acid). The acid does not remove water directly from the alcohol, but supplies hydrogen ions, which are the real catalyst for the dehydration. The part played by the hydrogen ions is explained in connection with the properties of alcohols in Chapter 8.

Ethene is obtained when ethanol is heated with excess of concentrated sulphuric(VI) acid at about 180°C. For demonstration purposes the preparation is carried out as shown in Fig. 6-1.

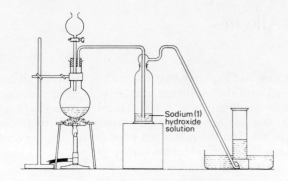

Fig. 6-1. Preparation of ethene from ethanol and concentrated sulphuric(VI) acid

Experiment

Set up the apparatus shown in Fig. 6-1. Into the 500-cm³ round-bottomed flask put 20 cm³ of industrial spirit. Add gradually with shaking 40 cm³ of concentrated sulphuric(VI) acid. To promote steady evolution of the gas and to reduce frothing add also one or two grams of clean sand. Heat the flask on a sand-tray until evolution of the gas becomes brisk, and then remove the flame. Begin to collect the gas when a sample, collected in a test-tube, burns quietly without explosion. The liquid in the flask darkens in colour, because carbon is liberated owing to oxidation of some of the alcohol by the concentrated acid (most organic compounds are reducing agents and 'charring' by concentrated sulphuric(VI) acid is common). Carbon

dioxide and sulphur dioxide are also formed, and these render the ethene impure. They are removed by passing the gas through a Dreschel bottle containing 10 per cent sodium(I) hydroxide solution. There is a tendency for the alkali to be sucked back into the flask owing to the rapidity with which it dissolves the sulphur dioxide. This tendency can be counteracted by momentarily opening the tap of the tap-funnel (which is included in the apparatus for this reason). If a dry sample of ethene is required it can be obtained by passing the gas through a U-tube containing anhydrous calcium(II) chloride and collecting the gas over mercury.

Alkenes can be obtained in a purer state by heating alcohols with syrupy phosphoric(V) acid. The latter method is often used in cases where concentrated sulphuric(VI) acid causes heavy charring. Thus, cyclohexene, a liquid cyclic alkene (b.p. 83°C), is usually prepared by distilling cyclohexanol with phosphoric(V) acid.

$$CH_2{-}CH_2$$
$$H_2C \quad\quad CHOH$$
$$CH_2{-}CH_2$$

cyclohexanol

$$\downarrow\ {-}H_2O$$

$$CH_2{-}CH$$
$$H_2C \quad\quad CH$$
$$CH_2{-}CH_2$$

cyclohexene

(*ii*) *With a Solid Catalyst.* In the second method the vapour of the alcohol is passed over aluminium(III) oxide or porous pot, which is heated to about 360°C. A convenient method of preparing ethene on a small scale is to use the 'powdered chalk' technique described below.

Experiment

Put about 1 cm³ of industrial spirit into a hard-glass test-tube, and add small amounts of powdered blackboard chalk until the liquid has been soaked up. Fill most of the remainder of the tube with a mixture of coarse and fine pieces of broken porous pot (or granular aluminium(III) oxide). Close the end of the tube with a rubber stopper and delivery-tube, and arrange to collect the gas evolved

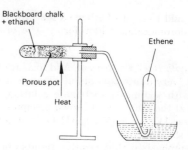

Fig. 6-2. 'Powdered chalk' method of preparing ethene

as shown in Fig. 6-2. Warm the middle of the tube by means of a Bunsen burner (preferably fitted with a 'spreader'), using a medium flame yellow at the tip. The alcohol is vaporized by heat conducted along the glass, and as the vapour passes through the catalyst it is dehydrated to ethene.

(*Note.* In the previous edition of this book the absorbent used for the alcohol was asbestos wool. The latter has been replaced by powdered blackboard chalk because of a possible health hazard in handling asbestos wool.)

Chemical Tests for Alkenes. (*i*) *Combustion.* Alkenes burn readily in air, forming carbon dioxide and water. The flames given by the lower members are more luminous than those of the lower alkanes owing to the higher proportion of carbon in the molecules. Mixtures of ethene with air or oxygen are highly explosive, but a test-tube full of the gas burns quietly.

(*ii*) *Decolorizing of Bromine.* Fill a gas-jar with bromine vapour and invert it over a jar of ethene. The brown colour of the bromine rapidly disappears, and oily streaks of 1,2-dibromoethane ($CH_2Br.CH_2Br$) are formed on the sides of the jars.

83

Alternatively, add to a jar of ethene a few drops of a solution of bromine in trichloromethane (chloroform). The brown solution is decolorized.

(iii) Reaction with Baeyer's Reagent. Add to a jar of ethene a little Baeyer's reagent (very dilute potassium(I) manganate(VII) solution made alkaline with sodium(I) carbonate solution). The liquid is decolorized, and at the same time a brown precipitate of a hydrated form of manganese(IV) oxide ($MnO_2 . xH_2O$) is obtained. This reaction is explained shortly.

It should be emphasized that positive results obtained with the above tests do not prove that a compound is an alkene. The reactions with bromine and Baeyer's reagent are tests for the carbon–carbon double bond, but this occurs in other types of compounds besides alkenes. The tests are also given by hydrocarbons, like ethyne (acetylene), $CH{\equiv}CH$, which contain a triple bond. Furthermore, positive results should be obtained with *both* bromine and Baeyer's reagent. Some saturated compounds such as phenylamine (aniline) rapidly decolorize bromine (in these cases hydrogen bromide is liberated), while others (e.g. trichloromethane) react with Baeyer's reagent.

Reactions of Alkenes

It will be seen shortly that the carbon–carbon double bond in alkenes is composed, not of two equally strong bonds, but of a stronger bond and a weaker bond. The latter makes alkenes extremely reactive. In particular, alkenes readily undergo addition reactions, in which their molecules add on complete molecules of other substances. Most of these reactions are of the type shown below. The weaker component of the double bond is broken, and different parts of the added molecule become linked to the two carbon atoms.

$$\begin{array}{c} X \quad Y \\ | \quad | \\ \diagdown C{=}C \diagup + XY \rightarrow -C{-}C{-} \\ \diagup \quad \diagdown \qquad\qquad | \quad | \end{array}$$

Because of their tendency to undergo addition reactions alkenes are described as *unsaturated* hydrocarbons. *An unsaturated compound is a compound which combines directly with other substances, forming in each case a single new compound.* In contrast, alkanes are saturated hydrocarbons, and their characteristic mode of reaction is by substitution.

1. Addition of Hydrogen. Alkenes combine directly with hydrogen, but only in the presence of a catalyst. Ethene and hydrogen yield ethane. Combination can be brought about by passing a mixture of the gases over finely divided platinum or palladium at ordinary temperatures or over finely divided nickel at 140°C.

$$CH_2{=}CH_2 + H_2 \rightarrow CH_3{-}CH_3$$

Direct combination of hydrogen with a substance is called *hydrogenation.* The catalytic effect of finely divided nickel in this type of reaction was discovered about 1900 by the French chemists Sabatier and Senderens. Nowadays the effect is utilized in the hydrogenation, or reduction, of a wide variety of compounds. The catalyst is prepared by reducing nickel(II) oxide in a stream of pure hydrogen. Reaction temperatures vary from 140°C to 300°C.

A more active form of nickel, known as *Raney nickel,* is prepared by treating an aluminium-nickel alloy with aqueous sodium(I) hydroxide. The aluminium dissolves, leaving the nickel in such a finely divided condition that it is pyrophoric (*i.e.* it sets on fire when exposed to the air). It is stored under water until required for use. Many hydrogenations which require heat when nickel from nickel(II) oxide is used as the catalyst can be carried out at room temperature with Raney nickel.

2. Addition of Halogens. The order of reactivity of halogens with alkenes is as follows:

$$F_2 > Cl_2 > Br_2 > I_2$$

Fluorine reacts explosively, forming chiefly tetrafluoromethane. With chlorine or bromine the halogen is added across the double bond, yielding 1,2-dichloroethane or 1,2-dibromoethane. The reaction is more rapid with chlorine than with bromine.

$$CH_2{=}CH_2 + Cl_2 \rightarrow CH_2Cl{-}CH_2Cl$$

Only slight combination occurs between ethene and iodine (dissolved in tetrachloromethane), the reaction being slow and incomplete.

When ethene is bubbled into *chlorine water*—which contains both free chlorine and chloric(I) acid (hypochlorous acid)—a mixture of 1,2-dichloroethane and 2-chloroethanol (ethylene chlorohydrin) is produced. The latter is not formed by direct addition of the acid (HOCl). Both chlorine and water are involved in the reaction, the mechanism of which is explained later in this chapter. The reaction can be represented simply by the following equation:

$$CH_2{=}CH_2 + H_2O + Cl_2 \rightarrow CH_2Cl{-}CH_2OH + HCl$$
$$\text{2-chloroethanol}$$

3. Addition of Hydrogen Halides. The order of reactivity of hydrogen halides with alkenes is

$$HI > HBr > HCl > HF$$

Hydrogen iodide and hydrogen bromide combine with ethene at ordinary temperatures, the reaction being more rapid with hydrogen iodide. Addition takes place similarly with hydrogen chloride, but only if a catalyst (anhydrous aluminium(III) chloride) is present. Hydrogen fluoride combines with alkenes only under pressure.

$$CH_2{=}CH_2 + HI \rightarrow CH_3{-}CH_2I$$
$$\text{iodoethane}$$

4. Hydration. Hydration is a reaction in which water is added chemically to a compound. Hydration of alkenes can be brought about indirectly by means of sulphuric(VI) acid, the overall reactions resulting in the formation of alcohols; e.g.

$$C_2H_4 + H_2O \rightarrow C_2H_5OH$$
$$\text{ethene} \qquad \text{ethanol}$$

This is clearly the converse of the reactions described earlier by which alkenes are prepared by dehydration of alcohols.

Fuming sulphuric(VI) acid (oleum) absorbs ethene at ordinary temperatures, but if concentrated sulphuric(VI) acid is used it is necessary to heat the acid to about 80°C. In both cases ethyl hydrogen sulphate(VI) is formed in solution. When the solution is diluted the ethyl hydrogen sulphate(VI) is largely hydrolysed to ethanol. The reaction is reversible, but since ethanol is the most volatile of the four compounds present it passes over most readily (together with some water) when the mixture is distilled. Ethanol and some of the other lower alcohols are manufactured from alkenes by this method.

$$C_2H_4 + H_2SO_4 \rightarrow C_2H_5HSO_4$$
ethyl hydrogen sulphate(VI)

$$C_2H_5HSO_4 + H_2O \rightleftharpoons C_2H_5OH + H_2SO_4$$
ethanol

5. Addition to Benzene. Under suitable conditions of temperature and pressure and in the presence of a catalyst (anhydrous aluminium(III) chloride) the lower alkenes combine with benzene to give homologues of the latter. Thus ethene and benzene yield ethylbenzene (Chapter 20). This is used in the manufacture of the plastic polystyrene.

$$C_6H_6 + C_2H_4 \rightarrow C_6H_5.C_2H_5$$
ethylbenzene

6. Oxidation. Alkenes can be oxidized in other ways besides combustion. Thus, if a mixture of ethene and oxygen is passed over a silver catalyst at 200°–300°C, epoxyethane (ethylene oxide) is produced. (The term 'epoxy-' is used when an oxygen atom is directly linked to two carbon atoms of a chain.) Epoxyethane is a colourless gas used in the manufacture of ethane-1,2-diol (ethylene glycol).

$$CH_2{=}CH_2 + \tfrac{1}{2}O_2 \rightarrow \overset{\displaystyle O}{\overset{\diagup \diagdown}{CH_2{-}CH_2}}$$
epoxyethane

As seen earlier, an alkaline solution of potassium(I) manganate(VII) (potassium permanganate) is decolorized by alkenes with simultaneous precipitation of brown hydrated manganese(IV) oxide. The reagent (Baeyer's reagent) oxidizes ethene to ethane-1,2-diol (ethylene glycol). It should be noted that in

$$\begin{array}{c} CH_2 \\ \| \\ CH_2 \end{array} + H_2O + O \rightarrow \begin{array}{c} CH_2OH \\ | \\ CH_2OH \end{array}$$
ethane-1,2-diol

testing organic compounds an alkaline solution of potassium(I) manganate(VII) is usually preferred to an acidic solution. The latter is such a strong oxidizing agent that it oxidizes many compounds completely to carbon dioxide and water.

Alkenes also combine directly with ozone to form *ozonides*. Ethene ozonide can be obtained by dissolving ethene in an inert solvent (e.g. tetrachloromethane) and passing ozonized oxygen into the solution, which is cooled in a freezing mixture of ice and salt.

$$CH_2{=}CH_2 + O_3 \rightarrow \begin{array}{c} O \\ \diagup \diagdown \\ CH_2 \quad CH_2 \\ | \qquad | \\ O{-}{-}O \end{array}$$
ethene ozonide

Ozonides are viscous liquids, which are unstable to heat and often explosive. In spite of this they are often used for establishing the position of a double bond in a carbon chain. The method, which is known as *ozonolysis*, is explained shortly by means of the isomeric butenes.

7. Polymerization. As stated in Chapter 4, polymerization is an addition reaction in which molecules of the same compound join together to form a single new compound. The lower alkenes polymerize when they are heated under pressure in the presence of a suitable catalyst. Ethene yields the well known plastic poly(ethene) (polyethylene, polythene) of formula $(C_2H_4)_n$. The manufacture of the polymer is described in Chapter 26. Poly(ethene) consists of long chain-like molecules containing over a thousand methylene ($-CH_2-$) units linked together as shown.

$$\cdots CH_2 \quad CH_2 \quad CH_2 \quad CH_2 \quad CH_2 \quad \cdots$$
$$CH_2 \quad CH_2 \quad CH_2 \quad CH_2 \quad CH_2$$

Evidence for the Existence of the Double Bond in Alkenes.

In describing the reactions associated with the alkene double bond its existence has been assumed in order to make the reactions clear. The evidence on which this assumption is based must now be examined.

If the carbon and hydrogen atoms have their usual valencies, there are three possible ways of representing the structure of the ethene molecule.

$$
\begin{array}{ccc}
\text{H} & \text{H} \;\; \text{H} & \text{H} \qquad \text{H} \\
| \quad \cdot & \cdot \;\; \cdot & \diagdown \qquad \diagup \\
\text{H--C--C--H} & \text{H--C--C--H} & \text{C=C} \\
| \quad \cdot & | \;\; | & \diagup \qquad \diagdown \\
\text{H} & \text{H} \;\; \text{H} & \text{H} \qquad \text{H} \\
\text{(A)} & \text{(B)} & \text{(C)}
\end{array}
$$

Formulæ A and B differ from C in representing free radicals, or molecules in which certain valency electrons of the carbon atoms are not paired with another electron. Now it has never

been found possible to prepare simple organic free radicals which can be kept under ordinary conditions. When free radicals such as methyl ($CH_3\cdot$) are produced in a reaction they either combine together or react with other molecules present. Thus, in the electrolysis of aqueous sodium acetate (Chapter 5) the free methyl radicals obtained at the anode combine in pairs to give ethane (C_2H_6). Since ethene does not polymerize spontaneously it is most unlikely that the molecule contains unpaired electrons. Thus formulæ A and B may be discarded, leaving formula C.

Structure C is supported by the fact that ethene can be obtained (although in poor yield) by electrolysis of an aqueous solution of potassium(I) butanedioate (potassium(I) succinate) $(CH_2.COOK)_2$. In the anion of this salt the two methylene groups are already joined. Formation of ethene at the anode therefore probably occurs in accordance with the following equation:

$$
\begin{array}{ccc}
CH_2.COO^- & & CH_2 \\
| & -2e \rightarrow & \| \quad + \; 2CO_2 \\
CH_2.COO^- & & CH_2
\end{array}
$$

A feature which distinguishes the C=C bond from the C—C bond is that free rotation cannot take place around the former. This results in certain alkenes existing in isomeric forms according to the relative positions in space of atoms or groups attached to the two carbon atoms. This type of isomerism, which is called *geometrical* isomerism, is described in Chapter 17.

The presence of the double bond in alkenes is also confirmed by physical evidence. Not only is the length of the C=C bond different from that of the C—C bond (Chapter 3) but, as will be seen shortly, the energies of formation of the two bonds also differ widely.

Table 6-2. SOME MANUFACTURES BASED ON ETHENE AND PROPENE

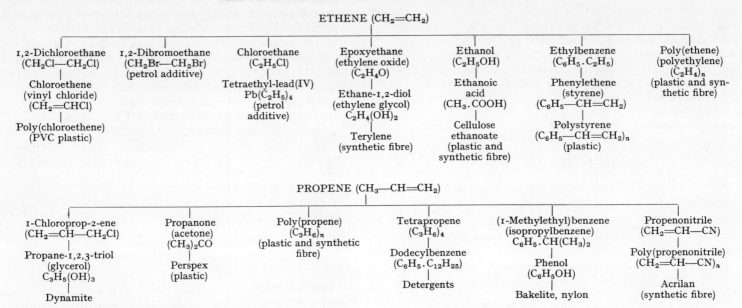

ETHENE (CH$_2$=CH$_2$)

1,2-Dichloroethane (CH$_2$Cl—CH$_2$Cl)	1,2-Dibromoethane (CH$_2$Br—CH$_2$Br) (petrol additive)	Chloroethane (C$_2$H$_5$Cl)	Epoxyethane (ethylene oxide) (C$_2$H$_4$O)	Ethanol (C$_2$H$_5$OH)	Ethylbenzene (C$_6$H$_5$.C$_2$H$_5$)	Poly(ethene) (polyethylene) (C$_2$H$_4$)$_n$ (plastic and synthetic fibre)
Chloroethene (vinyl chloride) (CH$_2$=CHCl)		Tetraethyl-lead(IV) Pb(C$_2$H$_5$)$_4$ (petrol additive)	Ethane-1,2-diol (ethylene glycol) C$_2$H$_4$(OH)$_2$	Ethanoic acid (CH$_3$.COOH)	Phenylethene (styrene) (C$_6$H$_5$—CH=CH$_2$)	
Poly(chloroethene) (PVC plastic)			Terylene (synthetic fibre)	Cellulose ethanoate (plastic and synthetic fibre)	Polystyrene (C$_6$H$_5$—CH=CH$_2$)$_n$ (plastic)	

PROPENE (CH$_3$—CH=CH$_2$)

1-Chloroprop-2-ene (CH$_2$=CH—CH$_2$Cl)	Propanone (acetone) (CH$_3$)$_2$CO	Poly(propene) (C$_3$H$_6$)$_n$ (plastic and synthetic fibre)	Tetrapropene (C$_3$H$_6$)$_4$	(1-Methylethyl)benzene (isopropylbenzene) C$_6$H$_5$.CH(CH$_3$)$_2$	Propenonitrile (CH$_2$=CH—CN)
Propane-1,2,3-triol (glycerol) C$_3$H$_5$(OH)$_3$	Perspex (plastic)		Dodecylbenzene (C$_6$H$_5$.C$_{12}$H$_{25}$)	Phenol (C$_6$H$_5$OH)	Poly(propenonitrile) (CH$_2$=CH—CN)$_n$
Dynamite			Detergents	Bakelite, nylon	Acrilan (synthetic fibre)

Manufacture of Alkenes. The lower alkenes are obtained on a large scale by several processes. One method (mentioned in Chapter 5) for ethene consists of pyrolysis of ethane. Ethene, propene, and the butenes are formed to some extent in the cracking of heavy petroleum oils. When alkenes are required as the main products (for chemical manufacture) the 'feedstock' used is a lighter alkane fraction, and cracking is carried out thermally (at 800°C) without a catalyst. Also, the vaporized oil is mixed with steam ('steam cracking'), which reduces the amount of carbon deposited. The process yields a mixture of alkanes and alkenes in the C$_1$—C$_4$ range, and these are separated by liquefaction under pressure, followed by fractional distillation.

Examples of petrochemicals made from ethene and propene are shown in Table 6-2. Some of these applications are explained in later chapters.

PROPENE, C_3H_6
(Propylene)

Propene can be prepared in the laboratory by catalysed dehydration of either propan-1-ol (normal propyl alcohol) or propan-2-ol (isopropyl alcohol). The methods used are those described earlier for ethene.

$$CH_3—CH_2—CH_2OH$$
propan-1-ol

$$CH_3$$
$$CHOH$$
$$CH_3$$
propan-2-ol

$$\longrightarrow CH_3—CH{=}CH_2 + H_2O$$
propene

Propene closely resembles ethene in its properties. Thus, it forms addition compounds with hydrogen, chlorine and bromine, hydrogen halides, concentrated sulphuric(VI) acid, etc. The equation for the addition reaction with chlorine is given below.

$$CH_3—CH{=}CH_2 + Cl_2 \rightarrow CH_3—CHCl—CH_2Cl$$
1,2-dichloropropane

At higher temperatures (500°–600°C) propene undergoes substitution with chlorine to give 1-chloroprop-2-ene (allyl chloride). The reaction, which is often described as the 'hot chlorination' of propene, is used in the manufacture of propane-1,2,3-triol (glycerol) (Chapter 8).

$$CH_3—CH{=}CH_2 + Cl_2 \rightarrow CH_2Cl—CH{=}CH_2 + HCl$$
1-chloroprop-2-ene

When hydrogen bromide or hydrogen iodide combines with propene the halogen atom unites with the middle carbon atom, which has one attached hydrogen atom, and not with the end carbon atom, which is joined to two hydrogen atoms. This is in accordance with *Markownikoff's rule*, which states that *when a hydrogen halide is added to an unsymmetrical alkene the halogen atom becomes attached to the carbon atom with the smaller number of hydrogen atoms*. The reason for this behaviour is explained shortly.

$$CH_3—CH{=}CH_2 + HBr \rightarrow CH_3—CH—CH_3$$
$$|$$
$$Br$$
2-bromopropane
(isopropyl bromide)

If propene is absorbed in warm concentrated sulphuric(VI) acid, the addition of the acid to the alkene also follows the Markownikoff pattern. In this case the —HSO_4 group is added to the carbon atom with the smaller number of hydrogen atoms, and 1-methylethyl hydrogen sulphate(VI) (isopropyl hydrogen sulphate) is formed in solution. When the solution is diluted with water and distilled an alcohol, propan-2-ol (isopropyl alcohol) is obtained. (Compare the corresponding reactions with ethene).

$$CH_3—CH{=}CH_2 + H_2SO_4 \rightarrow CH_3—CH—CH_3$$
$$|$$
$$HSO_4$$
1-methylethyl
hydrogen sulphate(VI)

$$CH_3—CH—CH_3 + H_2O \rightleftharpoons CH_3—CH—CH_3 + H_2SO_4$$
$$|\qquad\qquad\qquad\qquad\qquad |$$
$$HSO_4\qquad\qquad\qquad\qquad OH$$
propan-2-ol

89

Like ethene, propene can be polymerized with the aid of catalysts. The extent of polymerization depends on the nature of the catalyst and the conditions used. One process yields tetrapropene $(C_3H_6)_4$, from which detergents are made (Chapter 13). Another process produces poly(propene) (polypropylene) $(C_3H_6)_n$, a polymer of very high relative molecular mass. This is used both as a plastic and as a synthetic fibre (Chapter 26).

Propene is also the starting-point for obtaining propenonitrile (vinyl cyanide). A mixture of propene, ammonia, air, and steam is passed over a heated catalyst. The resulting propenonitrile is

$$CH_2{=}CH{-}CH_3 + NH_3 + 1\tfrac{1}{2}O_2 \rightarrow CH_2{=}CH{-}CN + 3H_2O$$

propenonitrile

polymerized to poly(propenonitrile) $(CH_2{=}CH{-}CN)_n$, which is the basis of 'Acrilan' fibres (Chapter 26).

BUTENES, C_4H_8
(Butylenes)

The butenes occur in three forms which are structurally different. These can be regarded as derived from propene $(CH_3{-}CH{=}CH_2)$ by substituting a hydrogen atom by a $CH_3{-}$ group in the three possible ways. The systematic names and structural formulæ of the butenes are given below.

$$CH_3{-}CH_2{-}CH{=}CH_2$$
But-1-ene

$$CH_3{-}CH{=}CH{-}CH_3$$
But-2-ene

$$CH_3{-}\underset{\underset{CH_3}{|}}{C}{=}CH_2$$
2-methylpropene

The ending -ene in the systematic names shows that the molecules contain an alkene double bond. The position of the latter is indicated (where necessary) by numbering the bonds in the carbon chain. Numbering starts from that end of the chain which gives the lower number for the position of the double bond. For branched-chain alkenes the name is based on that of the alkene with the longest straight chain of carbon atoms containing the double bond. Thus the third isomer is regarded as a methyl derivative of propene. In this case it is not necessary to specify the position of the double bond because the latter can occupy only the 1-position. The position of the substituting $CH_3{-}$ group is shown by numbering the carbon atoms of the chain, starting from that end of the chain already used for numbering the bonds.

The position of the double bond in the isomers of butene or in a higher alkene can be determined by means of the ozonides (ozonolysis). Formation of the ozonide (using an inert solvent, as described earlier) is attended by destruction of the double bond. When the resulting solution is reduced with hydrogen and a platinum catalyst the ozonide molecule undergoes fission at the position formerly occupied by the double bond. The products are aldehydes, ketones or mixtures of the two, depending on the nature of the original alkene. Identification of the products shows the location of the double bond in the parent alkene. Thus, in the case of the ozonide of but-1-ene two aldehydes are formed. These are propanal (propionaldehyde) and methanal (formaldehyde).

$$CH_3.CH_2.CH \underset{O \text{------} O}{\overset{O}{\diagup \diagdown}} CH_2 \xrightarrow[(Pt)]{H_2} CH_3.CH_2.CHO + CH_2O + H_2O$$

propanal   methanal

Reduction of the ozonide of but-2-ene yields ethanal as the only organic product, while similar treatment of the ozonide of 2-methylpropene gives a mixture of propanone (acetone) and methanal (formaldehyde).

$$CH_3.CH=CH.CH_3 \rightarrow 2CH_3.CHO$$
ethanal

$$(CH_3)_2C=CH_2 \rightarrow (CH_3)_2CO + CH_2O$$
propanone   methanal

The butenes occur in the gases obtained by cracking petroleum oils. After separation they are used to manufacture butanols (butyl alcohols), polybutenes (ingredients of lubricating oils), and buta-1,3-diene. The latter is a very important compound. It is made by catalytic dehydrogenation of but-1-ene and but-2-ene, and is used in the production of synthetic rubber (Chapter 26).

$$CH_3—CH_2—CH=CH_2 \xrightarrow{-H_2} CH_2=CH—CH=CH_2$$
but-1-ene   buta-1,3-diene

ALKENES AND ELECTRONIC THEORY

Electronic Structure of the Alkene Double Bond

Electron diffraction experiments show that the ethene molecule is flat and that the angles between the bonds are 120°. This is in accordance with the general rules (given in Chapter 4) that when

The flat ethene molecule

a carbon atom is joined to only three other atoms all four atoms lie in the same plane and the bonds tend to be as widely separated as possible.

The tetravalency of carbon in CH_4, C_2H_6, etc., was explained by hybridization of the one s and the three p atomic orbitals of the 'excited' carbon atom, giving four sp^3 hybrid orbitals with a tetrahedral distribution. When a carbon atom is attached to only three other atoms hybridization of the atomic orbitals takes place differently. The s orbital combines with only two of the p orbitals, so that the carbon atom has three sp^2 hybrid orbitals and one unchanged p orbital. The axes of the three hybrid orbitals are in the same plane and form angles of 120° with each other.

In ethene two of the hybrid orbitals are used in bonding with two hydrogen atoms, and the third in establishing a single bond with the other carbon atom. The unchanged p orbital has the form of two spheres, one above and one below the plane of the

Fig. 6-3. (a) Formation of a π bond by overlapping of p orbitals; (b) Shape of resulting molecular orbital

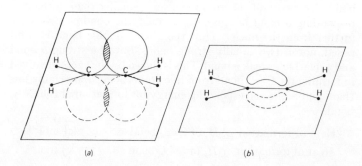

(a)   (b)

molecule (Fig. 6-3*a*). The second bond between the carbon atoms is formed by overlapping of the *p* orbital of one carbon atom with that of the other. The resulting molecular orbital takes the form of two banana-shaped figures, one being above, and one below, the plane of the six atoms (Fig. 6-3*b*).

In the hydrogen molecule (see Fig. 4-4) and the methane molecule (see Fig. 4-6) the H—H bond and the C—H bond are formed by overlapping of atomic orbitals in a direct line between the atomic nuclei. This is described as *collinear* overlapping of the orbitals, and is the usual way in which a normal covalent bond is produced. The resulting bond is called a σ *(sigma) bond*.

In ethane (CH_3—CH_3) the six C—H bonds and the C—C bond are all σ bonds. In ethene (CH_2=CH_2) the four C—H bonds are again σ bonds, but only one of the bonds forming the double bond is of this type. The other bond is produced by overlapping of electron orbitals *outside* the direct line of the atomic nuclei. This is known as *collateral* overlapping, and the resulting bond is called a π *bond*. In this kind of bond the attraction between the bond electrons and the atomic nuclei is smaller than in the corresponding σ bond because, on average, the bond electrons are further from the nuclei. Thus the alkene double bond is composed, not of two equally strong bonds, but of a strong σ bond and a relatively weak π bond. This is shown by the energy evolved in forming the C=C bond from free atoms being appreciably less than double the corresponding energy for formation of the C—C bond.

Heat of formation, $\triangle H$, of C=C bond = -607 kJ mol^{-1}

Heat of formation, $\triangle H$, of C—C bond = -335 kJ mol^{-1}

In addition reactions of alkenes it is the weaker π bond, and not the σ bond, which undergoes fission, making two electrons available for further combination.

Electrophilic Addition to Alkenes

Addition to alkenes normally takes place by an ionic mechanism. The substances added (halogens, hydrogen halides, sulphuric(VI) acid, etc.) are electrophilic reagents, that is, substances which make their attack in the first place at a centre of high electron density.

1. Addition of Hydrogen Halides. It has already been shown that the covalent bond in a hydrogen halide is polarized so that a partial positive charge is present on the hydrogen atom and a partial negative charge on the halogen atom. Since carbon and hydrogen are about equally electronegative, there is no appreciable polarization of the bonds in an alkene molecule. When, however, a molecule like HBr, which contains an electric dipole, approaches closely to an alkene molecule it can induce a dipole in the latter. The two electrons of the π bond are particularly susceptible to this effect. They are less strongly held by the carbon nuclei than those in the σ bond, and are therefore more mobile. Thus, when a hydrogen bromide molecule and an ethene molecule come close together both molecules possess electric dipoles, and, providing their combined energies are sufficient, addition occurs on collision.

Experiments show that the addition takes place in two steps. In the first step the hydrogen atom of the HBr molecule is added (as a proton) to the carbon atom with the partial negative charge, addition taking place by means of the two π electrons. As the other carbon atom loses its half-share in these two electrons it acquires a full positive charge, and becomes a carbonium ion. The addition of the proton is a relatively slow reaction. In the second step a negative bromide ion is added to the carbonium ion, this being a very rapid reaction.

$$(1) \quad \overset{\delta+}{H_2C}=\overset{\delta-}{CH_2} + \overset{\delta+}{H}-\overset{\delta-}{Br} \xrightarrow[\text{(slow)}]{} H_2\overset{+}{C}-CH_3 + Br^-$$

$$(2) \quad H_2\overset{+}{C}-CH_3 + Br^- \xrightarrow[\text{(fast)}]{} \underset{\underset{Br}{|}}{H_2C}-CH_3$$

bromoethane

The manner in which hydrogen bromide is added to propene is an illustration of Markownikoff's rule. The halogen atom becomes attached to the carbon atom with the smaller number of hydrogen atoms. This is readily understood from the two steps of the reaction mechanism.

$$(1) \quad CH_3-CH=CH_2 + H-Br \rightarrow CH_3-\overset{+}{CH}-CH_3 + Br^-$$

$$(2) \quad CH_3-CH-CH_3 + Br^- \rightarrow CH_3-CHBr-CH_3$$

2-bromopropane

As already noted, the order of reactivity of hydrogen halides with alkenes is HI > HBr > HCl > HF. Although this order depends on several factors, one of the most important is the energy required to stretch, and finally break, the bond between the hydrogen atom and the halogen atom. The amount of energy needed is least for hydrogen iodide and greatest for hydrogen fluoride.

2. **Addition of Chlorine and Bromine.** At first sight there is no indication that the reactions depend on ionic mechanisms. It is

$$CH_2=CH_2 + Br_2 \rightarrow CH_2Br-CH_2Br$$

found, however, that pure ethene and pure bromine do not combine appreciably unless some material with ionic character is present to act as a catalyst. This may be a trace of moisture or the glass walls of the reaction vessel. Thus it has been shown that if the glass is coated with paraffin wax the rate of combination is only one-seventeenth of the rate for uncoated glass.

It may be expected that in the presence of a polar catalyst partial charges will be produced in both the bromine and ethene molecules. Addition of bromine to ethene can therefore be represented by two steps, which are essentially similar to those for addition of hydrogen halides. In the first step a positive bromonium ion (Br^+) is added, and in the second a negative bromide ion (Br^-).

$$(1) \quad \overset{\delta+}{H_2C}=\overset{\delta-}{CH_2} + \overset{\delta+}{Br}-\overset{\delta-}{Br} \xrightarrow[\text{(slow)}]{} H_2\overset{+}{C}-CH_2-Br + Br^-$$

$$(2) \quad H_2\overset{+}{C}-CH_2-Br + Br^- \xrightarrow[\text{(fast)}]{} \underset{\underset{Br}{|}}{H_2C}-CH_2-Br$$

Experimental evidence of the two-step addition is provided by the reaction of ethene with bromine water containing a dissolved salt. With bromine water alone ethene yields a mixture of 1,2-dibromoethane, $CH_2Br.CH_2Br$, and 2-bromoethanol (ethylene bromohydrin), $CH_2Br.CH_2OH$. If sodium(I) chloride, which gives chloride ions, is also present, the products now include, besides those mentioned, a 'mixed' compound of formula $CH_2Cl.CH_2Br$. Similarly, if sodium(I) nitrate(V) is added, a compound of formula $CH_2NO_3.CH_2Br$ is obtained. Clearly the formation of these 'mixed' compounds involves addition of a Cl^- or NO_3^- ion. It is therefore probable that, at some stage in the formation of 1,2-dibromoethane, bromine is added as a Br^- ion. The production of 2-bromoethanol is explained by addition of the hydroxyl group as OH^-, following attack on the intermediate carbonium ion by a water molecule.

$$\overset{+}{H_2C}-CH_2Br + HOH \rightarrow H_2\overset{\overset{\displaystyle OH}{\displaystyle |}}{C}-CH_2Br + H^+$$

EXERCISE 6

1. In the preparation of ethene from ethanol and sulphuric(VI) acid state: (a) the concentration of the acid, (b) whether the alcohol or the acid is used in excess, (c) the approximate temperature of the reaction, (d) the names of two gaseous impurities in the product, (e) the reagent used to remove these impurities.

2. An alkene is obtained when the vapour of propan-1-ol $(CH_3.CH_2.CH_2OH)$ is passed over a catalyst at a suitable temperature. State (a) the name of the catalyst, (b) the approximate temperature, (c) the alkene produced, (d) the equation for the reaction, (e) the term which describes this type of reaction.

3. Name the organic compounds formed when ethene reacts with (a) bromine (at ordinary temperatures), (b) bromine water, (c) hydrogen bromide, (d) concentrated sulphuric(VI) acid, (e) Baeyer's reagent.

4. (a) Give the systematic names of the following:

(1) $CH_3-CH{=}CH_2$ (2) $CH_3-CH{=}CH-CH_3$

(3) $\underset{\underset{\displaystyle CH_3}{\displaystyle |}}{C_2H_5-C}{=}CH_2$

(b) On reduction with hydrogen and a platinum catalyst the ozonide of an alkene C_5H_{10} produced propanone (acetone), $(CH_3)_2CO$, ethanal (acetaldehyde), $CH_3.CHO$, and water. Give the structural formula and systematic name of the alkene.

5. Give the structural formulae of the following: (a) methylpropene, (b) hex-3-ene, (c) 3-methylpent-1-ene.

6. Give the terms used to describe the following: (a) addition of gaseous hydrogen to an organic compound, (b) addition of ethene molecules to each other, (c) organic molecules which can add on other molecules, (d) the type of bond formed by collateral overlapping of atomic orbitals, (e) the type of bond formed by collinear overlapping of atomic orbitals.

7. Supply the following information about the addition reaction between ethene and bromine: (a) Show the partial charges present just before combination on the two atoms of the bromine molecule and on the two carbon atoms; (b) Give the name and formula of the particle added to the ethene molecule in the first step of the reaction mechanism; (c) Write the equation for the second step in the reaction mechanism; (d) Why is bromine regarded as an electrophilic reagent in this reaction? (e) Give the formulæ of the three organic products formed if the reaction is carried out with bromine water containing dissolved sodium(I) chloride.

7 Alkynes

The alkynes (acetylenes) form an homologous series of general formula C_nH_{2n-2}, in which the functional group is $-C\equiv C-$. The systematic names of individual alkynes are derived by substituting the ending -*yne* for -*ane* in the names of the corresponding alkanes. Thus the first two members of the series are ethyne (C_2H_2, or $CH\equiv CH$) and propyne (C_3H_4 or $CH_3-C\equiv CH$). The trivial names for these are acetylene and methylacetylene respectively. The third member, butyne (dimethylacetylene), occurs in two isomeric forms, which are distinguished by numbering the position of the triple bond, as shown.

$$CH_3-CH_2-C\equiv CH \qquad CH_3-C\equiv C-CH_3$$
but-1-yne   but-2-yne

The first three members of the alkyne series are gases. Higher members are liquids, and eventually solids are formed. By far the most important member of the series is the first one, and this is the only one which will be considered in detail.

ETHYNE, C_2H_2
(Acetylene)

Laboratory Preparation. The first of the two methods now given for the preparation of ethyne is special to this hydrocarbon. The second is of general application.

(*i*) *From Calcium(II) Dicarbide* (CaC_2). This compound is a greyish-white solid containing the dicarbide ion C_2^{2-}. It is made

$$CaO + 3C \rightarrow CaC_2 + CO$$

by heating a mixture of calcium(II) oxide and coke in an electric furnace. When water is dropped on to calcium(II) dicarbide ethyne is given off and calcium(II) hydroxide remains.

$$CaC_2 + 2H_2O \rightarrow C_2H_2 + Ca(OH)_2$$

The gas obtained in this way is not pure, because calcium(II) dicarbide contains impurities which also react with water. The impurities are chiefly the sulphide, phosphide, and nitride of calcium(II), and these with water yield hydrogen sulphide, phosphine(PH_3), and ammonia. These gases can be removed by bubbling the ethyne through an aqueous solution of copper(II) sulphate (VI)-5-water, as shown in Fig. 7-1.

(*ii*) *From 1,2-Dibromoethane.* Ethyne of high purity can be prepared by dropping 1,2-dibromoethane (ethylene dibromide) into a boiling *ethanolic* solution of potassium(I) hydroxide. The

Fig. 7-1. Preparation of ethyne from calcium(II) dicarbide and water

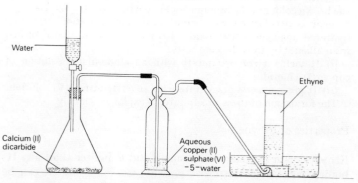

95

reaction involves elimination of two molecules of hydrogen bromide from each molecule of the halogen compound.

$$CH_2Br—CH_2Br + 2KOH \rightarrow CH{\equiv}CH + 2KBr + 2H_2O$$

An alternative method is to pass the vapour of 1,2-dibromo-ethane over heated soda-lime. This method is suitable for small-scale preparation of ethyne by the 'powdered chalk' procedure described earlier (see Fig. 6-2).

Chemical Tests for Ethyne. Mixtures of ethyne with air or oxygen are dangerously explosive. A test-tubeful of the gas, however, burns quietly with a luminous smoky flame, forming carbon dioxide and water. The smoky character of the flame is the result of incomplete combustion due to the high percentage (92·3) of carbon in the gas.

In testing for alkynes use is made of the two reagents used in the case of alkenes, namely, a solution of bromine in trichloromethane and an alkaline solution of potassium(I) manganate(VII) (Baeyer's reagent). There is an immediate reaction with the first, although the brown colour of the bromine is only partially destroyed. Baeyer's reagent is decolorized with simultaneous precipitation of brown hydrated manganese(IV) oxide. Ethyne is readily distinguished from alkenes by the following tests:

(*i*) It yields a *red* precipitate with an ammoniacal solution of copper(I) chloride;

(*ii*) It gives a *white* precipitate with silver(I) nitrate(V) solution. The formation of these precipitates is explained shortly.

Properties of Ethyne

Ethyne has a peculiar sweet smell, and is lighter than air. Its boiling point at atmospheric pressure is −84°C, but it can be liquefied at 0°C under a pressure of 26 atmospheres. The liquid is unstable and liable to decompose explosively into carbon and hydrogen when subjected to a mechanical shock. The instability is explained by the highly endothermic character of the hydrocarbon. 212 kilojoule are *absorbed* in the formation of one mole (26 g) of ethyne from carbon and hydrogen, and hence a large amount of energy is stored in its molecules.

Ethyne is slightly soluble in water, but is much more soluble in propanone (acetone), which dissolves 25 times its own volume of ethyne at 15°C. In welding the gas is obtained from steel cylinders, where it is stored in propanone under pressure. To reduce the risk of explosions the cylinders are packed with a porous material (such as asbestos) to absorb the solution.

Reactions of Ethyne

The previous chapter showed that the presence of the C≡C bond in alkenes confers high reactivity on these hydrocarbons. This is true also of the C≡C bond in alkynes. Thus the latter, like alkenes, are unsaturated and readily form addition compounds with electrophilic reagents. These reactions usually take place in two stages, the bond between the carbon atoms changing first into a double bond, and then into a single bond.

While most of the reactions of ethyne are due directly to the triple bond, one type of reaction is associated with the C—H bonds. The hydrogen atoms of these bonds possess a certain amount of acidic character, and undergo replacement by metal atoms with the formation of metal dicarbides (acetylides). This type of reaction is not found in alkenes.

1. **Action of Heat.** (*i*) *Polymerization to Benzene.* When ethyne is heated to about 400°C it polymerizes to the aromatic

hydrocarbon benzene.

$$3C_2H_2(g) \rightarrow C_6H_6(g)$$
$$\text{3 vols.} \qquad \text{1 vol.}$$

The polymerization is accompanied by a decrease in volume, three volumes of ethyne being converted into one volume of benzene vapour, which condenses to a liquid at ordinary temperatures.

Experiment

Collect about 4 cm³ of ethyne over water in a test-tube inverted in a large evaporating-dish, and mark the level of the water. Heat the ethyne as shown in Fig. 7-2, using a medium-size non-luminous flame. Continue the heating for about five minutes, and then allow the tube to cool to room temperature. There is a marked decrease in the volume of the remaining gas.

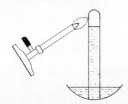

Fig. 7-2. Polymerization of ethyne

(Ethyne is slightly soluble in water, but, by using mercury instead of water, it can be shown that the contraction in volume in this experiment is not due to the ethyne dissolving.)

(*ii*) *Pyrolysis of Ethyne.* Because of the energy stored in its molecules ethyne is more easily decomposed by heat than ethane or ethene. Decomposition occurs at about 500°C, carbon and hydrogen being liberated.

$$C_2H_2 \rightarrow 2C + H_2$$

Experiment

Repeat the experiment described in (*i*), but this time heat the acetylene in the test-tube more strongly. There is a sudden flash of light, due to liberated energy, and soot is deposited in the tube.

2. Addition of Hydrogen. Ethyne combines directly with hydrogen when a mixture of the gases is passed over platinum black at ordinary temperatures or over finely divided nickel at 140°C (Sabatier and Senderens). Ethene and ethane are produced in turn.

$$CH \equiv CH + H_2 \rightarrow CH_2 = CH_2$$
$$CH_2 = CH_2 + H_2 \rightarrow CH_3 - CH_3$$

3. Reactions with Halogens. As in the case of alkenes, the order of reactivity of halogens with alkynes is: $F_2 > Cl_2 > Br_2 > I_2$ Fluorine again reacts explosively. An explosion also occurs if chlorine is mixed with ethyne under ordinary conditions. In this reaction carbon is liberated.

$$C_2H_2 + Cl_2 \rightarrow 2C + 2HCl$$

The reaction can be safely demonstrated as now described.

Experiment

Clamp a large boiling-tube in a stand. Put 1–2 g of bleaching-powder into the tube and add a little dilute hydrochloric acid. Wait until the tube is full of chlorine, and then drop one or two fragments of calcium(II) dicarbide into the tube. As the ethyne rises into the chlorine it explodes with a flash of light and a cloud of soot is liberated.

When separate streams of ethyne and chlorine are passed into a suitable solvent the two gases combine, forming firstly 1,2-dichloroethene (acetylene dichloride) and then 1,1,2,2-tetra-chloroethane (acetylene tetrachloride). These are oily liquids.

$$CH{\equiv}CH + Cl_2 \rightarrow CHCl{=}CHCl$$
1,2-dichloroethene

$$CHCl{=}CHCl + Cl_2 \rightarrow CHCl_2{-}CHCl_2$$
1,1,2,2-tetrachloroethane

The solvent employed in the manufacture of the final addition product is 1,1,2,2-tetrachloroethane itself, which increases in quantity as a result of the reaction. A catalyst (iron or antimony) is used to increase the rate of combination.

The reaction of ethyne with bromine (used in the form of vapour or as a solution in trichloromethane) follows a similar course. 1,2-Dibromoethene and 1,1,2,2-tetrabromoethane are produced in turn. The second stage of the addition, however, takes place only slowly. Ethyne and iodine scarcely react together, although a little 1,2-di-iodoethene is formed when ethyne is passed into an ethanolic solution of iodine.

4. Addition of Hydrogen Halides. The reactivities of hydrogen halides with alkynes (as with alkenes) are in the order: $HI > HBr > HCl > HF$. Addition of hydrogen iodide and hydrogen bromide occurs at ordinary temperatures, and again there are two stages, as represented below. It will be seen that the second stage of the addition of a hydrogen halide conforms to the Markownikoff pattern (Chapter 6), the halogen atom becoming attached to the carbon atom with the smaller number of hydrogen atoms.

$$CH{\equiv}CH + HBr \rightarrow CH_2{=}CHBr$$
bromoethene

$$CH_2{=}CHBr + HBr \rightarrow CH_3{-}CHBr_2$$
1,1-dibromoethane

Addition of hydrogen chloride to ethyne takes place very slowly under normal conditions, but is rapid at 150°C in the presence of mercury(II) chloride as a catalyst. The chief product is chloroethene (vinyl chloride), $CH_2{=}CHCl$. This is a colourless gas.

5. Hydration of Ethyne. A molecule of water can be added indirectly to the ethyne molecule by absorbing the gas in cold concentrated sulphuric(VI) acid, diluting the solution with water, and distilling. The distillate obtained is ethanal (acetaldehyde). The overall reaction is represented by the following equation:

$$C_2H_2 + H_2O \rightarrow C_2H_4O$$
ethanal

As in the hydration of ethene (Chapter 6) the overall reaction depends on preliminary addition of the acid to the hydrocarbon, but in this case two molecules of acid are added to one molecule of ethyne. The course of the reaction can be represented as follows:

$$CH \equiv CH \xrightarrow{H_2SO_4} CH_2 = CH(HSO_4) \xrightarrow{H_2SO_4} CH_3 - CH(HSO_4)_2$$

$$CH_3 - CH(HSO_4)_2 + H_2O \rightarrow CH_3 - CHO + 2H_2SO_4$$

6. Oxidation. As seen earlier, ethyne decolorizes Baeyer's reagent, giving a brown precipitate of hydrated manganese(IV) oxide. This is a complex redox reaction, the hydrocarbon being oxidized in stages to ethanedioic acid (oxalic acid).

$$\begin{matrix} CH \\ \| \\ CH \end{matrix} + 4O \rightarrow \begin{matrix} COOH \\ | \\ COOH \end{matrix}$$

7. Reactions Showing the Acidic Character of Ethyne. A hydrogen atom attached to a carbon atom usually has no acidic properties. There are exceptions, however, one being the hydrogen atom in the hydrogen cyanide ('prussic acid') molecule HCN. In this case a partial positive charge is produced on the hydrogen atom by relaying of the inductive $(-I)$ effect of the nitrogen atom to the H—C bond. The hydrogen atoms in ethyne also have a very weak acidic character, but the reason for this is uncertain.

Although ethyne does not yield hydrogen ions with water, its hydrogen atoms can be replaced by metal atoms with the formation of metal dicarbides (acetylides). Thus, if the gas is passed over heated sodium, ethyne is converted successively into monosodium(I) dicarbide (C_2HNa) and disodium(I) dicarbide (C_2Na_2), hydrogen being evolved at the same time.

Other metal dicarbides can be obtained by precipitation. For example, when ethyne is passed into an ammoniacal solution of copper(I) chloride a dark-red precipitate of *copper*(I) *dicarbide* is produced. With aqueous silver(I) nitrate(V) a white precipitate of *silver*(I) *dicarbide* is obtained. The dicarbides (acetylides) of copper(I) and silver(I) are unstable compounds, and are explosive in the dry state.

$$Cu_2Cl_2 + C_2H_2 \rightarrow Cu_2C_2 + 2HCl$$

$$2AgNO_3 + C_2H_2 \rightarrow Ag_2C_2 + 2HNO_3$$

Experiment

Bubble ethyne into each of two test-tubes, one containing an ammoniacal solution of copper(I) chloride and the other a solution of silver(I) nitrate(V). Filter the precipitates obtained. Open out the filter-papers containing the precipitates, and place them on a gauze over a tripod. Warm the filter-papers over a small flame. As the precipitates dry out there is a series of sharp, but harmless, explosions.

Manufacture and Uses of Ethyne. The chief method of obtaining ethyne on a large scale is still by treating calcium(II) dicarbide with water, but in recent years more economical processes have been developed. One of these consists of pyrolysis of methane derived from natural gas or from refinery processes such as 'cracking.' The methane is heated to a temperature of about 1500°C.

$$2CH_4 \rightarrow C_2H_2 + 3H_2$$

To avoid decomposition of the methane into carbon and hydrogen it is exposed to the high temperature for only a fraction of a second. The products also have to be rapidly cooled to prevent decomposition of the ethyne. The latter is separated from the hydrogen by dissolving it either in water under pressure or in an organic solvent.

Another process uses the light petroleum fraction naphtha as the starting point. When the naphtha is cracked under suitable conditions it yields both ethyne and ethene.

Ethyne (acetylene) is employed as a fuel in oxy-acetylene burners used for cutting and welding metals. Its main application, however, is in chemical manufacture, as now described.

(*i*) *Manufacture of Solvents.* A number of industrial solvents are made from ethyne with the help of chlorine. Because these are non-flammable they have largely replaced older industrial solvents such as benzene. Two of the best known are trichloroethene ($CHCl{=}CCl_2$) and tetrachloroethene ($CCl_2{=}CCl_2$). Both of these are produced from 1,1,2,2-tetrachloroethane, which is made from ethyne as described earlier. The first is used for extracting oil from seeds and for degreasing metal articles before they are electroplated. The second is chiefly employed in dry-cleaning.

(*ii*) *Manufacture of High Polymers.* Ethyne undergoes several addition reactions of the type shown below, in which addition of HX molecules under controlled conditions proceeds only to the half-way stage.

$$CH{\equiv}CH + HX \rightarrow CH_2{=}CHX$$

Typical addition products are chloroethene ($CH_2{=}CHCl$), propenonitrile ($CH_2{=}CH{-}CN$) and ethenyl ethanoate, or acetate ($CH_2{=}CH{-}OOC.CH_3$). These compounds can be obtained by passing a mixture of ethyne and hydrogen chloride, hydrogen cyanide, or ethanoic acid vapour over a heated catalyst. They all contain the ethenyl (vinyl) group $CH_2{=}CH{-}$, and under suitable conditions give rise to 'high polymers' (that is polymers of very high relative molecular mass). These are used as plastics or synthetic fibres, and are described in Chapter 26. A well known plastic of this group is poly(chloroethene), which is commonly called polyvinyl chloride, or PVC.

$$(-CH_2{-}CHCl{-})_n$$
poly(chloroethene)

Chloroethene and propenonitrile are now manufactured by cheaper processes than the addition process. The newer methods are based on ethene and propene respectively. Chloroethene is obtained by thermal decomposition of 1,2-dichloroethane, propenonitrile by the method given in Chapter 6.

$$CH_2Cl.CH_2Cl \rightarrow CH_2{=}CHCl + HCl$$
chloroethene

ALKYNES AND ELECTRONIC THEORY

Electronic Structure of the Alkyne Triple Bond. In Chapter 6 it was shown that the alkene double bond consists of two different single bonds. One is a σ (sigma) bond, formed by collinear overlapping of electron orbitals. The other, which is somewhat weaker, is a π bond, formed by collateral overlapping of electron orbitals (see Fig. 6-3). In the ethyne molecule ($H{-}C{\equiv}C{-}H$) the two C—H bonds are σ bonds, and so is one of the three bonds between the carbon atoms. The other bonds between the carbon atoms are both weaker π bonds. As in the case of the ethene

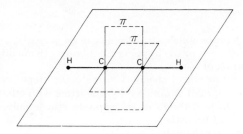

Fig. 7-3. The two π bonds in the ethyne molecule

molecule, they are produced by overlapping of electron orbitals outside the line connecting the carbon nuclei. For clarity only the axes of the overlapping electron orbitals are shown in Fig. 7-3, and the directions of the resulting π bonds are represented by broken lines. To illustrate the geometrical forms of the molecular orbitals of the electrons in the two π bonds four banana-shaped figures are required, one above the molecular axis, one below it, and one at each side.

As in methane and ethene, each carbon atom in the ethyne molecule had four electrons available for bond formation owing to 'promotion' of one of the two s electrons in the outer shell to the vacancy in the p sub-level. Hybridization of atomic orbitals again takes place, but in ethyne the orbital of the remaining s electron undergoes hybridization with only one of the three p orbitals. This results in two sp hybrid orbitals with axes at 180° to each other. One of the hybrid orbitals is used in bonding with a hydrogen atom, the other in bonding with the second carbon atom. Thus, all four atoms are in a straight line, and the bonds mentioned are σ bonds. The orbitals of the two remaining p electrons of each carbon atom have their axes

at right angles to each other and to the axis of the molecule. The two π bonds are formed by collateral overlapping of corresponding p orbitals, as in ethene.

Electrophilic Addition to Alkynes. Addition of electrophilic reagents to alkynes is thought to follow the same mechanism as that of addition to alkenes (Chapter 6). Thus, in the addition of the first hydrogen bromide molecule to the ethyne molecule partial charges are induced on the carbon atoms of the latter by the partial charges present in the former. As in the case of ethene, addition takes place in two stages.

$$(\text{I}) \qquad \overset{\delta+}{C}H \equiv \overset{\delta-}{C}H + \overset{\delta+}{H} - \overset{\delta-}{B}r \rightarrow \overset{+}{C}H = CH_2 + Br^-$$

$$(2) \qquad \overset{+}{C}H = CH_2 + Br^- \rightarrow CHBr = CH_2$$
$$\text{bromoethene}$$

The manner in which the second molecule of hydrogen bromide is added to bromoethene requires some explanation. The bromine atom in bromoethene has a fairly strong $-I$ effect, which would be expected to produce the partial charges now shown.

$$\overset{\delta+}{C}H_2 = CH - \overset{\delta-}{B}r$$

If electron displacement actually took place as represented, the carbon atom attached to the bromine atom would have the higher electron density. Hence on addition of the second HBr molecule the proton should be added to this carbon atom, and the Br^- ion to the other one. This would produce 1,2-dibromoethane ($CH_2Br.CH_2Br$) whereas in practice the compound obtained is 1,1-dibromoethane ($CH_3.CHBr_2$).

The reason for the formation of 1,1-dibromoethane is that bromoethene is a resonance hybrid with the two contributing structures given below.

$$CH_2 = CH - \overset{..}{\underset{..}{Br}}: \leftrightarrow \overset{..}{\overset{..}{C}H_2} - CH = \overset{+}{\underset{..}{Br}}:$$

It will be seen that owing to mesomerism the bromine atom tends to acquire a partial positive charge. Thus, the inductive effect and the mesomeric effect act in opposition. The formation of 1,1-dibromoethane shows that the mesomeric effect is the more powerful, and determines the actual electron displacements.

Evidence of resonance is provided by measurement of bond lengths. The presence of some double-bond character in the carbon-halogen bond is shown by this bond being appreciably shorter than usual, while the presence of some single-bond character in the C=C bond is indicated by an increase in the length of the bond. Another effect of resonance is to make the molecules more stable, or less reactive. Thus, it is difficult to remove the halogen atom of bromoethene by boiling with an aqueous alkali, whereas the halogen atom of an alkyl halide is readily removed in this way.

EXERCISE 7

1. Complete the following: Ethyne can be prepared by dropping water on to————, or by dropping 1,2-dibromoethane into a boiling ———— solution of ————.

2. Show by means of chemical formulæ the steps involved and the reagents required to obtain ethyne starting with ethanol.

3. State which of the following apply to ethene only, to ethyne only, to both, or to neither:

(*a*) It is lighter than air.

(*b*) It explodes when mixed with chlorine.

(*c*) It decolorizes Baeyer's reagent, giving a brown precipitate.

(*d*) It is rapidly absorbed by cold concentrated sulphuric(VI) acid.

(*e*) It forms a white precipitate with silver(I) nitrate(V) solution.

4. Give the structural formulæ of the compounds formed in each stage of the addition reactions between ethyne and (*a*) chlorine, (*b*) hydrogen bromide.

5. Give the systematic names of the organic compounds formed in question 4.

6. In which of the following molecules are all the carbon atoms in a straight line in space?

(*a*) $CH_3 - C \equiv CH$;   (*b*) $CH_3 - CH_2 - C \equiv CH$;

(*c*) $CH_3 - C \equiv C - CH_3$;   (*d*) $CH_2 = CH - C \equiv CH$.

7. Which of the following are true of the C≡C bond in alkynes?

(*a*) It is composed of one π bond and two σ bonds.

(*b*) It is shorter than the C=C bond in alkenes.

(*c*) It undergoes addition with electrophilic reagents.

(*d*) It adds on other molecules by means of its π electrons.

(*e*) In addition of hydrogen bromide the bromine atom is added first.

8 Alcohols and Ethers

MONOHYDRIC ALCOHOLS

The monohydric alcohols form an homologous series of compounds in which one hydroxyl group (—OH) is joined to an alkyl radical. They can be regarded as derived theoretically from the corresponding alkanes by substituting a hydrogen atom of the latter by a hydroxyl group. Their systematic names are obtained by replacing the final —*e* in the names of the alkanes by the ending —*ol*. Thus the first four members are as follows:

Alkane	*Alkanol* (alcohol)
Methane: CH_4	Methanol: CH_3OH (methyl alcohol)
Ethane: C_2H_6	Ethanol: C_2H_5OH (ethyl alcohol)
Propane: C_3H_8	Propanols: C_3H_7OH (propyl alcohols)
Butane: C_4H_{10}	Butanols: C_4H_9OH (butyl alcohols)

Propanol and butanol exist in isomeric forms (see Table 8-1). Easily the most important of the monohydric alcohols is ethanol, which is commonly known as 'alcohol.' Next in importance is the first member of the series, methanol.

METHANOL, CH_3OH

Boiling point: 64·6°C   *Density:* 0·80 g cm⁻³

(*Note:* Densities given here and later refer to 15°C.)

Occurrence and Manufacture

Alcohols do not occur free in Nature, but they are often found combined with acids in *esters*. A well-known ester of methanol is methyl salicylate ('oil of wintergreen'), which is extracted from the leaves of certain evergreen shrubs and used in embrocations.

Methanol is seldom prepared in the laboratory, although it can be obtained by the general methods given later for the preparation of alcohols. It is still manufactured to a small extent by destructive distillation of wood, which is chiefly carried out nowadays to make wood charcoal. The vapours given off in the distillation contain methanol, ethanoic, or acetic, acid, and propanone (acetone). Methanol extracted from the aqueous distillate is called 'wood spirit.' The modern method of manufacture is based either on the light petroleum fraction, naphtha, or on methane derived from refinery gas.

Synthetic Methanol Process. 'Synthesis gas' (a mixture of carbon monoxide and hydrogen) is obtained by mixing naphtha vapour or methane with steam and passing the mixture over a heated nickel catalyst (see Chapter 5). The synthesis gas is then converted to methanol by passing it at a suitable temperature and pressure over another catalyst. In the modern 'low-pressure process' the temperature is 250°C, the pressure 50 atmospheres, and the catalyst is a copper compound. The reaction is reversible.

$$CO + 2H_2 \rightleftharpoons CH_3OH(g)$$

Since the forward change is accompanied by decrease in volume the yield of methanol is favoured by increased pressure (in accordance with Le Chatelier's principle). Again, the reaction is exothermic in the forward direction, and therefore the best yield would be obtained by keeping the temperature low. The lower the temperature, however, the slower is the rate of reaction. In practice the temperature used is the 'optimum' one, that is, the temperature which, in conjunction with the pressure and the catalyst, gives the best yield in a given time.

Another factor of importance is the proportion of the reactant gases. The hydrogen is used in excess of the proportion (two to one by volume) indicated by the equation. In this way a bigger fraction of carbon monoxide is converted into methanol (in accordance with the law of Mass Action). After condensation of the methanol the small amount of carbon monoxide remaining is separated from hydrogen by absorption in an ammoniacal solution of copper(I) formate. The excess of hydrogen is then used to manufacture ammonia in an adjoining plant.

Properties and Uses of Methanol. Methanol is a colourless, pleasant-smelling liquid, which freezes at $-97°$C. It mixes with water in all proportions because of hydrogen-bonding between the $-OH$ groups of the two liquids. Methanol differs from ethanol in not forming a constant-boiling mixture with water.

Methanol burns in air with a pale-blue flame, forming carbon dioxide and water.

$$2CH_3OH + 3O_2 \rightarrow 2CO_2 + 4H_2O$$

With one or two exceptions the chemical reactions of methanol are similar to those of other monohydric alcohols (the general reactions are described shortly). The first member of an homologous series is often exceptional in some of its properties. One exception in the case of methanol is that it does not undergo dehydration to an alkene either with concentrated sulphuric(VI) acid or aluminium(III) oxide (see Chapter 6).

The chief use of methanol is as an intermediate in the manufacture of methanal (formaldehyde), which is used in making plastics. Methanol itself is employed in the manufacture of Perspex. Large amounts of methanol are also used in 'denaturing' ethanol in methylated spirit, that is, rendering it unfit for drinking. Methanol is poisonous, and when taken internally brings on blindness and insanity. Other applications of methanol are in the manufacture of various methyl compounds, such as chloromethane (methyl chloride), and as a solvent in the paint and varnish industry.

ETHANOL, C_2H_5OH

Boiling point: 78·4°C   *Density:* 0·79 g cm^{-3}

Manufacture of Industrial Ethanol. For many years the only method of obtaining ethanol on a large scale was by fermentation of molasses with yeast. Molasses is the viscous brown mother-liquor left when raw sugar has been crystallized from solution. The fermentation process is still used to a small extent, but has now been largely replaced by cheaper processes based on ethene. In Britain about 85 per cent of industrial ethanol is now derived from ethene. Two methods are used, but these are basically similar since both consist of adding one molecule of water to one molecule of the alkene. The 'hydration' can be carried out either directly or indirectly.

(*i*) *Direct Hydration of Ethene*. The reaction, which is reversible, is represented by the equation shown.

$$C_2H_4 + H_2O(g) \rightleftharpoons C_2H_5OH(g)$$

The same theoretical and practical considerations apply in this reaction as in the synthesis of methanol. The reaction is again exothermic in the forward direction.

In practice ethene is mixed with excess of steam. The mixture at 300°C and 65 atmospheres pressure is passed through a bed of silicon(IV) oxide (silica) impregnated with phosphoric(V) acid (the catalyst). About 30 per cent of the alkene undergoes combination. The issuing vapours are cooled, and a dilute solution of ethanol obtained. The unchanged ethene is recycled. The dilute ethanol is washed with sodium(I) hydroxide solution to remove traces of the acid catalyst (which is slightly volatile), and is then concentrated by fractional distillation. The limit of ethanol concentration which can be reached in this way is about 96 per cent by volume. The product is known as 'rectified spirit.'

(*ii*) *Indirect Hydration of Ethene*. In this process ethene is absorbed under pressure in hot concentrated sulphuric(VI) acid, giving a solution of ethyl hydrogen sulphate(VI). The latter is then hydrolysed to ethanol by diluting the solution with water. The ethanol is vaporized from the mixture by blowing in steam. Hydrolysis of ethyl hydrogen sulphate(VI) is reversible, but the equilibrium point is displaced to the right by removal of ethanol, which is more volatile than any of the other compounds present. The dilute solution of ethanol obtained is again concentrated by fractional distillation to give rectified spirit.

$$C_2H_4 + H_2SO_4 \rightarrow C_2H_5HSO_4$$

$$C_2H_5HSO_4 + H_2O \rightleftharpoons C_2H_5OH + H_2SO_4$$

(*iii*) *The Fermentation Process*. 'Fermentation' is the name given to chemical reactions brought about by certain lowly forms of life such as yeast, bacteria, and moulds. For many years, following the researches of Louis Pasteur about 1857, it was believed that the presence of living cells was necessary for fermentation to occur. In 1897, however, two German brothers called Buchner found that by grinding yeast cells with sand and subjecting the disrupted cells to high pressure a yellow liquid could be obtained which was as effective as the original yeast cells in fermenting sugar solutions. This showed that the chemical changes were due, not to any 'vital force' associated with living cells, but to some substance produced by the cells acting as a catalyst. Many biological catalysts of this kind have since been discovered. They are called *enzymes* (meaning literally 'in yeast'). Enzymes are highly complex proteins, and play an important part in nearly all the chemical changes which occur in plants and animals.

When sucrose (cane-sugar) is fermented by yeast the sugar is first hydrolysed to a mixture of the isomeric sugars glucose and fructose by the enzyme *invertase*, which is present in the yeast.

$$C_{12}H_{22}O_{11} + H_2O \rightarrow \underset{\text{glucose}}{C_6H_{12}O_6} + \underset{\text{fructose}}{C_6H_{12}O_6}$$

Both glucose and fructose are broken down by yeast into ethanol and carbon dioxide, this change being brought about by the enzyme (or group of enzymes) called *zymase*.

$$C_6H_{12}O_6 \rightarrow 2C_2H_5OH + 2CO_2$$

The main reactions are accompanied by various side-reactions, and the fermented liquid contains in addition to ethanol small amounts of other alcohols. The breaking down of complex sugar molecules into simpler ones is attended by evolution of energy. Part of this is used by the yeast organism for its own life processes, and part appears as heat.

Laboratory Preparation of Ethanol by Fermentation of a Sugar Solution. In the laboratory ethanol can be obtained by fermentation of cane-sugar or glucose, as now described.

Experiment

Into a 500 cm³ flat-bottomed flask put 100 cm³ of a 10 per cent solution of cane-sugar or glucose. Make a thin paste from about 3 g of baker's yeast and a little water, and add the paste to the flask. Close the flask with a stopper fitted with a tube leading to a Dreschel bottle containing lime-water (Fig. 8-1). Leave the apparatus in a warm place (preferably at about 25°C) for 24 hours. The contents of the flask soon begin to froth, and the lime-water is turned milky by the carbon dioxide evolved.

Decant about 20 cm³ of the remaining liquid into a 50 cm³ pear-shaped flask. Attach the latter to a fractionating column connected

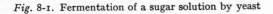

Fig. 8-1. Fermentation of a sugar solution by yeast

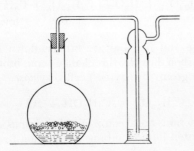

to a condenser (see Fig. 2-7). Distil the liquid, and collect the first 2–3 cm³ of distillate in a dry evaporating-dish. Warm the evaporating-dish for a few moments over a gauze and tripod, and apply a lighted taper to the liquid. The latter burns with a pale-blue flame characteristic of ethanol. Test a further portion of the distillate by means of the 'iodoform test' described shortly.

In some countries starch from potatoes, rice, corn, etc. is used for the manufacture of ethanol. Starch cannot be directly fermented by yeast, and must first be converted into a sugar. This is usually done by mixing the starch with water and malt. Malt contains an enzyme called *diastase*, which changes starch into maltose. Maltose ($C_{12}H_{22}O_{11}$) is isomeric with sucrose, and like the latter undergoes fermentation by yeast.

The manufacture of alcoholic beverages by fermentation is separate from that of industrial ethanol, but the two have much in common. Beer is made from malt as outlined above, wines are obtained by fermentation of grape-joice or other fruit-juices, and spirits by distilling the various dilute alcoholic solutions resulting from fermentation.

Absolute Ethanol. Ethanol is described as 'absolute' if its water content is less than 0·5 per cent by volume. As stated earlier, the percentage of ethanol in an ethanol–water mixture can be increased by fractional distillation to about 96 per cent by volume. It is impossible to reach a higher percentage by ordinary fractional distillation because the 96 per cent liquid is a constant-boiling, or *azeotropic*, mixture. It has a higher vapour-pressure, and therefore a lower boiling point, than any other mixture of ethanol and water. Thus, whereas pure ethanol boils at 78·4°C at standard pressure, the azeotropic mixture boils at 78·15°C.

In industry absolute ethanol is obtained by adding benzene to the constant-boiling mixture (rectified spirit) and distilling. The vapours given off are passed into a fractionating column. Benzene, ethanol, and water form a 'ternary mixture' of constant boiling point (68°C), which is lower again than the boiling point of the azeotropic binary mixture. Thus almost all of the water is removed from the rectified spirit by being incorporated in the ternary mixture. The latter passes out at the top of the column, while absolute ethanol of 99·9 per cent purity is obtained at the bottom of the column.

In the laboratory water is usually removed from rectified spirit by chemical treatment. Anhydrous calcium(II) chloride cannot be used for this purpose because it combines with ethanol to form a crystalline compound of formula $CaCl_2.4C_2H_5OH$. Methanol forms a similar compound. The combined alcohol is called 'alcohol of crystallization' because it is analogous to water of crystallization. The ethanol is dried in two stages. First it is refluxed with calcium(II) oxide for about six hours, after which it is distilled. Then it is refluxed with metallic calcium for about six hours, and again distilled. In this way the water content is reduced to less than 0·2 per cent by volume.

$$CaO + H_2O \rightarrow Ca(OH)_2$$
$$Ca + 2H_2O \rightarrow Ca(OH)_2 + H_2$$

Methylated Spirit. Manufactured ethanol is liable to a heavy excise duty, but relief from the duty is allowed in certain cases (e.g. school laboratories) providing the alcohol has been 'denatured,' or rendered unfit for drinking. The denatured liquid is called methylated spirit, and is produced in two main forms:

(*i*) *Industrial Methylated Spirit*. This consists of 95 per cent by volume of rectified spirit and 5 per cent of methanol.

(*ii*) *Mineralized Methylated Spirit*. This contains 90 per cent by volume of rectified spirit, 9·5 per cent of methanol, and small amounts of unpleasant-tasting substances such as kerosene and pyridine. The liquid is also coloured violet with a dye.

Owing to the presence of the poisonous methanol it is extremely dangerous to drink methylated spirit. There is no simple way of separating the ethanol from the methanol and other substances present.

Uses of Ethanol. Apart from its use in beverages ethanol has a wide variety of applications. The most important are given below.

(*i*) *As a Fuel*. Ethanol is used in methylated spirit burners, and is incorporated in some brands of motor spirit (its octane number is appreciably higher than those of petroleum hydrocarbons).

(*ii*) *As a Solvent*. Ethanol is an important solvent for dyes, drugs, oils, resins, plastics, soap, etc. With dyes it enters into the manufacture of coloured inks, stains, and indicator solutions (e.g. methyl orange). Solutions of drugs in ethanol are called 'tinctures' (e.g., tincture of iodine, quinine, etc.). Eau-de-Cologne and other perfumes are made by dissolving certain pleasant-smelling oils in ethanol. In 'French polishing' furniture the wood is rubbed with a solution of shellac in ethanol. Transparent soap is made by dissolving soap in ethanol and allowing the solution to evaporate.

(iii) *As an Antiseptic*. Ethanol is employed for the preservation of biological specimens. It is also the chief constituent of 'surgical spirit,' which is used for hardening skin, treating pimples, etc.

(iv) *In Chemical Manufacture*. Among the compounds manufactured from ethanol are ethoxyethane (diethyl ether), ethanal (acetaldehyde), ethanoic, or acetic, acid, and chloroethane.

Minor applications of ethanol include its use in ships' compasses, spirit-levels, and thermometers designed to measure low temperatures (ethanol freezes at $-114°C$).

The Homologous Series of Monohydric Alcohols

The alcohols which follow ethanol occur in isomeric forms, the number of which increases with the number of carbon atoms in the molecule. Table 8-1 shows the alcohols containing from one to four carbon atoms in the molecule. The two propanols correspond with the two ways in which an —OH group can replace a hydrogen atom from the propane molecule ($CH_3.CH_2.CH_3$). Two of the four butanols (the first and the third) are derived theoretically from butane ($CH_3.CH_2.CH_2.CH_3$), and the remaining two from 2-methylpropane.

The first column in Table 8-1 shows the systematic names and structures of the alcohols. The systematic names are based on the longest carbon chain containing the —OH group. The positions of the —OH group and any substituting alkyl radicals are indicated by numbering the carbon atoms of the chain, starting at the end of the chain which gives the lower number for the position of the —OH group.

Table 8-1. MONOHYDRIC ALCOHOLS
(General formulæ $C_nH_{2n+1}OH$, ROH)

Systematic name and structural formula	Common name	Type	Boiling point/°C
Methanol CH_3OH	Methyl alcohol	Primary	65
Ethanol $CH_3.CH_2OH$	Ethyl alcohol	Primary	78
Propanols, C_3H_7OH			
1. Propan-1-ol $CH_3.CH_2.CH_2OH$	*n*-Propyl alcohol	Primary	97
2. Propan-2-ol $CH_3.CH(OH).CH_3$	Isopropyl alcohol	Secondary	82
Butanols, C_4H_9OH			
1. Butan-1-ol $CH_3.CH_2.CH_2.CH_2OH$	*n*-Butyl alcohol	Primary	118

2. 2-Methylpropan-1-ol
$CH_3.CH.CH_2OH$
    |
    CH_3
	Isobutyl alcohol	Primary	108

3. Butan-2-ol $CH_3.CH_2.CH(OH).CH_3$	*s*-Butyl alcohol	Secondary	99·5

4. 2-Methylpropan-2-ol
$CH_3.C(OH).CH_3$
    |
    CH_3
	t-Butyl alcohol	Tertiary	82·5

Monohydric alcohols are classified as *primary* (*p*-), *secondary* (*s*-), or *tertiary* (*t*-), according to the groups present in their molecules. The characteristic groups are indicated below. As will be seen shortly, the properties of the different types of alcohol vary to some extent.

Primary: $-CH_2OH$

Secondary: $\diagdown\!CH(OH)\diagup$

Tertiary: $-\!\overset{\diagdown}{\underset{\diagup}{C}}(OH)$

As in the case of water molecules, hydrogen-bonding takes place between the —OH groups of alcohol molecules, and as a result the boiling points of alcohols are higher than those of alkanes of similar relative molecular mass. Thus, while the lower alkanes are gases, all the normal alcohols up to, and including, $C_{11}H_{23}OH$ are liquids. Higher members are solids. Secondary alcohols boil at a lower temperature than the corresponding primary alcohols, and tertiary alcohols at a still lower temperature. These effects are associated with the different shapes of the molecules and the different inductive effects of the various alkyl radicals.

Methanol, ethanol, and the two propanols are miscible with water in all proportions because of hydrogen-bonding to the water molecules. The hydrophobic character of the hydrocarbon chain, however, increases as it becomes longer, and solubility in water decreases. Butan-1-ol is only partially miscible with water,

while pentan-1-ol $(C_5H_{11}OH)$ and hexan-1-ol $(C_6H_{13}OH)$ dissolve only to a small extent. Higher alcohols are practically insoluble in water.

General Methods of Obtaining Alcohols. After methanol and ethanol the most important monohydric alcohols are the propanols and butanols. These are used as ·solvents and in the manufacture of other compounds.

(*i*) *From 'Synthesis Gas.'* The synthesis of methanol from a mixture of carbon monoxide and hydrogen has been described earlier. By modifying the conditions (temperature, pressure, and nature of the catalyst) other alcohols (e.g. propan-1-ol) can be made.

$$3CO + 6H_2 \rightarrow CH_3.CH_2.CH_2OH + 2H_2O$$

(*ii*) *Hydration of Alkenes.* Alcohols can be obtained by hydration of alkenes, but only in the case of ethene is a primary alcohol (ethanol) produced. Other alkenes yield secondary or tertiary alcohols. The only alcohol other than ethanol made by the direct method (phosphoric(V) acid catalyst) is propan-2-ol, which is obtained from propene.

$$CH_3.CH\!=\!CH_2 + H_2O \rightarrow CH_3.CH(OH).CH_3$$

The indirect method (involving formation of an intermediate alkyl hydrogen sulphate with concentrated sulphuric(VI) acid) is used to manufacture ethanol, propan-2-ol, butan-2-ol, and 2-methylpropan-2-ol. The last two are made from different butenes.

(*iii*) *Reduction of Aldehydes.* Certain alcohols are obtained on a large scale by reduction of the corresponding aldehyde. Reduction is brought about by hydrogen under pressure in the presence of a suitable catalyst. In this way butanal (*n*-butyral-dehyde) yields butan-1-ol.

$$CH_3.CH_2.CH_2.CHO + H_2 \rightarrow CH_3.CH_2.CH_2.CH_2OH$$
butanal   butan-1-ol

The aldehydes are produced from alkenes by the 'Oxo synthesis,' which is described in Chapter 10.

(*iv*) *From Esters.* A few of the higher alcohols, which are used to manufacture shampoos and detergents, are obtained by reduction or hydrolysis of naturally occurring esters. Thus dodecan-1-ol (lauryl alcohol), $C_{12}H_{25}OH$, is made from an ester present in coconut-oil. Hexadecan-1-ol (cetyl alcohol), $C_{16}H_{33}OH$, is a white solid used to stabilize emulsions such as hair-creams. It is prepared from an ester found in spermaceti wax, which is obtained from the head of the sperm whale.

In the laboratory alcohols can be prepared by reduction of aldehydes or ketones and by reduction or hydrolysis of esters. Details of the procedures are given later.

Reactions of Alcohols

The reactions of alcohols are largely determined by the functional hydroxyl group. Most of the reactions affect only this group, but some are more complicated and involve the alkyl radical as well. The —OH group in alcohols resembles that in water in showing *amphoteric* character; that is, it can behave as an acidic group towards bases and as a basic group towards acids. Since acids are electrophiles and bases are nucleophiles (Chapter 4), alcohols can act as either type of reagent with appropriate substances.

1. **Acidic Reactions.** The acidic character of alcohols is due to electron displacement in the O—H bond towards the strongly electronegative oxygen atom. The displacement leaves the hydrogen atom with a partial positive charge, so that a sufficiently strong electron donor can bring about heterolytic fission of the bond, the hydrogen atom being removed as a proton. Alcohols, however, are extremely weak acids in aqueous solution, and the solutions do not affect indicators. This is because alcohols are only feeble electrophiles, while water is only a weak nucleophile. Hence the ionization indicated below occurs to only a minute extent.

$$\overset{\delta+}{R}—\overset{\delta-}{O}—\overset{\delta+}{|H} + H_2O \rightleftharpoons R—O^- + H_3O^+$$
alkoxide ion

Alcohols show their acidic character in reactions with Group I metals and hydroxyl ions, both of which are strong nucleophiles, or electron donors. Thus, if sodium is added to methanol or ethanol in the cold, hydrogen is evolved and the sodium salt of the alcohol is left in solution in the form of ions. The reaction is much less vigorous than the one between sodium and water. Calcium, which readily liberates hydrogen from water, does not react with alcohols. This is because Group II metals are weaker electron donors than Group I metals.

$$CH_3OH + Na \rightarrow \underline{CH_3O^- + Na^+} + \tfrac{1}{2}H_2$$
sodium(I) methoxide

Experiment

Put about 2 cm³ of methanol into a dry test-tube and add in turn two small pieces of sodium. Test for hydrogen with a lighted taper. When the metal has dissolved pour the solution on to a clock-glass and evaporate it over a beaker of hot water. The solid residue is sodium(I) methoxide. When pure it is a white solid, but, prepared as described, it is usually coloured brown by impurities derived from the oil in which the sodium is kept.

It was seen previously (Chapter 3) that evolution of hydrogen when an organic compound is treated with sodium shows the presence of a hydroxyl group in the molecule.

In the reaction between an alcohol and hydroxyl ions a weak acid is reacting with a strong base, or a weak electrophile with a strong nucleophile. The reaction is reversible, the position of equilibrium depending on the concentrations of the substances present.

$$C_2H_5OH + OH^- \rightleftharpoons C_2H_5O^- + H_2O$$
$$\text{ethoxide ion}$$

When solid potassium(I) hydroxide is dissolved in absolute ethanol, the equilibrium point is well to the right. If, however, the proportion of water is high, it is far to the left. Thus when sodium(I) methoxide (prepared as described above) is added to excess of water it is almost completely converted into methanol and sodium(I) hydroxide.

$$CH_3O^- \, Na^+ + H_2O \rightleftharpoons CH_3OH + Na^+ + OH^-$$

The formation of OH^- ions is shown by the turning blue of red litmus paper.

2. Basic Reactions—Reactions with Acids. Alcohols react with acids (organic or inorganic) to form esters. The general reaction,

which is described as *esterification* of the alcohol, can be represented as follows:

$$ALCOHOL + ACID \rightleftharpoons ESTER + WATER$$

The reaction again is reversible and normally results in an equilibrium mixture of all four compounds. At ordinary temperatures esterification of the alcohol occurs very slowly, and equilibrium may not be reached for several weeks. It is therefore usual to heat the reactants in the presence of a catalyst, such as anhydrous zinc(II) chloride or concentrated sulphuric(VI) acid. These substances act not only as catalysts but also as dehydrating agents. They absorb the water formed in the reaction, thus displacing the position of equilibrium from left to right and increasing the yield of ester. Under certain conditions mineral acids and alcohols may yield other products (ethers and alkenes) as well as esters.

(*i*) *Hydrohalogen Acids.* Alcohols react with hydrohalogen acids to give alkyl halides. Thus bromomethane (methyl bromide) can be prepared by refluxing methanol with concentrated hydrobromic acid. The reaction takes place more rapidly if a little concentrated sulphuric(VI) acid is added to increase the concentration of hydrogen ions, which are the actual catalyst for the reaction.

$$CH_3OH + HBr \underset{}{\overset{H^+}{\rightleftharpoons}} CH_3Br + H_2O$$

Experiments show that the above reaction occurs by an ionic mechanism, in which a methanol molecule acts as a base by combining with a proton from the hydrobromic acid (or sulphuric(VI) acid) to give a *methyloxonium* ion. The combination takes place by the oxygen atom forming a co-ordinate covalent bond with the proton by means of one of its two lone pairs of electrons.

$$CH_3\overset{\cdot\cdot}{\underset{\cdot\cdot}{O}}\!-\!H + H^+ \rightleftharpoons CH_3\!-\!\overset{+}{\overset{\cdot\cdot}{O}}\!-\!H$$
$$\underset{\quad\; H}{\big|}$$

methyloxonium ion

Compare

$$H\!-\!O\!-\!H + H^+ \rightleftharpoons H_3O^+$$

The ester is produced from the methyloxonium ion by a one-step reaction between the latter and a Br^- ion. This step, involving a 'transition state,' is analogous to the reaction which occurs between an iodomethane molecule and an OH^- ion and which is described in detail in Chapter 4.

$$\bar{Br}: +CH_3\!-\!\overset{+}{O}H_2 \rightleftharpoons [\overset{\delta-}{Br}\cdots\cdots CH_3\cdots\cdots\overset{\delta+}{O}H_2]$$
$$\rightleftharpoons Br\!-\!CH_3 + H_2O$$

Practical details of the preparation of alkyl halides from alcohols and hydrohalogen acids are given in the next chapter.

(ii) *Nitric(V) Acid.* With concentrated nitric(V) acid monohydric alcohols yield small amounts of the alkyl nitrate(V) (RNO_3), but the main reaction consists of oxidation of the alcohol (see later). Considerable heat is evolved in the oxidation, which is liable to become violent and dangerous.

(iii) *Sulphuric(VI) Acid.* Concentrated sulphuric(VI) acid can react with alcohols in different ways according to the nature of the alcohol and the experimental conditions. When ethanol and concentrated sulphuric(VI) acid are refluxed together at about 80°C ethyl hydrogen sulphate(VI) is slowly formed.

$$C_2H_5OH + H_2SO_4 \rightleftharpoons C_2H_5HSO_4 + H_2O$$

At higher temperatures the alcohol undergoes *dehydration*. This may result in formation of an ether or an alkene, or both simultaneously. If excess of ethanol is heated with concentrated sulphuric(VI) acid at about 140°C the main product is ethoxyethane (diethyl ether), one molecule of water being eliminated from two molecules of the alcohol.

$$2C_2H_5OH \rightarrow (C_2H_5)_2O + H_2O$$

At 180°C and in the presence of excess of concentrated sulphuric(VI) acid dehydration takes place differently, an alkene being produced. Ethanol yields ethene. The use of this reaction for the laboratory preparation of ethene has been described in Chapter 6.

$$CH_3\!-\!CH_2OH \rightarrow CH_2\!\!=\!\!CH_2 + H_2O$$

As stated earlier, methanol does not give an alkene when heated with concentrated sulphuric(VI) acid (it forms methyl hydrogen sulphate(VI) or methoxymethane (dimethyl ether) according to the conditions). This is because the molecule of water eliminated in formation of an alkene is derived from the —OH group and a hydrogen atom attached to the carbon atom in the 2-position, that is, adjacent to the —CH_2OH group. Since the methanol molecule (CH_3OH) does not contain a hydrogen atom in this situation, it cannot form an alkene.

It is probable that all the mechanisms of the different reactions of primary alcohols with concentrated sulphuric(VI) acid depend on preliminary formation of alkyloxonium ions, as described earlier. Thus it is likely that ethyl hydrogen sulphate(VI) is produced by a substitution reaction between ethyloxonium ions and hydrogen-sulphate(VI) ions.

$$HSO_4^- + C_2H_5\overset{+}{-}OH_2 \rightleftharpoons HSO_4-C_2H_5 + H_2O$$

At about 140°C ethyloxonium ions dissociate into carbonium ions and water. A carbonium ion is very reactive, and if excess of ethanol is present it combines with another ethanol molecule to give an ether molecule.

$$C_2H_5\overset{+}{-}OH_2 \rightleftharpoons C_2H_5^+ + H_2O$$

$$C_2H_5^+ + \overset{..}{:O}-C_2H_5 \rightarrow C_2H_5-O-C_2H_5 + H^+$$
$$|$$
$$H$$

At about 180°C and in the presence of excess of sulphuric(VI) acid the carbonium ions decompose yielding ethene. Both this reaction and the previous one depend on a base (e.g. HSO_4^- ions) being present to remove the protons.

$$C_2H_5^+ \rightarrow C_2H_4 + H^+$$

(*iv*) *Carboxylic Acids*. Organic acids containing the carboxyl group (—COOH) are weak acids and react very slowly with alcohols even on boiling. In their case esters can be prepared by heating the alcohol and organic acid with a strong mineral acid, usually concentrated sulphuric(VI) acid. As explained earlier, the latter acts both as a catalyst (by supplying H^+ ions) and as a dehydrating agent. Ethanol and ethanoic, or acetic, acid yield the ester ethyl ethanoate, or acetate.

$$C_2H_5OH + CH_3.COOH \rightleftharpoons CH_3.COOC_2H_5 + H_2O$$
ethyl ethanoate

It should be noted that in writing the formula of an ester of a carboxylic acid the acid group is usually put first.

3. Reactions with Phosphorus Trichloride and Phosphorus Pentachloride. Both of these reagents bring about the replacement of the hydroxyl group in alcohols by a chlorine atom, the reactions taking place vigorously in the cold. With phosphorus trichloride ethanol forms chloroethane (ethyl chloride) and phosphonic acid (phosph*orous* acid).

$$3C_2H_5OH + PCl_3 \rightarrow 3C_2H_5Cl + H_3PO_3$$

Phosphorus pentachloride and ethanol yield chloroethane, phosphorus trichloride oxide (phosphorus oxychloride), and 'steamy' fumes of hydrogen chloride.

$$C_2H_5OH + PCl_5 \rightarrow C_2H_5Cl + POCl_3 + HCl$$

The evolution of hydrogen chloride when phosphorus pentachloride is added to an organic compound (which has been dried) is another method of showing the *presence of a hydroxyl group* in the compound.

4. Oxidation. Primary, secondary, and tertiary alcohols react differently with oxidizing agents. *Primary* alcohols are first converted to an aldehyde and then to a carboxylic acid. The group involved in the oxidation is the —CH$_2$OH group of the alcohol. The usual oxidizing agent used in the laboratory is an

(i) $$R—CH_2OH + O \rightarrow R—C\overset{H}{\underset{O}{\diagup}} + H_2O$$
aldehyde

(ii) $$R—CHO + O \rightarrow R—C\overset{OH}{\underset{O}{\diagup}}$$ carboxylic acid

acidified solution of sodium(I), or potassium(I), dichromate(VI). When ethanol is warmed with this oxidizing agent the vapour of ethanal (acetaldehyde), CH$_3$.CHO, is given off. If ethanol is refluxed with excess of the oxidizing agent, the ethanal is oxidized further to ethanoic acid, CH$_3$.COOH. Methanol similarly yields methanal (formaldehyde), HCHO, and methanoic, or formic, acid, HCOOH.

Another oxidizing agent which can be used is atmospheric oxygen in the presence of a suitable catalyst, such as platinum.

Experiment.

Put a small amount of methanol into a beaker, warm the beaker for a few moments over a gauze, and then remove the flame. Bend a clean 'flame-test' platinum wire so that the wire is at right angles to the holder. Grip the holder in a pair of tongs (in case the alcohol vapour ignites), and, after heating the wire in a Bunsen flame for a

Fig. 8-2. Catalytic oxidation of methanol vapour

few seconds, introduce it into the beaker (Fig. 8-2). As the wire is moved about in the beaker, it glows brightly owing to the heat evolved in oxidation of the methanol vapour at the surface of the platinum.

Secondary alcohols are also easily oxidized. In this case a ketone and water are produced. Unlike aldehydes, ketones are

$$\overset{R}{\underset{R'}{\diagup}}CHOH + O \rightarrow \overset{R}{\underset{R'}{\diagup}}C{=}O + H_2O$$
ketone

resistant to further oxidation. When propan-2-ol, $(CH_3)_2CHOH$, is heated with acidified potassium(I) dichromate(VI) it yields propanone (acetone), $(CH_3)_2CO$.

Tertiary alcohols are much more difficult to oxidize. When they are refluxed for some hours with acidified potassium(I) dichromate(VI) solution their molecules slowly break down owing to rupture of some of the C—C bonds. Mixtures of ketones and acids are produced.

5. **Decomposition with Solid Catalysts.** When the vapours of alcohols are passed over certain solid catalysts at suitable temperatures they decompose, giving different kinds of products according to the catalyst used.

With an *aluminium(III) oxide* catalyst the vapours of primary, secondary, and tertiary alcohols are *dehydrated* to alkenes. As seen in Chapter 6, ethene can be prepared from ethanol by passing ethanol vapour over aluminium(III) oxide at 360°C, using the 'powdered chalk method.'

$$C_2H_5OH \xrightarrow{Al_2O_3} C_2H_4 + H_2O$$

Tertiary alcohols undergo dehydration very readily, alkenes being produced in some cases by the action of heat alone.

With a *copper* catalyst different types of products are obtained from primary, secondary, and tertiary alcohols. If the vapour of a primary alcohol is passed over copper turnings at 250–300°C *dehydrogenation* occurs, and an aldehyde is formed. In similar circumstances a secondary alcohol is dehydrogenated to a ketone. A tertiary alcohol does not undergo dehydrogenation, but instead loses water and gives an alkene. These reactions are illustrated by the examples given.

$$CH_3 \cdot CH_2OH \xrightarrow{Cu} CH_3 \cdot CHO + H_2$$
$$\text{ethanol} \qquad\qquad \text{ethanal}$$

$$(CH_3)_2CHOH \xrightarrow{Cu} (CH_3)_2CO + H_2$$
$$\text{propan-2-ol} \qquad\qquad \text{propanone}$$

$$(CH_3)_3COH \xrightarrow{Cu} (CH_3)_2C{=}CH_2 + H_2O$$
$$\text{2-methylpropan-2-ol} \qquad \text{2-methylpropene}$$

Chemical Tests for Lower Aliphatic Alcohols.

Note. In describing a chemical test by which a compound is recognized it is not enough merely to state that a certain product is obtained. For purposes of identification it is necessary to produce a characteristic colour, smell, or precipitate. A common procedure is to prepare a solid derivative, which can be readily purified and then identified by taking its melting point

(*i*) *Distinction Between Methanol and Ethanol.* A quick method of distinguishing between methanol and ethanol is by means of the *tri-iodomethane (iodoform) reaction* This is given by ethanol, but not by methanol. In the presence of an alkali iodine converts ethanol into tri-iodomethane (CHI_3), a yellow solid with a characteristic smell. The reaction is carried out as now described.

To 0·5 cm³ of ethanol add 5 cm³ of a 10 per cent solution of potassium(I) iodide, and heat the mixture until it begins to boil. Now add one or two drops of a concentrated solution of sodium(I) chlorate(I) (sodium hypochlorite). The solution is alkaline, and also liberates iodine from potassium(I) iodide. Tri-iodomethane is at once precipitated.

The above reaction is not specific to ethanol. It is also given by propan-2-ol and certain compounds like ethanal (acetaldehyde) and propanone (acetone) which are not alcohols.

(ii) *Ester Formation*. Alcohols can be recognized as a class by the esters which they yield when heated with organic acids (usually in the presence of concentrated sulphuric(VI) acid as a catalyst). The esters have pleasant fruity odours.

Put into a test-tube 1 cm³ of ethanol, 1 cm³ of glacial ethanoic acid, and a few drops of concentrated sulphuric(VI) acid. Warm the tube over a *small* flame for a few minutes, keeping it well shaken. The ester ethyl ethanoate is formed. Pour the contents of the tube into a beaker containing water. The liquid ester floats on the surface of the water, and its fruity odour is readily detected.

Individual alcohols can be identified by preparing a solid ester and taking its melting point. A convenient, although rather complex, ester is the alkyl 3,5-dinitrobenzoate. This can be precipitated even from a dilute aqueous solution of an alcohol by means of 3,5-dinitro-benzoyl chloride (see Chapter 25). Methyl 3,5-dinitrobenzoate melts at 109°C, and the ethyl ester at 93°C.

Distinction Between Primary, Secondary, and Tertiary Alcohols. The different classes of alcohols can be distinguished by their different behaviour with oxidizing agents. A tertiary alcohol does not react with a warm *alkaline* solution of potassium(I) manganate(VII) (potassium permanganate), whereas both a primary alcohol and a secondary alcohol decolorize the solution and precipitate brown hydrated manganese(IV) oxide.

To distinguish between a primary alcohol and a secondary alcohol each may be distilled in turn with acidified potassium(I) dichromate(VI) solution: The distillate obtained from the primary alcohol contains an aldehyde, while that from the secondary alcohol contains a ketone. The distillate is tested with *Schiff's reagent* (this is prepared as described in Chapter 10).

The reagent gives an immediate purple colour with an aldehyde, but with a ketone the colour develops only after some time.

An alternative method of distinction is to pass the vapours of the alcohols over gently heated copper turnings. This can be done by the 'powdered chalk' method given in Chapter 6. As seen earlier, a primary alcohol yields an aldehyde and a secondary alcohol a ketone, while a tertiary alcohol produces an alkene. The latter decolorizes a solution of bromine in trichloromethane.

The different isomers of an alcohol (e.g. the four butanols) can be distinguished by means of the different melting points of their 3,5-dinitrobenzoates, as described earlier.

POLYHYDRIC ALCOHOLS

A *polyhydric alcohol* is an alcohol which contains more than one hydroxyl group in the molecule. The two most important polyhydric alcohols have the trivial names ethylene glycol and glycerol. Ethylene glycol is a *dihydric* alcohol. It is derived theoretically from ethane by substituting two hydroxyl groups for two hydrogen atoms attached to different carbon atoms in the ethane molecule. Hence the systematic name of the compound is ethane-1,2-diol. Glycerol is a *trihydric* alcohol, and is a trisubstitution derivative of propane. Its systematic name is propane-1,2,3-triol. Thus we have:

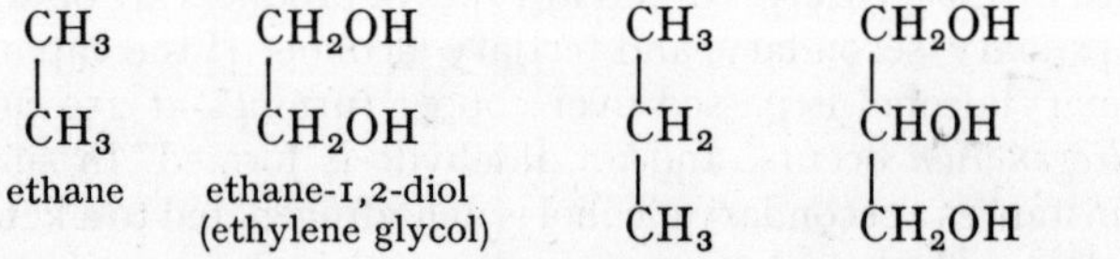

$$
\begin{array}{cccc}
CH_3 & CH_2OH & CH_3 & CH_2OH \\
| & | & | & | \\
CH_3 & CH_2OH & CH_2 & CHOH \\
 & & | & | \\
 & & CH_3 & CH_2OH
\end{array}
$$

ethane ethane-1,2-diol propane propane-1,2,3-triol
(ethylene glycol) (glycerol)

Ethane-1,2-diol, $C_2H_4(OH)_2$. This compound is commonly called 'glycol'. It is a colourless syrupy liquid with a sweet taste. It freezes at $-17°C$, boils at $197.5°C$, and has a density of 1.12 g cm^{-3}. Its high boiling point and viscosity are due to relatively strong hydrogen-bonding between the molecules caused by the presence of two hydroxyl groups. It is chiefly used in the manufacture of Terylene (Chapter 26) and as an 'anti-freeze' in motor-car radiators.

Ethane-1,2-diol is made on a large scale by hydrolysis of epoxyethane (ethylene oxide) at $200°C$ and 20 atmospheres pressure.

$$\begin{array}{c} H_2O \\ | \\ H_2C \end{array} \!\!\! \rangle O + H_2O \rightarrow \begin{array}{c} CH_2OH \\ | \\ CH_2OH \end{array}$$

In the laboratory the dihydric alcohol can be obtained by refluxing 1,2-dibromoethane with aqueous potassium(I) carbonate for about 8 hours.

$$C_2H_4Br_2 + K_2CO_3 + H_2O \rightarrow C_2H_4(OH)_2 + 2KBr + CO_2$$

The reactions of ethane-1,2-diol are those expected of a compound with two primary alcohol groups in the molecule. The hydrogen atoms of both —OH groups can be displaced by sodium to give the derivative $CH_2ONa.CH_2ONa$. Phosphorus pentachloride replaces the two hydroxyl groups by chlorine atoms, forming 1,2-dichloroethane ($CH_2Cl.CH_2Cl$). Reactions with acids take place in stages. Thus when heated with ethanoic acid and concentrated sulphuric(VI) acid the alcohol yields firstly the monoethanoate and then the diethanoate (ethane-1,2-diyl diethanoate). Both primary alcohol groups undergo oxi-dation with nitric(V) acid, giving eventually ethanedioic acid (oxalic acid), $(COOH)_2$.

Propane-1,2,3-triol, $C_3H_5(OH)_3$. Glycerol, or 'glycerine,' is also a syrupy liquid with a sweet taste. Again there is a strong tendency to form intermolecular hydrogen bonds, this being reflected in the high boiling point ($290°C$) and complete miscibility with water. The alcohol undergoes 'supercooling' very readily; that is, it remains in the liquid state at temperatures well below its freezing point. If it is cooled to $0°C$ it forms colourless crystals which, on warming, melt at $18°C$.

Propane-1,2,3-triol occurs naturally in animal fats and vegetable oils mainly in the form of esters of higher carboxylic acids. These esters form the raw materials for the manufacture of both soap and the alcohol (see Chapter 13). The latter is also made on a large scale (chiefly in the U.S.A.) from propene.

An important derivative of propane-1,2,3-triol is the ester propane-1,2,3-triyl trinitrate(V) (glyceryl trinitrate), $C_3H_5(NO_3)_3$. This is commonly called 'nitroglycerine.' It is made by treating the alcohol with a mixture of concentrated nitric(V) acid and concentrated sulphuric(VI) acid, and is used in the manufacture of dynamite (Chapter 13) and cordite (Chapter 18). Other uses of the alcohol are as an anti-freeze and in the manufacture of synthetic resins, printing-ink, shoe-polish, and pharmaceutical products.

ETHERS

While monohydric alcohols are theoretically derived from water by substituting one hydrogen atom of the water molecule by an alkyl radical, aliphatic ethers are obtained by replacing both hydrogen atoms by alkyl radicals.

$$H—O—H \qquad R—O—H \qquad R—O—R'$$

water $\qquad\qquad$ alcohol $\qquad\qquad$ ether

If the alkyl radicals in the ether are similar, the latter is called a *simple ether*. If they are different, it is known as a *mixed ether*.

Under I.U.P.A.C. rules ethers may be named in two ways. A particular ether may be specified by means of the alkyl radicals attached to the central oxygen atom. In the case of a mixed ether the alkyl radicals are placed in alphabetical order. Thus dimethyl ether is an example of a simple ether, and ethyl methyl ether typefies a mixed ether.

$$CH_3—O—CH_3 \qquad CH_3—O—C_2H_5$$

methoxymethane $\qquad$ methoxyethane
(dimethyl ether) $\qquad$ (ethyl methyl ether)

In the substitutive system of nomenclature an ether is regarded as formed by substituting a hydrogen atom of an alkane by an alkoxy (RO—) group. For mixed ethers the higher alkane is chosen as the parent compound. As the substitutive system is now the more favoured one, the names given first for the ethers cited are those based on this system.

Since ethers do not contain a hydroxyl group they are unable to form hydrogen bonds. This is reflected in their properties. Although ethers are isomeric with monohydric alcohols their boiling points are considerably lower than those of the corresponding alcohols. Thus methoxymethane, $(CH_3)_2O$, is a gas (b.p. $-23°C$), while its isomer ethanol, C_2H_5OH, is a liquid (b.p. $78°C$). Again, whereas the lower alcohols are completely miscible with water because of hydrogen bonding, the lower ethers are only partially miscible with water. Ethers are excellent solvents for fats and for some non-metallic elements.

The most important aliphatic ether is ethoxyethane (diethyl ether), which is commonly called 'ether.' This is used as a solvent and as an anæsthetic. It is a colourless mobile liquid with a sweet smell. It freezes at $-125°C$, boils at $34\cdot6°C$, and has a density of $0\cdot72$ g cm^{-3}. Ethoxyethane is very flammable, and this fact, combined with its high volatility, necessitates great care in its handling.

Preparation of Ethers

(*i*) *Williamson's Continuous Ether Process.* Ethoxyethane is prepared both in the laboratory and industrially by heating ethanol with concentrated sulphuric(VI) acid at about $140°C$. The overall reaction consists of dehydration of the alcohol under the catalytic influence of hydrogen ions supplied by the acid (the mechanism of the reaction has been given earlier).

$$2C_2H_5OH \xrightarrow{\text{H}^+} (C_2H_5)_2O + H_2O$$

Other acids (e.g. hydrochloric acid) can be used instead of sulphuric(VI) acid, but are less efficient.

Theoretically it should be possible for the acid to continue changing ethanol into the ether indefinitely, and for this reason the method of preparation became known as the 'continuous ether process.' In practice there is a limit to the amount of ethanol which can be converted by a given quantity of acid owing to reduction of the latter by the organic compounds The industrial process is made to operate continuously for several months by passing ethanol vapour into a hot mixture of ethanol and concentrated sulphuric(VI) acid, vaporizing off the ether and water, and adding more acid from time to time.

The laboratory preparation of ethoxyethane is carried out by means of the apparatus shown in Fig. 8-3. A mixture of ethanol (12 cm³) and concentrated sulphuric(VI) acid (10 cm³) is put into a small flask attached to a condenser and receiver. The latter consists of a small distilling-flask, the bulb of which is immersed in iced water. A length of rubber tubing is connected to the side-arm of the receiver, the free end being hung over the side of the bench to lead away any vapour which escapes condensation. The oil-bath is kept at 140°–145°C while ethanol (12 cm³) is added from the dropping-funnel at the same rate as the ether distils over. The crude ether collected contains ethanol, water, and sulphur dioxide. The latter is removed by stirring with a little anhydrous sodium(I) carbonate, and the liquid is then left in contact with anhydrous calcium(II) chloride which removes the water and alcohol. Finally the ether is redistilled, the fraction passing over at 34°–38°C being collected in a cooled receiver.

This method of preparing ethers from alcohols is successful only with the lower primary alcohols. With secondary and tertiary alcohols and with higher primary alcohols the competing reaction of alkene formation largely takes over. Secondary alcohols give poor yields of ethers, and tertiary alcohols practically none. In these cases the alternative method of preparation now given can be used.

(ii) *Williamson's Synthesis of Ethers*. This method consists of heating an alkyl halide (usually an iodo-compound) with a

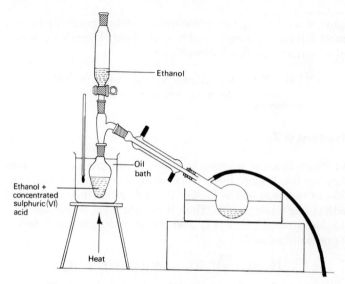

Fig. 8-3. Preparation of ethoxyethane (diethyl ether)

sodium(I) alkoxide in ethanolic solution. The general reaction is expressed by the following equation:

$$RONa + R'I \rightarrow R\text{—}O\text{—}R' + NaI$$

Williamson's synthetic method can be used for the preparation of either simple or mixed ethers. Thus, to obtain methoxyethane (ethyl methyl ether) sodium is dissolved in excess of methanol in a flask attached to a reflux condenser and iodoethane (ethyl iodide) is added. The mixture is gently refluxed for about two hours, when a precipitate of sodium(I) iodide is gradually formed (most salts are sparingly soluble in alcohols). The ether is separated by fractional distillation.

$$CH_3ONa + C_2H_5I \rightarrow CH_3\!-\!O\!-\!C_2H_5 + NaI$$

sodium   iodoethane   methoxyethane
methoxide

Reactions of Ethers

As ethers lack the reactive hydroxyl group present in alcohols they are much more inert. Their most important reaction consists of cleavage of the C—O—C bonds when they are heated with strong mineral acids. Thus, if ethoxyethane is refluxed with concentrated sulphuric(VI) acid, ethyl hydrogen sulphate(VI) is gradually formed.

$$(C_2H_5)_2O + 2H_2SO_4 \rightarrow 2C_2H_5HSO_4 + H_2O$$

In the same way a concentrated solution of hydriodic acid converts ethers into alkyl iodides on heating.

$$R\!-\!O\!-\!R' + 2HI \hookrightarrow RI + R'I + H_2O$$

This reaction is used to identify ethers. A simple ether like ethoxyethane yields only one alkyl iodide, while a mixed ether gives two. By separating and identifying the products the nature of the original ether can be established.

On exposure to light and air ethers absorb oxygen and form unstable peroxides of the type $C_4H_{10}O_4$ These may be present in old samples of ether and cause an explosion if the ether is distilled.

Their presence can be detected by the blue colour which they give with an acidified solution of potassium(I) iodide and starch. Their formation can be avoided by storing ether in dark bottles and adding a small amount of ethanol.

Like alcohols ethers can act as weak bases. They are absorbed by concentrated mineral acids to give substituted oxonium salts in solution. Ethoxyethane and hydrochloric acid form diethyloxonium chloride.

$$C_2H_5\!-\!\overset{..}{\underset{|}{O}}: \; + HCl \rightarrow \left[C_2H_5\!-\!\overset{..}{\underset{|}{O}}\!-\!H \right]^+ + Cl^-$$
$$\quad\; C_2H_5 \qquad\qquad\qquad C_2H_5$$

The salts exist only in concentrated acid solution. If the solutions are diluted with water, the ether separates out again.

EXERCISE 8

1. In the reactions.

    (A) $CO + 2H_2 \rightleftharpoons CH_3OH$ (g)
    (B) $C_2H_4 + H_2O$ (g) $\rightleftharpoons C_2H_5OH$ (g)

which of the following apply to (A) only, to (B) only, to both, or to neither?

(a) The reaction is exothermic in the forward direction.
(b) The higher the temperature the greater is the yield of the alcohol.
(c) The higher the pressure the greater is the yield of the alcohol.
(d) Use of a catalyst increases the yield of the alcohol

2. Give the terms used to describe: (a) ethanol concentrated by fractional distillation to about 96 per cent by volume, (b) ethanol

which is at least 99·5 per cent pure, (c) a mixture of two liquids which boils at a constant temperature (at a fixed pressure), (d) organic catalysts which bring about fermentation.

3. State which of the following apply to methanol only, to ethanol only, to both, or to neither:

(a) It is present in methylated spirit.

(b) When the vapour is passed over heated aluminium(III) oxide an alkene is obtained.

(c) If it is warmed with acidified potassium(I) dichromate(VI) solution the latter turns green.

(d) It gives a yellow precipitate when it is boiled with potassium(I) iodide solution and sodium(I) chlorate(I) (sodium hypochlorite) solution is added.

4. Give the systematic names of the following alcohols:

(a) $CH_3.CH(OH).CH_3$,   (b) $C_2H_5.CH(OH).CH_3$,

(c) $C_2H_5.CH_2OH$,   (d) $(CH_3)_2 CH.CH_2OH$.

5. Write the structural formulae of (a) butan-1-ol, (b) pentan-2-ol, (c) hexan-1-ol, (d) 2-methylpropan-2-ol.

6. Name the organic compounds formed in the reactions between propan-1-ol and (a) sodium, (b) concentrated sulphuric(VI) acid at 80°C, (c) concentrated sulphuric(VI) acid at 180°C, (d) ethanoic acid in the presence of concentrated sulphuric(VI) acid, (e) phosphorus pentachloride.

7. (a) Write equations for the mechanism by which ethanol forms bromoethane with concentrated hydrobromic acid, (b) state whether the alcohol acts as an electrophile or a nucleophile in this reaction.

8. Which of the following statements are true?

(a) Williamson's 'continuous ether process' is not continuous.

(b) Methoxyethane can be synthesized by refluxing an alcoholic solution of sodium ethoxide with iodomethane.

(c) The formula C_3H_8O can represent one ether and two alcohols.

(d) Ethoxyethane has a lower boiling point than ethanol because its molecules are larger.

9 Halogen Derivatives of Alkanes

It was shown in Chapter 5 that chlorine and bromine bring about slow substitution in alkanes. Thus methane gives rise to the following series of substitution compounds with chlorine: chloromethane (methyl chloride), CH_3Cl, dichloromethane (methylene dichloride), CH_2Cl_2; trichloromethane (chloroform), $CHCl_3$, and tetrachloromethane (carbon tetrachloride), CCl_4. Ethane yields a similar series. Although the compounds are classified as halogen derivatives of alkanes, only in a few special cases are they made directly from the hydrocarbons. This is partly because the reactions are slow and partly because substitution usually results in a mixture of products which require separation. Iodo-derivatives are always prepared indirectly.

MONOHALOGEN DERIVATIVES

Substitution of one hydrogen atom of an alkane molecule by a halogen atom produces an *alkyl halide*. Alkyl halides are represented by the general formula $C_nH_{2n+1}X$, or RX, in which R stands for the alkyl radical and X for the halogen atom. It has been customary to name alkyl halides according to the nature of the alkyl radical and halogen atom, but they are now named systematically as substitution derivatives of alkanes. Like alcohols, alkyl halides are classified as primary, secondary, and tertiary. These have the types of structure shown below.

Primary	*Secondary*	*Tertiary*
	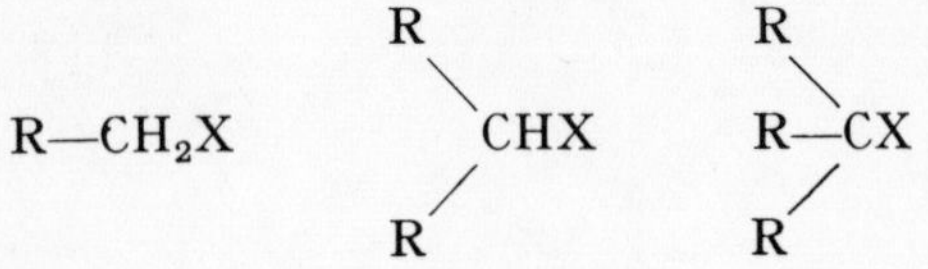 R	R
	\	\
$R—CH_2X$	CHX	$R—CX$
	/	/
	R	R

The most important alkyl halides on the chloro-, bromo-, and iodo-derivatives of methane and ethane. The fluoro- derivatives are of little consequence and will be omitted in this discussion.

Melting points and boiling points of alky halides are higher than those of the corresponding alkanes. This is partly due to their greater molecular mass and partly to mutual attraction between the electric dipoles present in the molecules (Chapter 4) The boiling points of the lower alky halides are shown in Table 9-1. It will be seen that there is a progressive increase in boiling point with increase in molecular mass from the chloro-derivatives to the iodo-derivatives and from methyl to butyl.

Table 9-1. BOILING POINTS OF ALKYL HALIDES (°C)

Alkyl radical	Cl	Br	I
Methyl (CH_3—)	−24	4·5	42
Ethyl (C_2H_5—)	12·5	38	72
Propyl (C_3H_7—)	46	71	102
Butyl (C_4H_9—)	78	102	129

Chloromethane, bromomethane, and chloroethane are gases at ordinary temperatures and pressures. The other compounds are colourless pleasant-smelling liquids, some of which have low boiling points and are very volatile. Their densities increase in the order: RCl < RBr < RI. Thus the densities of (liquid) chloroethane (C_2H_5Cl), bromoethane (C_2H_5Br), and iodoethane (C_2H_5I) are 0·91, 1·43, and 1·93 g cm^{-3} respectively. Alkyl halides are only slightly soluble in water, but dissolve readily in ethanol, ether, and other organic solvents.

Alkyl halides, like other halogen derivatives of alkanes (e.g. chloroform), are narcotics, and their vapours can cause permanent damage to the liver and kidneys. The vapours are also flammable. The bromo- and iodo- compounds slowly turn brown on keeping owing to liberation of bromine or iodine under the influence of light.

Laboratory Preparation of Alkyl Halides

The most convenient alkyl halides to prepare are those which have boiling points in the range $70°$–$120°C$. These are less volatile than the compounds of lower boiling point, and losses by evaporation are smaller.

(i) *From the Alcohol and the Hydrohalogen Acid.* The chloro-, bromo-, and iodo-compounds can be obtained by refluxing the alcohol with hydrochloric acid, hydrobromic acid, or hydriodic acid. The reactions are reversible, but under the conditions used the equilibrium point is well to the right.

$$ROH + HX \rightleftharpoons RX + H_2O$$

The order of reactivity of the hydrohalogen acids with alcohols is: $HI > HBr > HCl$. Primary alcohols react only slowly with hydrochloric acid, and it is necessary to use a catalyst (anhydrous zinc(II) chloride) to increase the rate of reaction. 1-Chlorobutane $(CH_3.CH_2.CH_2.CH_2Cl)$ can be prepared by refluxing butan-1-ol for about an hour with concentrated hydrochloric acid and anhydrous zinc(II) chloride.

In the preparation of a primary alkyl bromide from the alcohol and hydrobromic acid concentrated sulphuric(VI) acid is used as a catalyst (the mechanism of the reaction has been described in the last chapter). Sulphuric(VI) acid, however, cannot be used as a catalyst with hydriodic acid because the latter is oxidized to iodine. A catalyst is not required in this case because of the greater reactivity of hydriodic acid.

A better method of obtaining primary alkyl bromides is to reflux the alcohol with a concentrated aqueous solution of potassium(I) bromide and concentrated sulphuric(VI) acid. Hydrogen bromide is formed *in situ*, and this reacts with the alcohol. Details for the preparation of 1-bromobutane (*n*-butyl bromide) are given below.

$$C_4H_9OH + KBr + H_2SO_4 \rightleftharpoons C_4H_9Br + KHSO_4 + H_2O$$

Experiment

As fumes of hydrogen bromide are given off in the reaction the experiment should be performed in a fume-cupboard, or steps should be taken to absorb the fumes in water. A suitable arrangement for this purpose is shown in Fig. 9-1.

Dissolve 8 g of potassium(I) bromide in 10 cm³ of water and add the solution to 7 cm³ of butan-1-ol in the 50 cm³ pear-shaped flask. Add also 10 cm³ of concentrated sulphuric(VI) acid in portions of 1–2 cm³. Swirl the flask round to mix the liquids and cool it under the tap. Attach the flask to the remainder of the apparatus. It is easier to control the refluxing if the flask is placed on a gauze and heated by means of a small Bunsen flame, the latter being adjusted to keep the contents of the flask gently boiling.

Continue the refluxing for half-an-hour and then change over the apparatus for direct distillation. Distil the liquid in the flask, using a small direct flame from a Bunsen burner from which the chimney has been removed. Stop the distillation when no more oily drops are obtained.

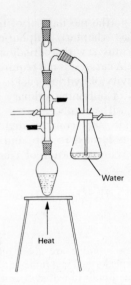

Water

Heat

Fig. 9-1. Preparation of 1-bromobutane

Purification. The distillate consists of two layers, an upper aqueous layer and a lower layer of the alkyl bromide. Impurities in the lower layer include unchanged butan-1-ol, water, hydrogen bromide, bromine, and sulphur dioxide. Decant off most of the top layer and transfer the remaining liquid to a separating-funnel. Shake the bromo-compound firstly with 10 cm³ of water and then with 10 cm³ of 'bench' sodium(I) hydroxide solution (this removes the acidic impurities and bromine). In each case allow the liquid to separate into two layers and discard the upper layer. On the second occasion run the lower layer into a small dry conical flask, and add some pieces of anhydrous calcium(II) chloride. The liquid has a

cloudy appearance owing to suspended droplets of water. Leave it to dry for ten minutes, swirling the flask round occasionally.

Filter the alkyl halide, which should now be clear, through a funnel containing a plug of cotton wool into a clean dry 50 cm³ pear-shaped flask. Add a little powdered pumice and distil the liquid, collecting the fraction which passes over at 99°–103°C. The yield is about 7 g (5·5 cm³), which is 65 per cent of the theoretical yield.

(*ii*) *From the Alcohol and Sulphur Dichloride Oxide.* Sulphur dichloride oxide ($SOCl_2$) is commonly called thionyl chloride. It is a colourless fuming liquid (b.p. 78°C). The alcohol is added slowly to this liquid, and the mixture is refluxed for one hour. The alkyl chloride produced is then distilled off. A typical preparation is that of 1-chlorohexane (*n*-hexyl chloride) from hexan-1-ol.

$$C_6H_{13}OH + SOCl_2 \rightarrow C_6H_{13}Cl + SO_2 + HCl$$
hexan-1-ol   1-chlorohexane

The use of sulphur dichloride oxide for making alkyl chlorides has two particular advantages. It gives excellent yields and the products are easily purified from the two gaseous by-products. The method is limited to the preparation of alkyl chlorides because sulphur dibromide oxide and sulphur di-iodide oxide are not available as commercial substances.

(*iii*) *From the Alcohol, Red Phosphorus, and Bromine (or Iodine).* Alkyl halides can be obtained from alcohols and phosphorus halides. This type of reaction is used to prepare alkyl bromides and iodides, but not chlorides because yields of the latter are poor. In practice the phosphorus halides are not used directly. Instead, the alcohol is refluxed with *red* phosphorus and bromine or iodine (yellow phosphorus reacts with these too vigorously).

Phosphorus tribromide or phosphorus tri-iodide is first formed, and this reacts with the alcohol. Refluxing is carried out on a water-bath and continued for about an hour. The alkyl halide is then distilled off and purified as described earlier for 1-bromo-butane.

$$3C_2H_5OH + PBr_3 \rightarrow 3C_2H_5Br + H_3PO_3$$

Manufacture and Uses of Alkyl Halides

Large-scale production of alkyl halides is mostly limited to the derivatives of methane and ethane.

Chloromethane (CH_3Cl) is manufactured in Britain by *Groves's process*, in which hydrogen chloride under pressure is passed into methanol in the presence of anhydrous zinc(II) chloride as a catalyst. In the U.S.A. chloromethane is obtained by chlorinating methane from natural gas. The reaction temperature is 400°C, and as a safeguard against explosions a large excess of methane is used. A mixture of all four chloromethanes is produced. The mixture is liquefied by cooling under pressure, and then separated by fractional distillation. Chloromethane is chiefly used as a 'methylating' agent (a substance which introduces the CH_3— group into a molecule).

Chloroethane (C_2H_5Cl) is made by a two-stage process, involving chlorination of ethane at 400°C and addition of the resulting hydrogen chloride to ethene at 200°C in the presence of anhydrous aluminium(III) chloride as a catalyst.

$$C_2H_6 + Cl_2 \rightarrow C_2H_5Cl + HCl$$
$$C_2H_4 + HCl \rightarrow C_2H_5Cl$$

Chloroethane is chiefly used as an 'ethylating' agent and in the manufacture of the petrol additive tetraethyl-lead(IV).

In spite of their greater reactivity the monobromo- and mono-iodo- derivatives of methane and ethane are used on a much smaller scale than the chloro-compounds. This is because of their greater cost. They are made by treating the alcohols with red phosphorus and the appropriate halogen, and are chiefly employed as methylating and ethylating agents in laboratories.

Reactions of Alkyl Halides

Most of the important reactions of alkyl halides are of the *substitution* type, in which the functional halogen atom is replaced by another atom or group. Substitution can be brought about by water, ammonia, aqueous or alcoholic solutions of alkalies, potassium(I) cyanide, and silver(I) salts. The actual reagents in these cases are either neutral molecules (e.g. H_2O or NH_3) or negative ions (e.g. OH^- or CN^-). In both cases the reactants are nucleophilic reagents (Chapter 4), and therefore the reactions are described as *nucleophilic* substitutions. Alkyl halides themselves are electrophilic reagents.

Under suitable conditions alkyl halides also give rise to *elimination* reactions, in which a molecule of hydrogen halide is removed from the alkyl halide molecule.

Nucleophilic Substitutions

Owing to the strongly electronegative character of the halogen atom there is a relatively large electron displacement in the carbon-halogen bond of an alkyl halide. As a result the carbon atom has a partial positive charge, or low electron density, which makes it vulnerable to attack by a nucleophilic reagent. The latter donates a share in a lone pair of electrons to form a co-ordinate covalent bond with the carbon atom.

1. Hydrolysis. The term 'hydrolysis' means literally 'splitting apart by water.' Generally, however, it is used to describe reactions in which an atom or group is substituted by a hydroxyl group by means of water, aqueous alkalies, or aqueous acids.

Water alone hydrolyses primary alkyl halides only slowly, even on boiling. This is partly due to the small mutual solubility of the compounds and partly to water being a relatively weak nucleophile.

$$R-X + H_2O \rightleftharpoons R-OH + HX$$

The OH^- ion, on the other hand, is a strong nucleophile (its oxygen atom has a full negative charge instead of the partial negative charge on the oxygen atom of the water molecule). Thus hot aqueous alkalies cause rapid hydrolysis of primary alkyl halides, particularly if ethanol is added to act as a common solvent.

$$\overset{-}{HO}: + \overset{\delta+}{CH_3}-\overset{\delta-}{I} \rightleftharpoons HO-CH_3 + I^-$$

For alkyl halides containing the same halogen rates of hydrolysis are in the order: primary < secondary < tertiary. The mechanisms of hydrolysis of primary and tertiary alkyl halides are different, and have been described in detail in Chapter 4. There is evidence that in the case of secondary alkyl halides both mechanisms operate simultaneously.

2. Reaction with Alkalies in Ethanolic Solution (Alcoholysis). If a primary alkyl halide is refluxed with a concentrated ethanolic solution of sodium(I) hydroxide (or potassium(I) hydroxide) an ether is produced. Here the nucleophilic reagent which brings about substitution is the ethoxide ion. Thus the formation of ethoxyethane (diethyl ether) from iodoethane and an ethanolic solution of sodium(I) hydroxide may be represented as shown.

$$C_2H_5\overset{-}{O}: + \overset{CH_3}{\underset{|}{CH_2}}-I \rightleftharpoons C_2H_5-O-C_2H_5 + I^-$$

The reaction is described as an *alcoholysis* because of its similarity to hydrolysis. Thus alcoholysis consists of substitution of an atom or group by the ethoxide group (C_2H_5O-) by means of either ethanol or ethoxide ions. Like water, ethanol itself is only a weak nucleophile, and does not react appreciably with alkyl halides. However, it brings about alcoholysis of strong electrophiles such as acyl halides (Chapter 12).

3. Reaction with Potassium(I) Cyanide. Cyanide ions substitute in alkyl halides in the same way as hydroxyl ions and ethoxide ions. Thus, when potassium(I) cyanide is refluxed with iodomethane in aqueous-ethanolic solution methyl cyanide is produced. The mixed solvent is necessary because of the small solubilities of potassium(I) cyanide in ethanol and iodomethane in water. The methyl cyanide is isolated by fractional distillation.

$$CN^- + CH_3I \rightarrow CH_3CN + I^-$$

This is an important general reaction of alkyl halides because of the additional carbon atom incorporated in the molecule.

Since the —CN group can be readily converted by hydrolysis into a —COOH group, the reaction can be used to synthesize a carboxylic acid from an alcohol through the alkyl halide. The resulting acid contains one more carbon atom in the molecule than the alcohol. For example, ethanoic, or acetic acid may be obtained from methanol by the following steps:

$$CH_3OH \xrightarrow{HI} CH_3I \xrightarrow{KCN} CH_3CN \xrightarrow{H_2O} CH_3.COOH$$
(methanol)   (ethanoic, or acetic acid)

4. Reactions with Silver Salts. Alkyl halides form *esters* with the silver salts of both inorganic and organic acids. The reactants are usually dissolved in ethanol and heated together under reflux. Alkyl iodides react more rapidly than the bromides or chlorides. Iodomethane gives a precipitate of silver(I) iodide merely by shaking in the cold with an ethanolic solution of silver(I) nitrate(V). The ester produced is methyl nitrate(V).

$$AgNO_3 + CH_3I \rightarrow CH_3NO_3 + AgI \downarrow$$
methyl nitrate(v)

If iodomethane is refluxed with silver(I) ethanoate, or acetate, in ethanolic solution, silver(I) iodide is again precipitated and the ester methyl ethanoate, or acetate, is formed. This can be isolated by fractional distillation.

$$CH_3.COOAg + CH_3I \rightarrow CH_3.COOCH_3 + AgI \downarrow$$
methyl ethanoate

5. Reaction with Ammonia (Ammonolysis). 'Ammonolysis' is analogous to hydrolysis and alcoholysis, and results in the halogen atom of an alkyl halide being substituted by an amino (—NH₂) group. The product is an *amine*. The reaction is carried out by heating the alkyl halide with excess of an ethanolic solution of ammonia in a sealed vessel (to prevent escape of ammonia). In this way iodomethane yields *methylamine*.

$$CH_3I + 2NH_3 \rightarrow CH_3NH_2 + NH_4I$$
methylamine

The reaction probably takes place in two stages. In the first stage a positively charged methylammonium ion is formed, and in the ethanolic solution this remains in close association with the expelled halide ion, giving a salt, methylammonium iodide. The positive sign of the methylammonium ion is usually written outside the square bracket, so that the ion is represented by the formula $[H_3N—CH_3]^+$, or $[CH_3—NH_3]^+$.

In the second stage of the reaction a proton is removed from the methylammonium ion by an ammonia molecule. This is an acid-base reaction, and is reversible. The methylammonium ions are thus partially converted into methylamine.

(Stage I)   *Ammonolysis*

$$H_3N: + CH_3—I \rightleftharpoons [H_3\overset{+}{N}—CH_3]I^-$$

methylammonium iodide

(Stage II)

$$[CH_3—NH_3]^+ I^- + NH_3 \rightleftharpoons CH_3—NH_2 + [NH_4]^+ I^-$$

6. Reduction. As seen in Chapter 5, reducing agents convert alkyl halides into alkanes. The reducing agent may be zinc and a mineral acid, sodium and ethanol, or an aluminium-mercury couple and ethanol. On reduction iodoethane gives ethane.

$$C_2H_5I + 2H \rightarrow C_2H_6 + HI$$

The reduction mechanism is complex and not fully understood. The reaction is classified as nucleophilic substitution because a reducing agent is essentially an electron donor. The electrons are supplied by the metal atoms, which are oxidized to cations.

The Elimination Reaction

This is an alternative to substitution as a method of reaction between an alkyl halide and the stronger nucleophilic reagents. The latter, which include OH^-, $C_2H_5O^-$, and CN^-, are also strong bases. Under suitable conditions they may cause an alkyl halide to lose not only its halogen atom (as an anion), but also a hydrogen atom (as a proton). This is equivalent to removal, or elimination, if one molecule of hydrogen halide. The hydrogen atom removed is one attached to a carbon atom in the 2-position, that is, the second carbon atom from the halogen atom. The reaction produces an alkene, as now shown.

$$R-CH_2-CH_2X \xrightarrow{-HX} R-CH=CH_2$$

To understand why the elimination reaction occurs one must bear in mind that a nucleophilic reagent attacks an atom with a partial positive charge, or low electron density. In the substitution reactions this is the carbon atom of the $-CH_2X$ group. This, however, is not the only site of low electron density in an alkyl halide molecule. The $-I$ (electron-attracting) effect of

the halogen atom X is relayed beyond the C—X bond to neighbouring C—H bonds, producing electron displacements which confer partial positive charges on the hydrogen atoms. In this way the hydrogen atoms acquire some acidic character and become vulnerable to attack by a strong base like OH^- or $C_2H_5O^-$. The base is able to remove one of the hydrogen atoms as H^+, while at the same time it expels the halogen atom X as X^-. The two electrons released by heterolytic fission of the C—H bond are used to form an extra bond (a π bond) with the carbon atom originally joined to X. Thus the elimination reaction between an ethanolic solution of sodium(I) hydroxide and an alkyl halide may be represented as follows:

$$C_2H_5\overset{-}{O}: + \overset{\delta+}{H} \quad \underset{\underset{|}{|}}{\overset{\underset{|}{|}}{-C-C}}-\overset{\delta-}{X} \rightarrow C_2H_5OH + \overset{\diagup}{\underset{\diagup}{C}}=\overset{\diagdown}{\underset{\diagdown}{C}} + X^-$$

With certain exceptions alkyl halides must be regarded as capable of reacting by substitution and elimination simultaneously. The exceptions are the methyl halides, which do not possess a carbon atom in the 2-position and therefore cannot react by elimination. In other cases the relative extents to which the competing reactions occur depend on their respective rates. These in turn are governed by the nature of the alkyl halides and by the experimental conditions. The tendency for substitution to give way to elimination increases from primary alkyl halides to secondary, and from secondary to tertiary. Substitution is favoured by dilute aqueous alkalies (OH^- ions), elimination by concentrated ethanolic alkalies ($C_2H_5O^-$ ions).

Dilute aqueous sodium(I) hydroxide brings about substitution in all three classes of alkyl halides to give alcohols. Only in the case of the tertiary compounds is a small amount of alkene also formed by elimination.

With concentrated *ethanolic sodium(I) hydroxide* and *primary* alkyl halides the main reaction is again substitution (but in this case to give ethers). The yield of alkene by elimination is only 1–10 per cent of the total organic product. For *secondary* alkyl halides, however, elimination preponderates, relatively little ether being formed. Thus with 2-bromopropane (isopropyl bromide), CH_3—$CHBr$—CH_3, 70–80 per cent of the product is the alkene propene. For *tertiary* alkyl halides elimination is outstandingly the chief mode of reaction, the yield of alkene approaching 100 per cent of the total organic product.

Chemical Tests for Alkyl Halides

Alkyl halides can be recognized by means of the following tests:

(*i*) Beilstein's test (Chapter 3) can be used to show that a halogen element is present. A positive result does not necessarily mean, however, that the compound is an alkyl halide.

(*ii*) With ethanolic silver(I) nitrate(V) solution a precipitate of silver(I) halide is formed at once or on warming in a water-bath. The halogen can be identified by tests on the silver(I) halide precipitated.

(*iii*) By refluxing the alkyl halide with aqueous potassium(I) hydroxide, the corresponding alcohol is liberated. This can be distilled off and identified. The halogen is left as the potassium(I) halide and can be identified, after acidifying the solution with nitric(V) acid, by means of silver(I) nitrate(V) solution.

DIHALOGEN DERIVATIVES

The only important dihalogen derivatives of alkanes are dichloromethane, dichloroethane, and dibromoethane.

Dichloromethane (CH_2Cl_2). Dichloromethane (methylene dichloride) is a colourless oily liquid, which boils at 41°C. It is manufactured by chlorination of either methane from natural gas or chloromethane (methyl chloride) obtained by the Groves process. In the second method a mixture of chloromethane and chlorine is passed up a tower maintained at a temperature of about 350°C by heat given out in the reaction. A mixture of vapours, containing unchanged chloromethane as well as di-, tri-, and tetrachloromethane, passes out at the top of the tower. The vapours are led into a series of fractionating towers, which condense and separate the individual compounds.

Dichloromethane is chiefly used as a paint 'stripper' and as a solvent for extracting oils.

1,2-Dichloroethane ($CH_2Cl.CH_2Cl$). This compound, which is commonly called ethylene dichloride, is formed by direct combination of ethene (ethylene) and chlorine. It is made on a large scale by passing the two gases into a solvent (1,2-dichloroethane itself) at 25°C, a catalyst being used to assist the reaction.

1,2-Dichloroethane is a heavy colourless liquid (b.p. 84°C) with a pleasant smell. It is almost insoluble in water, but dissolves readily in organic solvents.

If 1,2-dichloroethane is refluxed with an *aqueous* solution of sodium(I) hydroxide, the two halogen atoms are substituted by hydroxyl groups and ethane-1,2-diol (ethylene glycol) is formed

$$CH_2Cl.CH_2Cl + 2OH^- \rightarrow CH_2OH.CH_2OH + 2Cl^-$$

With a boiling *ethanolic* solution of sodium(I) hydroxide two molecules of hydrogen chloride are eliminated, and ethyne is produced.

$$CH_2Cl.CH_2Cl - 2HCl \rightarrow CH \equiv CH$$

1,2-Dichloroethane is used to some extent as a petrol additive (for 'scavenging' lead), as a fumigant for dried fruits and cereals, and as an intermediate in the manufacture of synthetic rubber.

1,1-*Dichloroethane* (ethylidene dichloride), $CH_3.CHCl_2$, is isomeric with 1,2-dichloroethane. It is a colourless liquid (b.p. 57°C) which is obtained by adding phosphorus pentachloride to ethanal (acetaldehyde). Phosphorus trichloride oxide is also formed. 1,1-Dichloroethane has little practical importance.

$$CH_3.CHO + PCl_5 \rightarrow CH_3.CHCl_2 + POCl_3$$

1,2-Dibromoethane ($CH_2Br.CH_2Br$). By far the largest proportion of bromine produced industrially is used in the manufacture of 1,2-dibromoethane for addition to 'leaded' petrol. This application of the halogen compound was described in Chapter 5. The dibromo-compound is manufactured in the same way as the dichloro-compound, ethene being passed into a solution of bromine in 1,2-dibromoethane itself.

In the laboratory 1,2-dibromoethane is usually prepared by passing ethene into liquid bromine covered by a layer of water. The preparation is dangerous, partly because of the highly corrosive character of liquid bromine and partly because excess of ethene, which is flammable, escapes from the apparatus.

1,2-Dibromoethane is also a heavy colourless liquid (b.p. 131·5°C). It reacts in the same way as the dichloro-compound with aqueous and alcoholic alkalies, but is more reactive. The use of 1,2-dibromoethane for preparing ethyne in the laboratory has been described in Chapter 7.

OTHER HALOGEN DERIVATIVES

Trichloromethane ($CHCl_3$). Trichloromethane (chloroform) can be prepared in the laboratory by adding propanone (acetone) to a suspension of bleaching-powder in water and distilling the mixture. Bleaching-powder may be regarded as composed of calcium(II) chlorate(I) (calcium hypochlorite), $Ca(OCl)_2$, and calcium(II) hydroxide, $Ca(OH)_2$. The propanone is first chlorinated by the chlorate(I) ions to trichloropropanone, which then reacts with hydroxyl ions to give trichloromethane and ethanoate ions.

$$\begin{array}{c} CH_3 \\ \diagdown \\ CO + 3OCl^- \rightarrow \\ \diagup \\ CH_3 \end{array} \quad \begin{array}{c} CCl_3 \\ \diagdown \\ CO + 3OH^- \\ \diagup \\ CH_3 \end{array}$$

$$\begin{array}{c} CCl_3 \\ \diagdown \\ CO + OH^- \rightarrow CHCl_3 + CH_3.COO^- \\ \diagup \\ CH_3 \end{array}$$

On a large scale trichloromethane is obtained by the process described for the production of dichloromethane, in which dichloromethane, trichloromethane, and tetrachloromethane are produced simultaneously by chlorination of monochloromethane.

Trichloromethane is a colourless liquid with a sweet sickly

smell. It boils at 61°C and has a density of 1.50 g cm^{-3}. When the vapour is inhaled it rapidly produces unconsciousness. The liquid is immiscible with water, but miscible with ethanol and ether. Trichloromethane is an excellent solvent for oils and fats. The liquid is non-flammable, but when it is heated in air the vapour burns with a green-edged flame.

Trichloromethane does not give a precipitate of silver(I) chloride with ethanolic silver(I) nitrate(V) solution, even on warming. It is not hydrolysed by water, but hydrolysis occurs with boiling aqueous alkalies. With sodium(I) hydroxide solution it yields sodium(I) formate and sodium(I) chloride.

$$CHCl_3 + 4NaOH \rightarrow HCOONa + 3NaCl + 2H_2O$$

One of the best-known reactions of trichloromethane is the *isocyano reaction*. When trichloromethane is warmed with phenylamine (aniline) and an ethanolic solution of potassium(I) hydroxide the compound isocyanobenzene (phenyl isocyanide) is formed. This has a nauseating smell.

$$\underset{\text{phenylamine}}{C_6H_5NH_2} + CHCl_3 + 3KOH \rightarrow \underset{\text{isocyanobenzene}}{C_6H_5NC} + 3KCl + 3H_2O$$

Experiment

Put into a test-tube 1 cm^3 of trichloromethane, $2-3 \text{ cm}^3$ of an ethanolic solution of potassium(I) hydroxide, and 2 or 3 drops of phenylamine. Warm the tube gently, and shake it. The characteristic smell of isocyanobenzene is soon noticed. (*On no account must the remaining liquid be poured into the sink.* It must be either discarded out-of-doors, or, when cool, treated carefully with excess of concentrated hydrochloric acid, which destroys the isocyano-compound.)

The isocyano reaction serves as a chemical test for trichloromethane or, conversely, for the class of compounds called 'primary amines' (Chapter 15). Phenylamine is a primary aromatic amine.

Tri-iodomethane (CHI_3) corresponds chemically to trichloromethane, but is a light-yellow crystalline solid. It is chiefly of interest because of the *tri-iodomethane (iodoform) reaction*, which is used to distinguish between ethanol and methanol (Chapter 8).

Tetrachloromethane (CCl_4). Tetrachloromethane (carbon tetrachloride) is well known for its use in fire-extinguishers of the 'Pyrene' type. It is also employed as a cleaning-agent, as a solvent for oils and fats, and in the manufacture of 'Arctons' (see chlorofluorocarbons). It is a heavy colourless liquid (b.p. 77°C) with a peculiar sweet smell. It is immiscible with water. The vapour is toxic and when inhaled causes unconsciousness and death.

Tetrachloromethane is chiefly manufactured in Britain by passing chlorine into carbon disulphide at 40°C in the presence of iron as a catalyst. Disulphur dichloride is also formed.

$$CS_2 + 3Cl_2 \rightarrow CCl_4 + S_2Cl_2$$

The products are separated by fractional distillation, and a further yield of tetrachloromethane is obtained by heating the disulphur dichloride with more carbon disulphide at 80°C, iron again being used as a catalyst.

$$CS_2 + 2S_2Cl_2 \rightarrow CCl_4 + 6S \downarrow$$

Tetrachloromethane is the final product obtained by substituting the hydrogen in methane by chlorine. The substitution can be reversed by treating tetrachloromethane in ethanolic solution with zinc and hydrogen chloride. Trichloromethane, dichloromethane, chloromethane, and methane are formed in turn.

$$CCl_4 \xrightarrow{2H} CHCl_3 \xrightarrow{2H} CH_2Cl_2 \xrightarrow{2H} CH_3Cl \xrightarrow{2H} CH_4$$

Like trichloromethane, tetrachloromethane is not hydrolysed by water, and no precipitate is formed when the liquid is shaken with silver(I) nitrate(V) solution (aqueous or ethanolic). Hydrolysis can be brought about by boiling the compound with sodium(I) hydroxide solution.

$$CCl_4 + 6NaOH \rightarrow 4NaCl + Na_2CO_3 + 3H_2O$$

Fluorocarbons

A fluorocarbon is a compound formed by replacing all the hydrogen atoms of a hydrocarbon by fluorine atoms. Normally fluorine reacts explosively with hydrocarbons, but if the reactants are diluted with nitrogen and cobalt(II) fluoride is present as a catalyst, substitution proceeds smoothly. Ethane is converted into hexafluoroethane.

$$C_2H_6 + 6F_2 \xrightarrow{CoF_2} C_2F_6 + 6HF$$

Fluoroalkanes from CF_4 to C_4F_{10} are colourless, odourless gases. Higher members are oily liquids, and these have properties which make them useful for special purposes in spite of their high cost. They are non-flammable, are not decomposed by heat, and have low surface tensions and freezing points. They can therefore be used as lubricants for machinery which operates at either low or high temperatures. Their high electrical resistance makes it possible to use them in this way in electrical machinery. They are immune to chemical attack by all reagents except fused caustic alkalies, so that they can be employed as solvents in cases where a particularly inert liquid is required.

Another important fluorocarbon is tetrafluoroethene (C_2F_4), which is used to manufacture poly(tetrafluoroethene), $(C_2F_4)_n$. The latter is described in Chapter 26.

Chlorofluorocarbons are derivatives containing chlorine and fluorine. They are known by the trade-name of 'Arctons' ('Freons' in the U.S.A.). A typical compound is dichlorodifluoromethane, which is manufactured by passing hydrogen fluoride into tetrachloromethane in the presence of antimony pentachloride as a catalyst.

$$CCl_4 + 2HF \xrightarrow{SbCl_5} CCl_2F_2 + 2HCl$$

Dichlorodifluoromethane is a colourless gas with no smell. It is non-flammable, non-toxic, and easily liquefied by pressure.

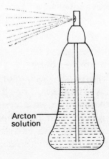

Arcton solution

Fig. 9-2. An aerosol spray

These and other properties make it suitable for use as a refrigerant, and it is utilized in this way in household refrigerators. Another use is as the propellant in aerosol sprays for hair lacquers, insecticides, etc. A solution of the material in the liquefied gas is held under pressure in a container (Fig. 9-2). When a valve is opened the pressure of the vapour expels the solution as a fine spray, from which the solvent immediately evaporates.

GRIGNARD REAGENTS

When magnesium is suspended in a solution of an alkyl halide in dry ethoxyethane (diethyl ether) a vigorous exothermic reaction occurs, and the metal dissolves.

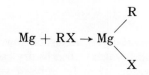

In the case of iodomethane the resulting solution contains methylmagnesium(II) iodide (CH_3—Mg—I).

Although the reaction is applicable to alkyl halides in general, the bromo- and iodo-compounds are preferred to the less reactive chloro-compounds. The solutions obtained are called *Grignard reagents* after the French chemist who discovered them in 1901. Grignard reagents are important as intermediate compounds in the preparation of several classes of organic compounds. They are particularly valuable as a means of organic synthesis. In carrying out these preparations it is unnecessary to isolate the intermediate compounds, as the ethereal solutions can be used directly. Methylmagnesium(II) iodide may be taken as representative.

1. Preparation of Alkanes. An alkane is obtained by reaction between a Grignard reagent and water. Thus when water is added to an ethereal solution of methylmagnesium(II) iodide methane is produced. A white precipitate of magnesium(II) iodide hydroxide (basic magnesium iodide) is also formed.

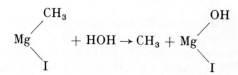

It is important that the substances and the apparatus used for preparation of a Grignard reagent should be rigorously dried. Otherwise, the above type of reaction begins as soon as the magnesium starts dissolving in the ethereal solution of alkyl halide. The metal becomes covered with a layer of the 'basic' salt, and further dissolving is prevented. The substances are dried with sodium, and the apparatus is left for some time in a warm oven.

2. Preparation of Primary Alcohols. If dry oxygen is passed into the ethereal solution of a Grignard reagent the gas is absorbed. When the resulting solution is hydrolysed a primary alcohol is obtained. This contains the same number of carbon atoms in the molecule as the parent alkyl iodide.

$$2\,\underset{\underset{I}{|}}{\overset{\overset{CH_3}{|}}{Mg}} + O_2 \rightarrow 2\,\underset{\underset{I}{|}}{\overset{\overset{OCH_3}{|}}{Mg}}$$

$$\underset{\underset{I}{|}}{\overset{\overset{OCH_3}{|}}{Mg}} + HOH \rightarrow CH_3OH + \underset{\underset{I}{|}}{\overset{\overset{OH}{|}}{Mg}}$$

Hydrolysis is carried out either with ice-cold water or with a mixture of ice and dilute sulphuric(VI) acid. Cooling is necessary to slow down, and thus minimize, the further reaction which takes place between a Grignard reagent and an alcohol (an alcohol, like water, gives an alkane). The acid prevents precipitation of the 'basic' magnesium salt, which therefore does not require separation.

3. Synthesis of Secondary Alcohols. An ethereal solution of a Grignard reagent reacts additively with an aldehyde by means of the carbonyl group of the latter. When the addition compound is hydrolysed a secondary alcohol is obtained. Methylmagnesium(II) iodide and ethanal (acetaldehyde) yield propan-2-ol (isopropyl alcohol).

$$\underset{\underset{I}{|}}{\overset{\overset{CH_3}{|}}{Mg}} \xrightarrow{CH_3.CHO} \underset{\underset{I}{|}}{\overset{\overset{OCH{<}^{CH_3}}{|}}{Mg}} \xrightarrow{HOH} \underset{CH_3}{\overset{CH_3}{CHOH}} + \underset{\underset{I}{|}}{\overset{\overset{OH}{|}}{Mg}}$$

Methanal (formaldehyde), HCHO, differs from other aldehydes in giving a *primary* alcohol (ethanol) in the above case.

4. Synthesis of Tertiary Alcohols. A ketone also forms an addition compound with a Grignard reagent by means of its carbonyl group. The addition compound on hydrolysis forms a tertiary alcohol. With methylmagnesium(II) iodide and propanone (acetone) the alcohol obtained is 2-methylpropan-2-ol (tertiary butyl alcohol).

$$\underset{\underset{I}{|}}{\overset{\overset{CH_3}{|}}{Mg}} \xrightarrow{(CH_3)_2CO} \underset{\underset{I}{|}}{\overset{\overset{OC(CH_3)_3}{|}}{Mg}} \xrightarrow{HOH} (CH_3)_3COH + \underset{\underset{I}{|}}{\overset{\overset{OH}{|}}{Mg}}$$

5. Synthesis of Carboxylic Acids. Below 0°C an ethereal solution of a Grignard reagent absorbs dry carbon dioxide, and the addition compound when hydrolysed yields a carboxylic acid. The addition compound is usually made by adding the ethereal solution to solid carbon dioxide. In this way ethanoic, or acetic, acid can be synthesized from carbon dioxide and methylmagnesium(II) iodide.

$$\underset{\underset{I}{|}}{\overset{\overset{CH_3}{|}}{Mg}} \xrightarrow{CO_2} \underset{\underset{I}{|}}{\overset{\overset{O\overset{\|}{C}.CH_3}{|}}{Mg}} \xrightarrow{HOH} CH_3.C\overset{\overset{O}{\diagup}}{\underset{\underset{OH}{}}{}} + \underset{\underset{I}{|}}{\overset{\overset{OH}{|}}{Mg}}$$

1. Give the systematic names of the following alkyl halides (that is, name them as halogen derivatives of alkanes):
(a) $CH_3.CH_2.CH_2Cl$; (b) $(CH_3)_2 CHBr$; (c) $(CH_3)_3 CBr$.

2. Which of the following could be used for the laboratory preparation of 1-iodopropane from propan-1-ol; (a) concentrated hydriodic acid alone, (b) concentrated hydriodic acid with concentrated sulphuric(VI) acid as a catalyst, (c) potassium(I) iodide and concentrated sulphuric(VI) acid, (d) red phosphorus and iodine?

3. In the reaction $HO^- + CH_3I \rightarrow HO{-}CH_3 + I^-$:
(a) Which reactant is the nucleophile?
(b) Which reactant supplies electrons to form the new C—O bond?
(c) Why does the oxygen atom lose its negative charge?
(d) Bearing in mind that a CH_3—radical has a $+I$ effect, would you expect the reaction with iodoethane to be faster or slower than the one with iodomethane under similar conditions?

4. Name the chief organic product (or products) formed in the substitution reactions between bromoethane and the following in ethanolic solution:
(a) potassium(I) hydroxide, (b) silver (I) acetate, (c) ammonia.

5. 2-bromopropane reacts with an ethanolic solution of potassium (I) hydroxide by substitution and elimination simultaneously. Give the formulæ and names of (a) the nucleophilic reagent in these reactions, (b) the organic product (or products) formed in the substitution reaction, (c) the organic product (or products) formed in the elimination reaction.

6. Name the chief organic compounds formed when (1) aqueous potassium(I) hydroxide is heated with (a) 1,2-dibromoethane, (b) trichloromethane; and when (2) ethanolic potassium(I) hydroxide is heated with (c) 1,2-dibromoethane, (d) trichloromethane and phenylamine.

7. Give the structural formulæ and the names of the organic compounds obtained when an ethereal solution of the Grignard reagent C_2H_5 MgI is combined with each of the following and the addition products are hydrolysed: (a) methanal (formaldehyde), (b) ethanal (acetaldehyde), (c) carbon dioxide.

10 Aldehydes and Ketones

Aldehydes and ketones are two closely related homologous series of compounds. Both contain the *carbonyl* group, and this results in many chemical similarities. At the same time differences arise from the fact that in aldehydes, except methanal (formaldehyde), the carbonyl group is joined to an alkyl radical and a hydrogen atom, whereas in ketones the carbonyl group is attached to two alkyl radicals.

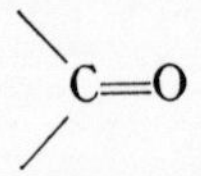

$$\begin{array}{c} \diagdown \\ \diagup \end{array} C = O$$

carbonyl group

$$\begin{array}{c} R \\ \diagdown \\ \diagup \\ H \end{array} C = O \qquad \begin{array}{c} R \\ \diagdown \\ \diagup \\ R' \end{array} C = O$$

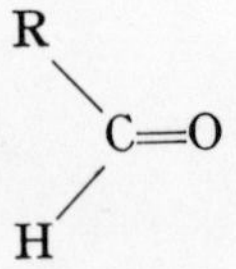

aldehyde 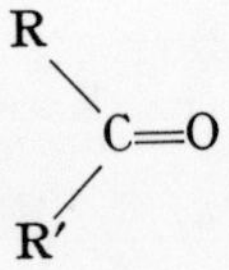   ketone

As seen in Chapter 8, aldehydes represent an intermediate stage in the oxidation of *primary* alcohols to carboxylic acids.

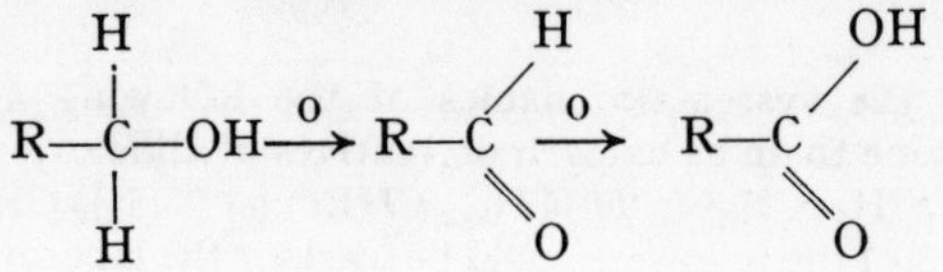

$$\begin{array}{ccc}
H & H & OH \\
| & | & | \\
R-C-OH \xrightarrow{O} & R-C \xrightarrow{O} & R-C \\
| & \| & \| \\
H & O & O \\
(-1) & (+1) & (+3)
\end{array}$$

The numbers below the formulæ are the oxidation numbers (Chapter 4) of the carbon atoms in the —CH_2OH, —CHO, and —COOH groups when these are attached to an alkyl radical. It will be seen that, in accordance with the general principle of oxidation processes, there is a progressive increase in the oxidation numbers.

Ketones are formed by oxidation of *secondary* alcohols. Unlike aldehydes they are difficult to oxidize further.

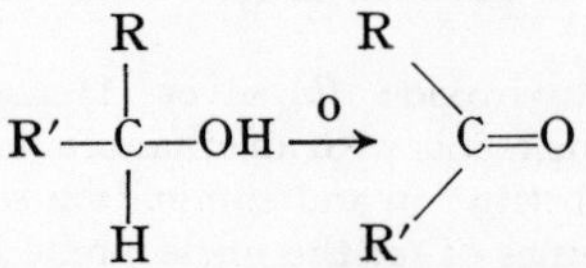

$$\begin{array}{cc}
R & R \\
| & \diagdown \\
R'-C-OH \xrightarrow{O} & C=O \\
| & \diagup \\
H & R' \\
(0) & (+2)
\end{array}$$

ALDEHYDES

The systematic names of *straight-chain* aldehydes are obtained by substituting the ending -*al* for the terminal -*e* in the names of the corresponding alkanes. Thus, the systematic names of the first four straight-chain aldehydes shown in Table 10-1 are methanal, ethanal, propanal, and butanal. *Branched-chain* aldehydes are given names based on the longest straight carbon chain which contains the —CHO group. The positions of substituting alkyl radicals are then indicated by numbers starting at the end —CHO group. This is illustrated by the systematic name of isobutyraldehyde.

Table 10-1. ALIPHATIC ALDEHYDES
(General formulæ: $C_nH_{2n}O$, R.CHO)

Systematic name	Common name	Structural formula	Boiling point/°C
Methanal	Formaldehyde	HCHO	-21
Ethanal	Acetaldehyde	$CH_3.CHO$	21
Propanal	Propionaldehyde	$CH_3.CH_2.CHO$	49
Butanals			
1. Butanal	*n*-Butyraldehyde	$CH_3.CH_2.CH_2.CHO$	75
2. 2-Methyl-propanal	Isobutyraldehyde	$CH_3.CH.CHO$ $\overset{|}{CH_3}$	61

The trivial names of aldehydes are derived from the trivial names of the carboxylic acids which they yield on oxidation. Thus formaldehyde oxidizes to formic acid, acetaldehyde to acetic acid, etc.

Aldehydes show the usual increase in boiling point with increase in molecular mass. The first member, methanal, is a gas with a strong pungent smell. When inhaled it has a lachrymatory, or tear-producing, effect. The other members shown in Table 10-1 are liquids with unpleasant oily odours. The lower aldehydes dissolve readily in water, but the solubility decreases with increase in length of the carbon chain. The vapours of aldehydes are flammable.

The carbon atom of the —CHO group is attached to only three other atoms. In accordance with the general rule given in Chapter 4, the carbon atom and the three other atoms lie in the same plane and the bond angles are approximately 120°C. Thus methanal is one of the few examples of a flat organic molecule.

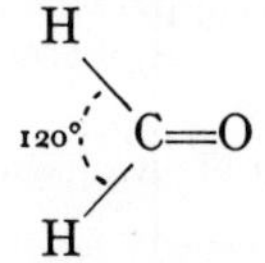

Laboratory Preparation of Aldehydes

(*i*) *By Oxidation of Primary Alcohols.* In the laboratory primary alcohols are usually oxidized to aldehydes with a mixture of sodium(I) dichromate(VI) and medium sulphuric(VI) acid. The sodium salt ($Na_2Cr_2O_7$) is preferred to the potassium salt because of its greater solubility in alcohols. Potassium(I) manganate(VII) (potassium permanganate)—either in alkaline or acidified solution—also oxidizes a primary alcohol to aldehyde, but in this case it is difficult to avoid further oxidation to the carboxylic acid. Even with acidified sodium(I) dichromate(VI) a small amount of the acid is formed. The production of ethanal from ethanol is represented by the following equations:

$$Cr_2O_7{}^{2-} + 8H^+ \rightarrow 2Cr^{3+} + 4H_2O + 3O$$

$$CH_3.CH_2OH + O \rightarrow CH_3.CHO + H_2O$$

The reaction is exothermic, and no heating is required except at the beginning and end of the experiment. The method now described yields an aqueous solution of ethanal.

Experiment

Put 15 cm³ of water into a 50 cm³ pear-shaped flask and add carefully, in small portions, 5 cm³ of concentrated sulphuric(VI) acid, swirling the flask round well to mix the liquids. Add also a little powdered pumice. Set up the apparatus shown in Fig. 10-1, including a trough of cold (preferably, iced) water to cool the receiver.

Fig. 10-1. Preparation of an aqueous solution of ethanal.

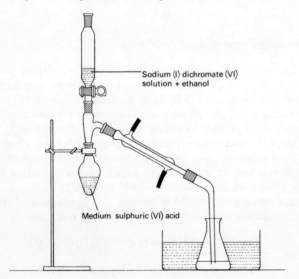

Sodium (I) dichromate (VI) solution + ethanol

Medium sulphuric (VI) acid

Prepare in a beaker a solution of 10 g of sodium(I) dichromate(VI) in 10 cm³ of water, and mix the solution with 8 cm³ of industrial spirit. Put the mixture into the dropping-funnel. Heat the acid in the flask until the liquid begins to boil (using a Bunsen burner from which the chimney has been removed), and then take away the flame. Run in the mixture from the dropping-funnel drop by drop, so that the heat evolved in the reaction is just sufficient to keep the contents of the flask boiling. The liquid in the flask turns green owing to reduction of orange dichromate(VI) ions to green chromium(III) ions. When about four-fifths of the mixture in the funnel have been added to the flask heat the latter with a small flame to maintain gentle boiling. The distillate consists of impure aqueous ethanal.

The chief impurities in the product are water, ethanol, and acetic acid. Pure ethanal can be obtained from the impure liquid by fractional distillation.

Note. If methanal is prepared from methanol by the above method it is necessary to have a layer of water in the receiver to absorb the gas. In this case the condenser serves to cool the gas and thus assist absorption.

(ii) By Dehydrogenation of Primary Alcohols. The term 'aldehyde' is a condensed form of '*al*cohol *dehyd*rogenated,' and signifies that an aldehyde can be derived from an alcohol by removal of hydrogen. Actually this applies only to primary alcohols.

$$R.CH_2OH—H_2 \rightarrow R.CHO$$

As seen in Chapter 8, ethanol can be dehydrogenated to ethanal by passing the vapour of the alcohol over a catalyst of copper turnings at 250°–300°C. This can be done on a small scale by the 'powdered chalk' method.

Chemical Tests for Aldehydes

Aldehydes produce an immediate purple colour with *Schiff's reagent*. This is an aqueous solution of magenta dye (rosaniline hydrochloride) which has been decolorized by passing sulphur dioxide through the solution. The last traces of colour are most quickly removed with carbon (animal charcoal).

Other reagents used to recognize aliphatic aldehydes are Fehling's solution, Tollens's reagent, aqueous sodium(I) hydroxide, and 2:4-dinitrophenylhydrazine. These reagents and the reactions which they give are described shortly.

Manufacture and Uses of Aldehydes

Methanal is made on a large scale by passing a mixture of methanol vapour and air over a silver catalyst at 600°C. Both oxidation and dehydrogenation of the alcohol take place.

$$CH_3OH + \tfrac{1}{2}O_2 \rightarrow HCHO + H_2O$$

$$CH_3OH - H_2 \rightarrow HCHO$$

Methanal is the most important aldehyde, large amounts being used in the manufacture of synthetic resins and plastics. A 40 per cent solution is sold under the name of *formalin* for use as a disinfectant. It is employed in sterilizing soil, embalming bodies, and preserving biological specimens. Other uses are in the manufacture of the drug hexamine (see later) and the high explosive cyclonite.

Ethanal is made by catalysed oxidation of ethene (*Wacker process*). A mixture of ethene and oxygen is passed into an aqueous solution of palladium(II) chloride and copper(II) chloride at 40°C. The overall reaction is represented by the following equation:

$$C_2H_4 + \tfrac{1}{2}O_2 \rightarrow C_2H_4O$$

Ethanal is chiefly used as an intermediate in the manufacture of other compounds (e.g. ethanoic acid).

Certain higher aldehydes are obtained by the *Oxo synthesis*, in which alkenes are converted into aldehydes containing an extra carbon atom in the molecule. The alkene is dissolved in an inert solvent (e.g. octane), and a mixture of carbon monoxide and hydrogen ('synthesis gas') is passed into the solution. The reaction takes place at 150°C and a pressure of 300 atmospheres, and is assisted by a cobalt(II) salt as catalyst.

$$R.CH=CH_2 + CO + H_2 \rightarrow R.CH_2.CH_2.CHO$$

One aldehyde produced in this way is butanal (from propene). The aldehydes are usually converted by reduction into the corresponding alcohols for use in making esters.

Reactions of Aldehydes

The functional group in aldehydes is the *aldehydic* group, which is composed of a carboxyl group and a hydrogen atom.

aldehydic group

Nearly all the reactions of aldehydes are due, directly or indirectly, to the carbonyl group. The double bond in this group resembles the alkene double bond (Chapter 6) in being composed of a σ (sigma) bond and a weaker π (pi) bond. In both cases the double bond tends to revert to two single bonds by addition of suitable reagents. Thus aldehydes, like alkenes, are unsaturated compounds.

As was seen in Chapter 6, the substances which add to alkenes are *electrophilic* reagents (halogens, hydrogen halides, etc.). In contrast, those which combine with aldehydes are mostly *nucleophilic* reagents. The difference is caused by a different electron distribution in the double bond. In alkenes the double bond joins two similar atoms, and there is no permanent displacement of the bond electrons towards either atom. In aldehydes there is a large displacement of the bond electrons towards the strongly electronegative oxygen atom. The more mobile electrons of the π bond are particularly affected, and the banana-shaped molecular orbital of these electrons (see Fig. 6-2) becomes distorted, as indicated in Fig. 10-2. Thus the oxygen atom of the

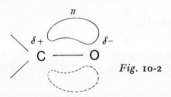

Fig. 10-2

carbonyl group has a high partial negative charge, while the carbon atom has a high partial positive charge (the existence of these charges is confirmed by the large electric dipole moments of aldehydes). The partial positive charge on the carbon atom is responsible for the tendency of the latter to combine with nucleophilic reagents. In these reactions the aldehydes themselves act as electrophilic reagents.

The carbonyl group influences the properties of aldehydes indirectly in 'activating' the hydrogen atoms of neighbouring C—H bonds. This is due to relaying of the —I effect of the oxygen atom to the C—H bonds, so that partial positive charges are created on the hydrogen atoms. (A similar effect was noted in Chapter 9 for the halogen atom in an alkyl halide). One result of this 'activation' is that the hydrogen atom of the —CHO group can be oxidized readily to a hydroxyl group, C—H becoming C—OH. Thus aldehydes are reducing agents.

Methanal differs from other members of the series in not having an alkyl radical attached to the —CHO group. Consequently some of its reactions are different.

1. Oxidation. As stated earlier, oxidizing agents such as acidified sodium(I) dichromate(VI) convert aldehydes to carboxylic acids. Methanal yields methanoic, or formic, acid (HCOOH), while ethanal gives ethanoic, or acetic, acid ($CH_3.COOH$).

Aldehydes are slowly oxidized by atmospheric oxygen, and hence laboratory samples of aldehydes often contain the corresponding acids as impurities. The ease with which even mild oxidizing agents bring about oxidation is utilized in chemical tests for aldehydes. Two mild oxidizing agents used in this way are Fehling's solution and Tollens's reagent.

Fehling's solution is an alkaline solution of copper(II) tartrate. As the reagent deteriorates on keeping, it is usually prepared in two parts which are added together when a test is being carried out. 'Fehling's solution A' is a solution of copper(II) sulphate(VI), while 'Fehling's solution B' is a solution of sodium(I) potassium(I) tartrate ('Rochelle salt') made alkaline with sodium(I) hydroxide. When equal amounts of the two liquids are added to each other a deep-blue solution, containing copper(II) tartrate, is produced. If this solution is heated with an aldehyde a yellow precipitate of copper(I) oxide is obtained. The precipitate turns red on standing. The change of valency of the copper from the divalent state to the monovalent state is represented simply by the equation given.

$$2CuO + CH_3.CHO \rightarrow Cu_2O + CH_3.COOH$$

Experiment

To 1 cm^3 of a dilute aqueous solution of ethanal add a few drops of 10 per cent sodium(I) carbonate solution to neutralize any ethanoic acid present. Now add two or three drops of Fehling's solution and boil the mixture. A yellow precipitate of copper(I) oxide is formed.

Repeat the experiment, using the solution obtained by boiling a few small pieces of apple with water. Apples contain glucose, a sugar in which the aldehydic group is present.

Tollens's reagent is an ammoniacal solution of silver(I) oxide. When the solution is treated in the cold with an aldehyde a precipitate of silver is obtained in the form of a silver 'mirror.'

$$Ag_2O + CH_3.CHO \rightarrow 2Ag + CH_3.COOH$$

Experiment

A *new* test-tube should be used for this experiment, as it is essential for the glass to be clean. To 5 cm^3 of 'bench' silver(I) nitrate(V) solution add a few drops of dilute sodium(I) hydroxide solution. A brown precipitate of silver(I) oxide is produced. Dilute some 'bench' ammonia solution to about three times its volume with water, and add this solution drop by drop to the test-tube until the precipitate of silver(I) oxide just dissolves. Now add two drops of very dilute ethanal solution. A deposit of silver is slowly formed as a 'mirror' on the sides of the tube.

2. Nucleophilic Addition. The lower aldehydes form addition compounds with water, the lower alcohols, hydrogen cyanide, hydrogensulphate(IV) (bisulphite) ions, ammonia, and 'substituted ammonias' such as hydroxylamine (NH_2OH). They also add on hydrogen with reducing agents. In some cases the addition compounds are unstable and eliminate water. These cases are considered later.

Most of the reagents mentioned above can be represented by the formula HX in which X stands for —OH, —OC_2H_5, etc. In the addition reactions HX is added across the double bond of the aldehyde molecule, as shown below. It should be mentioned, however, that the hydrogen atom added to the oxygen atom is not necessarily derived from HX (it may come from the solvent or from an acid catalyst).

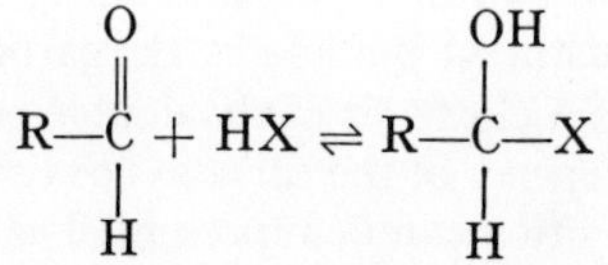

$$
\begin{array}{c}
\text{O} \\
\parallel \\
\text{R—C} \\
\mid \\
\text{H}
\end{array}
+ \text{HX} \rightleftharpoons
\begin{array}{c}
\text{OH} \\
\mid \\
\text{R—C—X} \\
\mid \\
\text{H}
\end{array}
$$

The mechanisms of the addition reactions have not been definitely established in all cases, so that some of the mechanisms which follow must be regarded only as probable ones in the light of the experimental evidence available.

(i) *Water*. When the lower aldehydes (e.g. ethanal) are dissolved in water they are partially converted into 'diols' by addition of water.

$$CH_3 . CHO + H_2O \rightleftharpoons CH_3 . CH(OH)_2$$

The diol addition compounds cannot be isolated owing to their instability (molecules containing two —OH groups attached to the same carbon atom are usually unstable). Evidence of their formation, however, is obtained in several ways. For example, when ethanal is dissolved in water enriched with the heavy isotope ^{18}O of oxygen, some of the isotope becomes incorporated in the aldehyde. This can only be explained by a reversible reaction between the water and the aldehyde. About 60 per cent of ethanal exists in the diol form in water at ordinary temperatures.

In neutral solution the attack on the aldehyde molecule is probably made by a water molecule at the carbonyl carbon atom with its partial positive charge. The oxygen atom of the water molecule donates a share in a lone pair of electrons to form a co-ordinate covalent bond with the carbon atom. The latter, however, already has eight electrons in its outer quantum shell, and cannot acquire two more without losing a similar number. The necessary adjustment is made by the carbon atom giving up its share in the two π electrons of the double bond, so that these become the sole property of the carbonyl oxygen atom. Thus the first step in the addition can be represented as shown.

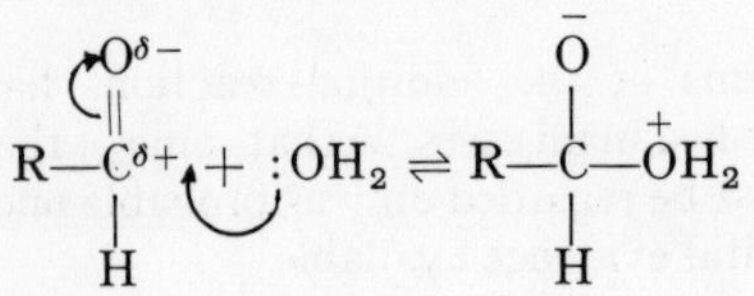

The next two steps probably occur simultaneously. These are addition of a proton from a water molecule at the negatively charged oxygen atom and removal of a proton by another water molecule from the positively charged —OH_2 group.

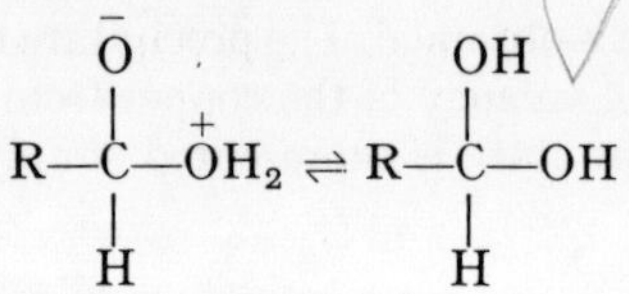

Addition of water to aldehydes takes place more rapidly if a trace of alkali is present to act as a catalyst. This is because the OH^- ion is a stronger nucleophile than a water molecule. For the alkaline solution the mechanism is thought to have the following form:

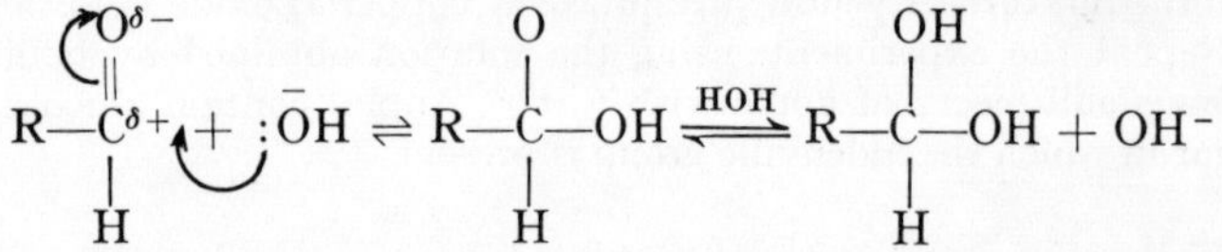

(ii) *Alcohols*. The lower alcohols are probably added to aldehydes in the same manner as water. With ethanal and ethanol a simple addition compound called 1-ethoxyethanol (hemi-acetal) is formed. This cannot be isolated. Under the influence of a strong acid catalyst, however, a further reaction takes place. Thus, if dry hydrogen chloride is passed into a mixture of ethanal and ethanol, the —OH group of the first addition compound is substituted by an ethoxy group ($-OC_2H_5$). The product is 1,1-diethoxyethane (acetal). This is a stable liquid which boils at $104°C$.

$$CH_3-\overset{\overset{\displaystyle OH}{|}}{\underset{\underset{\displaystyle H}{|}}{C}}-OC_2H_5$$

1-ethoxyethanol

$$CH_3-\overset{\overset{\displaystyle OC_2H_5}{|}}{\underset{\underset{\displaystyle H}{|}}{C}}-OC_2H_5$$

1,1-diethoxyethane

(iii) Hydrogen Cyanide. Aldehydes combine with hydrogen cyanide in aqueous solution to form *cyanohydrins*. The reactions are very slow with hydrogen cyanide alone, but take place rapidly if a little alkali is added as a catalyst. The alkali liberates CN^- ions from HCN molecules. Addition of hydrogen cyanide to ethanal yields 2-hydroxypropanonitrile (acetaldehyde cyanohydrin). The suffix *-onitrile* indicates the presence of a $-CN$ group, the carbon atom of which forms part of the carbon chain in the molecule. The two steps in the addition can be represented as follows:

$$CH_3-\overset{\overset{\displaystyle O^{\delta-}}{||}}{\underset{\underset{\displaystyle H}{|}}{C^{\delta+}}} \xrightarrow{CN^-} CH_3-\overset{\overset{\displaystyle \bar{O}}{|}}{\underset{\underset{\displaystyle H}{|}}{C}}-CN \xrightarrow{HOH} CH_3-\overset{\overset{\displaystyle OH}{|}}{\underset{\underset{\displaystyle H}{|}}{C}}-CN + OH^-$$

2-hydroxypropanonitrile

In the addition of hydrogen cyanide to aldehydes an **extra** carbon atom is incorporated in the molecule. Since the $-CN$ group is easily hydrolysed to a carboxyl ($-COOH$) group, the addition reaction provides a means of synthesizing hydroxycarboxylic acids. Thus 2-hydroxypropanonitrile, when hydrolysed, forms 2-hydroxypropanoic acid (lactic acid),

$$CH_3.CH(OH).COOH.$$

(iv) Hydrogensulphate(IV) (Bisulphite) Ions. When an aldehyde is added to a saturated solution of sodium(I) hydrogensulphate(IV) (sodium bisulphite) combination occurs with the anions of the acid salt. The addition product is still an anion, the 1-hydroxyethylsulphate(IV) ion, so-called because the 1-hydroxyethyl group, $CH_3.CH(OH)-$, has replaced the hydrogen atom of the original ion. The trivial name of the new ion is the acetaldehyde-bisulphite ion.

$$CH_3-\overset{\overset{\displaystyle O^{\delta-}}{||}}{\underset{\underset{\displaystyle H}{|}}{C^{\delta+}}} + HSO_3^- \rightleftharpoons CH_3-\overset{\overset{\displaystyle OH}{|}}{\underset{\underset{\displaystyle H}{|}}{C}}-SO_3^-$$

Although the mechanism of the above addition has not been definitely established, it probably depends on dissociation of the anions of the sodium(I) salt in aqueous solution ($HSO_3^- \rightleftharpoons H^+ + SO_3^{2-}$). A sulphate(IV) ion (sulphite ion) is added in the first step, and a proton in the second.

In the case of some aldehydes the ions formed by addition combine with sodium(I) ions present to give a crystalline precipitate of the sodium(I) salt. This does not happen, however, with the ions formed from methanal and ethanal because the sodium(I) salts are too soluble to be precipitated in aqueous solution. With benzaldehyde ($C_6H_5.CHO$) crystals of the sodium(I) salt are readily obtained.

(v) *Polymerization of Aldehydes*. In the presence of alkalies or acids as catalysts aldehyde molecules combine with each other and give polymers. Under laboratory conditions ethanal (C_2H_4O) forms three simple polymers. These are ethanal dimer, $(C_2H_4O)_2$, ethanal trimer, $(C_2H_4O)_3$, and ethanal tetramer, $(C_2H_4O)_4$. All three compounds are commercially important.

Ethanal dimer is commonly called *aldol*, but its systematic name is 3-hydroxybutanal. The manner in which the last name is derived will be seen shortly. The polymer is a colourless syrupy liquid used in the manufacture of synthetic rubber. It can be made by dissolving ethanal in very dilute sodium(I) hydroxide solution and allowing the mixture to stand at ordinary temperature. The polymer is isolated by fractional distillation under reduced pressure.

The mechanism of the 'aldol reaction' is thought to depend on activation of the hydrogen atoms attached to the carbon atom in the 2-position of the carbon chain. Owing to relaying of the $-I$ effect of the carbonyl oxygen atom the hydrogen atoms joined to the carbon atom in the 2-position have a partial positive charge, which gives them a certain amount of acidic character. One of the hydrogen atoms can be removed as a proton by the strongly basic OH^- ion.

$$CH_3.CHO + OH^- \rightleftharpoons \bar{C}H_2.CHO + H_2O$$

The resulting negative ion (which is called a *carbanion*) is strongly nucleophilic, and undergoes addition to a second molecule of ethanal as described earlier for OH^-. In the final step a proton is added from a water molecule. Thus the polymer has the following structure (which is that of 3-hydroxybutanal):

$$CH_3-\overset{\overset{\displaystyle OH}{|}}{\underset{\underset{\displaystyle H}{|}}{C}}-CH_2.CHO$$

Aldehydes like methanal (formaldehyde), $HCHO$, which do not have a hydrogen atom attached to a carbon atom in the 2-position do not give the 'aldol reaction'.

The formation of more complex molecules from ethanal with an alkali catalyst does not stop at the dimer. The latter still possesses an aldehyde group at one end of its molecule, and this is capable of bringing about further combination with ethanal molecules. Thus by a repetition of the combination very long chain-like molecules can be built up. This occurs when ethanal is heated with a more concentrated solution of sodium(I) hydroxide, the product in this case being a brown resinous mass called *aldehyde resin*. The exact constitution of the latter is unknown, but its relative molecular mass is very high (about 50 000).

Experiment

To two drops of ethanal add 5 cm^3 of 'bench' sodium(I) hydroxide solution. Shake the tube well and heat it. The contents turn yellow and then brown, while a smell resembling that of rotten apples is given off. When the tube is cool a dark viscous liquid remains.

Ethanal Trimer and Ethanal Tetramer. Since the carbon and oxygen atoms of the unsaturated carbonyl group carry opposite partial charges, an addition reaction might be expected to take place between different ethanal molecules by the oxygen atom of one molecule uniting with the carbon atom of another. This type of addition actually does occur under the catalytic influence

$$CH.CH_3$$

$$O \quad \quad O$$

$$CH_3.CH \quad CH.CH_3$$

$$O$$

ethanal trimer
(paraldehyde)

of strong acids. Two different polymers are obtained according to the conditions. In ethanal trimer (paraldehyde), $(C_2H_4O)_3$, three molecules of ethanal are joined together, forming a six-membered ring structure. Ethanal tetramer (metaldehyde), $(C_2H_4O)_4$, has a similar type of ring structure, but this is eight-membered.

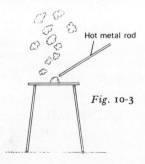

Hot metal rod

Fig. 10-3

Ethanal trimer is a colourless liquid formed when concentrated sulphuric(VI) acid is carefully added to ethanal cooled in a freezing-mixture (the reaction is strongly exothermic). The trimer is used in medicine as a soporific.

Ethanal tetramer is produced as a white solid when hydrogen chloride is passed into an ethereal solution of ethanal cooled in a freezing-mixture. It is sold in tablet form for killing slugs and for use as a solid fuel by campers. If the solid is touched with a hot metal rod it sublimes and fluffy masses of the condensed vapour ('chemical snow') are formed in the air (Fig. 10-3).

(*vi*) *Reduction of Aldehydes.* Aldehydes can be reduced to primary alcohols by zinc and dilute hydrochloric acid, sodium amalgam and water, or molecular hydrogen with a nickel catalyst at 140°C. Methanal is reduced to methanol, and ethanal to ethanol.

$$R.CHO \xrightarrow[\text{or } H_2]{2H} R.CH_2OH$$

Two reagents often used to reduce the carbonyl group in aldehydes and ketones are lithium(I) tetrahydridoaluminate(III) (lithium aluminium hydride), $LiAlH_4$, and sodium(I) tetrahydridoborate(III) (sodium borohydride), $NaBH_4$. These are selective in action; that is, they react with some reducible groups, but do not affect others.

Reduction of aldehydes is carried a stage further—to alkenes—if zinc amalgam and concentrated hydrochloric acid are used as the reducing agent (*Clemmensen's method* of reduction). For example, propanal is reduced to propane.

$$CH_3.CH_2.CHO + 4H \rightarrow CH_3.CH_2.CH_3 + H_2O$$

propanal   propane

3. Addition—elimination Reactions. As stated earlier, some of the addition compounds formed by aldehydes with nucleophilic reagents are unstable and split off water. The overall reaction is then described as a *condensation*. (In some condensations ammonia or some other simple substance is eliminated instead of water).

(*i*) *Ammonia*. Combination occurs between ammonia and the lower aldehydes. Thus, if ammonia gas is passed into a cooled ethereal solution of ethanal, a white precipitate of 'ethanal-ammonia' is formed. The composition of the precipitate, however, never corresponds with that of the simple addition compound. Gradual elimination of water converts the latter firstly into iminoethane and then into more complex condensation products.

$$CH_3-\overset{\displaystyle H}{\underset{\displaystyle O}{C}}$$

$$\downarrow NH_3$$

$$CH_3-\overset{\displaystyle H}{\underset{\displaystyle NH_2}{C}}-OH$$

'ethanal-ammonia'

$$\downarrow -H_2O$$

$$CH_3-C\overset{\displaystyle H}{\underset{\displaystyle NH}{}}\quad \text{iminoethane}$$

The mechanism by which the addition compound is formed is probably similar to that described earlier for the addition of water to an aldehyde in neutral solution.

(*ii*) *'Substituted Ammonias.'* These can be represented by the formula NH_2X, in which X stands for a group which has replaced a hydrogen atom from the ammonia molecule. The condensation reactions of substituted ammonias with aldehydes are much more important than that of ammonia itself. The products are stable crystalline substances, which are usually precipitated from aqueous or ethanolic solution. They are easily purified by recrystallization, and after purification their melting points can be used to identify particular aldehydes. As in the case of ammonia the condensation reactions probably depend on an unstable addition compound being formed first. The condensations can be represented in general by the equation shown.

$$R.CHO + H_2NX \rightarrow H_2O + R.CH:NX$$

Some examples of substituted ammonias and the reactions which they undergo with ethanal are given below.

Hydrazine ($NH_2.NH_2$) gives rise to *hydrazones*; e.g.

$$CH_3.CHO + H_2N.NH_2 \rightarrow CH_3.CH:N.NH_2 + H_2O$$

ethanal
hydrazone

Phenylhydrazine ($C_6H_5.NH.NH_2$) yields *phenylhydrazones*; e.g.

$$CH_3.CHO + H_2N.NH.C_6H_5$$
$$\rightarrow CH_3.CH:N.NH.C_6H_5 + H_2O$$

ethanal phenylhydrazone

Hydroxylamine (NH_2OH) produces *aldoximes*; e.g.

$$CH_3.CHO + H_2NOH \rightarrow CH_3.CH:NOH + H_2O$$
ethanal oxime

The above reagents are unsuitable for the identification of methanal and ethanal. In these cases the condensation products are too soluble for precipitation to occur. A common reagent used nowadays is 2,4-dinitrophenylhydrazine, which gives precipitates with nearly all aldehydes (including methanal and ethanal). The use of the reagent is described below.

Experiment

Dissolve about 0·2 g of 2,4-dinitrophenylhydrazine in 2 cm³ of glacial ethanoic acid. Add two drops of ethanal, and dilute the solution to four or five times its volume with water. A yellow precipitate of the 2,4–dinitrophenylhydrazone is formed. If this is recrystallized from ethanol and dried, it is found to have a melting point of 168°C.

4. Chlorination and Bromination. Chlorine and bromine slowly bring about substitution of hydrogen in the alkyl radical of an aldehyde. If chlorine is passed into ethanal the hydrogen atoms of the methyl radical are replaced in turn. Antimony(III) chloride acts as a catalyst for the reaction.

$$CH_3.CHO + 3Cl_2 \rightarrow CCl_3.CHO + 3HCl$$
trichloroethanal

A similar reaction occurs with bromine, but takes place more slowly.

The substitutions are assisted by the carbonyl group activating the hydrogen atoms attached to the carbon atom in the 2-position. Activation not only confers some acidic character on the hydrogen atoms, but also facilitates their replacement by chlorine or bromine. Thus, when propanal is chlorinated, the hydrogen atoms of the methylene group are substituted in preference to those of the methyl group.

$$CH_3.CH_2.CHO + 2Cl_2 \rightarrow CH_3.CCl_2.CHO + 2HCl$$
2,2-dichloropropanal

Trichloroethanal (commonly called *chloral*) is a colourless oily liquid, but is usually encountered in the form of its 'hydrate,' which consists of colourless crystals. These have the formula $CCl_3.CHO.H_2O$, or, more correctly, $CCl_3.CH(OH)_2$, the systematic name being 2,2,2-trichloroethanediol. The 'hydrate' is used in medicine as a soporific and in the manufacture of the insecticide 'D.D.T.'

Trichloroethanal gives most of the reactions of ordinary aldehydes, but differs in its greater ability to form stable addition compounds. Thus the 'hydrate' is one of the few examples of a stable compound containing two —OH groups attached to the same carbon atom. Trichloroethanal also forms stable addition compounds with ammonia and hydroxylamine.

METHANAL, HCHO

While the reactions of methanal (formaldehyde) resemble in general those of other aliphatic aldehydes, they differ in certain respects. In other members of the series the partial positive charge on the carbonyl carbon atom is decreased to some extent by the $+I$ (electron-repelling) effect of the alkyl radical. In methanal, which does not contain an alkyl radical, this effect is absent, and therefore the partial positive charge on the carbon atom is higher. As a result methanal reacts more readily with nucleophilic reagents, and some of the reactions follow a different course.

1. Polymerization. Methanal differs from other aldehydes in polymerizing spontaneously when its aqueous solution is allowed to stand. Thus a white deposit of poly(methanal), $(CH_2O)_n$, often appears in a stock-bottle containing 'formalin.' The polymer can be made simply by evaporating a little of the aqueous solution in a clock-glass over a water-bath. The polymer molecule is a linear polyoxymethylene of variable length, and can be represented as shown.

$$-CH_2-O-CH_2-O-CH_2-O-$$
poly(methanal)

If poly(methanal) is heated to about 200°C it undergoes depolymerization, and methanal is given off.

When a concentrated aqueous solution of methanal is distilled with concentrated sulphuric(VI) acid the distillate contains *methanal trimer*, $(CH_2O)_3$. This can be extracted and obtained as colourless crystals. The molecules have a six-membered ring structure consisting of alternate methylene groups and oxygen atoms.

If methanal solution is made weakly alkaline with calcium(II) hydroxide solution, the aldehyde is converted in the course of a week or so into a complex mixture of sugars called *formose*. Some of these sugars are of the glucose type, and their formation may be represented by the equation given.

$$6CH_2O \rightarrow C_6H_{12}O_6$$

The reaction has no practical importance.

2. Reaction with Sodium(I) Hydroxide. As mentioned earlier, methanal does not undergo the 'aldol reaction' because its molecule does not contain a hydrogen atom linked to a carbon atom in the 2-position. Neither does it resinify when heated with aqueous sodium(I) hydroxide. If, however, methanal solution is shaken in the cold with a concentrated solution of sodium(I) hydroxide, simultaneous oxidation and reduction of the aldehyde occur. The latter is partly reduced to methanol and partly oxidized to methanoic acid (which yields sodium methanoate, or formate).

$$2CH_2O + NaOH \rightarrow CH_3OH + HCOONa$$

The type of reaction described is known as *Cannizzaro's reaction*. It is given by aldehydes which do not contain a suitably situated hydrogen atom (as specified above) in the molecule. Other examples besides methanal are trimethylethanal, $C(CH_3)_3.CHO$, and benzaldehyde, $C_6H_5.CHO$.

3. Reaction with Ammonia. With ammonia methanal forms a stable condensation product called hexamethylenetetramine (usually known as *hexamine*).

$$6CH_2O + 4NH_3 \rightarrow (CH_2)_6N_4 + 6H_2O$$
hexamine

Experiment

Put on to a clock-glass a few drops of 'formalin' solution and about the same amount of concentrated ammonia solution. Place the clock-glass over a beaker containing water, and boil the water. A white residue of hexamine is left on the clock-glass.

Hexamine is used in medicine and in the manufacture of phenolic resins and plastics (e.g. Bakelite). It is also an intermediate in the production of the high-explosive Cyclonite.

KETONES

As noted at the beginning of this chapter, the carbonyl group in ketones is joined to two alkyl radicals. There may be similar or different. When they are similar the compound is described as a *simple* ketone; if they are different, it is called a *mixed* ketone. The first three members of the series are listed in Table 10-2.

Table 10-2. ALIPHATIC KETONES

(General formulæ: $C_nH_{2n}O$, R.CO.R')

Systematic name	Common name	Structural formula	Boiling point/°C
Propanone	Dimethyl ketone, acetone	$CH_3.CO.CH_3$	56·1
Butanone	Ethyl methyl ketone	$CH_3.CO.C_2H_5$	79·6
Pentanone	Diethyl ketone	$C_2H_5.CO.C_2H_5$	102

The systematic names of aliphatic ketones are based on the longest straight chain of carbon atoms (including that of the carbonyl group) in the molecule. The name of the corresponding alkane is then modified by replacing the terminal -*e* by -*one*.

Thus we obtain the names propanone and butanone. In the case of many higher ketones it is necessary to specify the positions of the carbonyl group and any substituting alkyl radicals by numbering the carbon atoms in the chain. This is done in the same way as described for the monohydric alcohols in Chapter 8. An illustration is provided by ethyl isopropyl ketone, which has the following structure and systematic name:

$$CH_3.CH_2.CO.\overset{\overset{\displaystyle CH_3}{\displaystyle |}}{CH}.CH_3$$

2-methylpentan-3-one

As usual, melting points and boiling points rise with increase in molecular mass. The lower ketones are colourless flammable liquids with pleasant smells, while the higher ones are solids. The liquids are lighter than water, the densities of the first three members of the series being 0·79, 0·80, and 0·81 g cm^{-3} respectively. The first two ketones are miscible with water in all proportions, but the other liquids are only partially miscible.

The general formula, $C_nH_{2n}O$, of ketones is the same as that of aldehydes. Hence ketones are isomeric with aldehydes which contain the same number of carbon atoms in the molecule. For example, propanone ($CH_3.CO.CH_3$) is isomeric with propanal ($C_2H_5.CHO$).

Laboratory Preparation of Ketones

Ketones are prepared either by oxidation of secondary alcohols in the liquid phase or by dehydrogenation of secondary alcohols in the vapour phase; e.g.

$$(CH_3)_2CHOH + O \rightarrow (CH_3)_2CO + H_2O$$

propan-2-ol   propanone

$$\begin{array}{c} CH_3 \\ | \\ CHOH - H_2 \rightarrow \\ | \\ C_2H_5 \end{array} \qquad \begin{array}{c} CH_3 \\ | \\ CO \\ | \\ C_2H_5 \end{array}$$

butan-2-ol   butanone

Oxidation is carried out with sodium(I) dichromate(VI) and medium sulphuric(VI) acid, as described earlier for the preparation of ethanal. Dehydrogenation is effected by passing the alcohol vapour over copper turnings at 250°–300°C.

Propanone is sometimes prepared in the laboratory by dry distillation of anhydrous calcium(II) ethanoate or barium(II) ethanoate, but the yield is poor.

$$(CH_3.COO)_2Ca \rightarrow (CH_3)_2CO + CaCO_3$$

Manufacture and Uses of Propanone

Propanone is by far the most important ketone. It is made on a large scale by catalysed oxidation of propene, using the Wacker process described earlier for ethanal. A mixture of propene and oxygen is passed through a warm aqueous solution containing palladium(II) chloride and copper(II) chloride.

$$CH_3.CH{=}CH_2 + \tfrac{1}{2}O_2 \rightarrow CH_3.CO.CH_3$$

Propanone is also obtained as a by-product in the manufacture of phenol by the *cumene process* (see Chapter 24).

Propanone is widely used as a solvent, e.g. in the manufacture of cordite (Chapter 18) and cellulose ethanoate rayon (Chapter 26). Ethyne used in cutting and welding metals is stored by dissolving it in propanone under pressure. Propanone is also an intermediate in the manufacture of the plastic Perspex (Chapter 26).

Reactions of Ketones

As ketones contain the carbonyl group they have many similarities to aldehydes. Thus they undergo addition or addition-elimination with many of the same nucleophilic reagents as aldehydes. Since, however, the hydrogen atom of the —CHO group in aldehydes is replaced in ketones by a second alkyl radical, the two classes also differ in some ways. In particular, the reducing reactions associated with the hydrogen atom (or C—H bond) in aldehydes are absent in ketones.

Again, the inductive ($+I$) effect of the second alkyl radical reinforces that of the first one, decreasing still further the partial positive charge on the carbonyl carbon atom. This reduces the attraction of the carbon atom for nucleophilic reagents. Thus, ketones have a much more limited capacity than aldehydes for forming stable addition compounds.

Finally, the carbonyl group in ketones again activates the hydrogen atoms which are attached to carbon atoms adjacent to the carbonyl group. As a result these hydrogen atoms have a slight acidic character. They also undergo substitution by halogens more readily.

1. Oxidation. Ketones are not oxidized by exposure to air, neither do they reduce Fehling's solution or Tollens's reagent. (Commercial samples of ketones often contain traces of aldehydes as impurities, and these produce a silver mirror slowly with Tollens's reagent). Ketones can be oxidized only by prolonged heating with strong oxidizing agents. Thus when propanone is refluxed with acidified sodium(I) dichromate(VI) solution the molecules are slowly broken down and oxidized to ethanoic, or acetic acid, carbon dioxide, and water.

$$CH_3.CO.CH_3 + 4O \rightarrow CH_3.COOH + CO_2 + H_2O$$

2. Nucleophilic Addition. The weaker electrophilic character of the lower ketones as compared with the lower aldehydes is shown by the failure of the ketones to combine appreciably with weak nucleophiles like water and ethanol. Also, while nearly all aldehydes combine with hydrogen cyanide and hydrogensulphate(IV) ions (bisulphite ions), addition of these to ketones is confined to those ketones which have a CH_3— radical attached to the carbonyl group. Thus addition occurs with propanone and butanone, but not with pentanone (diethyl ketone). The addition compounds have structures analogous to those of the addition compounds formed by aldehydes; e.g.

$$CH_3-\underset{\underset{CH_3}{|}}{\overset{\overset{OH}{|}}{C}}-CN \qquad CH_3-\underset{\underset{CH_3}{|}}{\overset{\overset{OH}{|}}{C}}-SO_3^-$$

The addition compound formed with hydrogen cyanide is a colourless liquid, commonly called *acetone-cyanohydrin*. Its systematic name is 2-hydroxy-2-methylpropanonitrile. The other addition product shown, the *acetone-bisulphite* ion, is named systematically as the 1-hydroxy-1-methylethylsulphate(IV) ion. The sodium salt is obtained as a colourless crystalline precipitate when propanone is shaken with a saturated solution of sodium(I) hydrogensulphate(IV) (sodium bisulphite).

Ketones polymerize much less readily than aldehydes. Only one polymer of propanone is known. This is a liquid dimer usually known as *diacetone alcohol*, which is formed in small amounts under the catalytic influence of OH^- ions. Actually the

$$2C_3H_6O \rightleftharpoons (C_3H_6O)_2$$

dimer is both an alcohol and a ketone with the structure $(CH_3)_2C(OH).CH_2.CO.CH_3$. Its systematic name is 4-hydroxy-4-methyl-pentan-2-one. It is produced by the aldol mechanism described earlier, an activated hydrogen atom being removed by the basic catalyst to give the nucleophilic ion $\overset{-}{C}H_2.CO.CH_3$.

Ketones are reduced by the same reagents as aldehydes, secondary alcohols or alkanes being formed according to the reagent used. With zinc and dilute hydrochloric acid, lithium(I) tetrahydridoaluminate (III), or sodium (I) tetrahydridoborate (III) propanone yields propan-2-ol.

$$(CH_3)_2CO + 2H \rightarrow (CH_3)_2CHOH$$
$$\text{propan-2-ol}$$

Reduction of propanone can be carried a stage further (to propane) by using zinc amalgam and concentrated hydrochloric acid (Clemmensen's method).

$$(CH_3)_2CO + 4H \rightarrow C_3H_8 + H_2O$$
$$\text{propane}$$

3. Addition-elimination Reactions. Ketones undergo a large number of condensation reactions, but only a few can be mentioned here. The reactions probably depend on preliminary formation of an addition compound, from which water is subsequently eliminated.

Propanone molecules condense together when propanone is distilled with medium-concentrated sulphuric(VI) acid. The product is a colourless liquid, trimethylbenzene, which is an aromatic compound.

$$3(CH_3)_2CO \rightarrow C_6H_3(CH_3)_3 + 3H_2O$$

Ketones react more slowly than aldehydes with ammonia, but the results are similar. Thus, if ammonia gas is dissolved in propanone and the solution is allowed to stand in a closed vessel for two or three weeks, a mixture of complex condensation products is obtained.

Ketones also condense in the same way as aldehydes with 'substituted ammonias'. Thus with a solution of phenylhydrazine in ethanoic acid propanone forms propanone-phenylhydrazone. This, however, remains in solution because of its high solubility.

$$(CH_3)_2CO + H_2N.NH.C_6H_5$$
$$\rightarrow (CH_3)_2C:N.NH.C_6H_5 + H_2O$$

If 2,4-dinitrophenylhydrazine is used, the corresponding 2,4-dinitrophenylhydrazone is obtained as a yellow crystalline precipitate.

With hydroxylamine ketones form crystalline condensation products called *ketoximes*. Propanone gives propanone oxime.

$$(CH_3)_2CO + H_2NOH \rightarrow (CH_3)_2C:NOH + H_2O$$
propanone oxime

4. Substitution by Chlorine and Bromine. Chlorine and bromine bring about progressive substitution in the alkyl radicals of ketones. In the case of propanone the first chloro-derivative is chloropropanone ($CH_2Cl.CO.CH_3$). If chlorination is continued, all six hydrogen atoms in the molecule are eventually replaced, giving hexachloropropanone ($CCl_3.CO.CCl_3$).

Chemical Tests for the Lower Ketones

The reactions most frequently used for the identification of the lower ketones are the following:

(*i*) The production of a colourless crystalline precipitate of the addition compound with a saturated solution of sodium(I) hydrogensulphate(IV) (sodium bisulphite). The test is given only by ketones of the type $CH_3.CO.R$.

(*ii*) The formation of a yellow crystalline precipitate of the 2,4-dinitrophenylhydrazone. Individual ketones can be identified by recrystallizing the derivatives from ethanol and taking their melting points (the propanone derivative melts at 128°C). Another derivative used in the same way is the ketoxime.

(*iii*) Ketones can be distinguished from aldehydes by their not giving an immediate violet colour with Schiff's reagent. Also, they fail to reduce either Fehling's solution or Tollens's reagent.

EXERCISE 10

1. Give the systematic names of the following aldehydes:
(a) $C_2H_5.CHO$, (b) $C_2H_5.CH_2.CHO$, (c) $(CH_3)_2 CH.CHO$.

2. Name the alcohols which, on oxidation, yield the aldehydes specified in question 1.

3. Which of the following statements are true of aldehydes in general?

(a) Their functional group is the carbonyl group.

(b) The double bond in the carbonyl group is polarized.

(c) The carbonyl carbon atom tends to react with electrophilic reagents.

(d) They can act as reducing agents.

(e) They can act as oxidizing agents.

4. State what you *see* (if anything) when ethanal is added to (a) Schiff's reagent, (b) Fehling's solution (which is then boiled), (c) a relatively small amount of a warm acidified solution of potassium(I) manganate(VII) (potassium permanganate), (d) a saturated solution of sodium(I) hydrogensulphate(IV) (sodium bisulphite), (e) a solution of 2,4-dinitrophenylhydrazine in ethanoic, or acetic, acid.

5. Name the organic compounds formed in (b), (c), (d), and (e) in question 4.

6. With which of the following (if any) does methanal (form-aldehyde) react differently from ethanal (acetaldehyde):

(a) Tollens's reagent, (b) aqueous sodium(I) hydroxide, (c) ammonia, (d) 2,4-dinitrophenylhydrazine, (e) lithium(I) tetrahydrido-aluminate(III)?

7. Which of the reagents given in question 6 have no reaction with propanone?

8. Name the following compounds: (a) $(CH_3)_2 C(OH)CN$, (b) $(CH_3)_2 C(OH)SO_3^- Na^+$, (c) $(CH_3)_2 C:NOH$, (d) $CH_2Br.CO.CH_3$.

11　Alkanoic Acids

Alkanoic acids (fatty acids) are the final products of oxidation of primary alcohols by sodium(I) dichromate(VI) and sulphuric(VI) acid. Their functional group is the carboxyl (—COOH) group, which is attached to an alkyl radical (or a hydrogen atom in the case of the first member, methanoic acid). Alkanoic acids can therefore be represented by the following general formula:

$$R-C\overset{\displaystyle O}{\underset{\displaystyle OH}{\big\langle}}$$

The first six members of the homologons series, together with two important higher members, are shown in Table 11-1 opposite.

The homologous series was originally called the 'fatty acid' series because the last two members, which were two of the earliest known, were found as esters in fats like beef-fat and mutton-fat. The term 'fatty acid' has now been replaced by 'alkanoic acid.' The acids are derived from alkanes by replacing the end CH_3— radical in the main carbon chain by a —COOH group. They are named systematically by substituting -oic acid for the final -e in the names of the alkanes. Branched-chain members are named as alkyl-substitution derivatives of the acids with the longest straight chain of carbon atoms. The positions of substituents are shown by numbering the carbon atoms in the chain, starting at the carboxyl carbon atom. The common names formic acid and acetic acid have been retained as 'recommended' names, and rank equally with methanoic acid and ethanoic acid.

The lower alkanoic acids are liquids with penetrating, and often unpleasant, smells. Thus the odours of perspiration and rancid butter are chiefly caused by butanoic acid, while that of goats comes from hexanoic acid (caproic acid), a skin secretion of these animals (Latin caper = a goat). From C_{10}— upwards the members are solids.

Melting points and boiling points increase with molecular mass, although higher members decompose before their boiling points are reached. The boiling points are considerably higher than those of the corresponding alkanes or alcohols. This is mainly because hydrogen-bonding of a special kind takes place in the liquid state.

Table 11-1. ALKANOIC ACIDS
(General formulæ: $C_nH_{2n}O_2$, R.COOH)

Systematic name	Common name	Structural formula	Boiling point/°C
Methanoic acid	Formic acid	HCOOH	100·5
Ethanoic acid	Acetic acid	$CH_3.COOH$	118
Propanoic acid	Propionic acid	$CH_3.CH_2.COOH$	141
Butanoic acids			
1. Butanoic acid	n-Butyric acid	$CH_3.(CH_2)_2.COOH$	163
2. 2-Methylprop-anoic acid	Isobutyric acid	$(CH_3)_2.CH.COOH$	154
Pentanoic acid	n-Valeric acid	$CH_3.(CH_2)_3.COOH$	187
Hexanoic acid	n-Caproic acid	$CH_3.(CH_2)_4.COOH$	205
Hexadecanoic acid	Palmitic acid	$(C_{15}H_{31}.COOH)$	Decomposes
Octadecanoic acid	Stearic acid	$(C_{17}H_{35}.COOH)$	Decomposes

Two molecules of acid are linked together by *two* hydrogen bonds, giving 'associated' molecules. This is illustrated by the following molecular structure of ethanoic acid in the liquid state:

$$CH_3.C \begin{array}{c} O—H \cdots O \\ \diagup \qquad \diagdown \\ \diagdown \qquad \diagup \\ O \cdots H—O \end{array} C.CH_3$$

The first four alkanoic acids are miscible with water in all proportions. Freezing-point experiments show that their molecules are not 'associated' in aqueous solution. Solubility in water decreases as the length of the hydrophobic hydrocarbon chain increases. Thus the liquid members which follow butanoic acid are only partially miscible with water. Solid members like hexadecanoic acid are insoluble in water.

Owing to the absence of an alkyl radical from its molecule methanoic acid has some exceptional properties. Ethanoic acid may be taken as a typical member of the series. The pure acid is called *glacial* ethanoic acid, because in winter it often freezes into ice-like crystals. Actually the freezing point of the acid is 17°C, but, as often happens with organic liquids, supercooling readily occurs, and ethanoic acid can be cooled well below its freezing point without solidification taking place. The pure acid is very hygroscopic, and has a marked blistering action on skin. It should therefore be handled with care. It is also highly flammable.

Laboratory Preparation of Alkanoic Acids

(*i*) *By Oxidation of Primary Alcohols*. Alkanoic acids are usually obtained by oxidizing primary alcohols with sodium(I) dichromate(VI) and medium sulphuric(VI) acid. The materials are the same as those used for the preparation of aldehydes (Chapter 10), but in preparing acids the conditions are such that the aldehydes first formed undergo further oxidation. The preparation of ethanoic acid from ethanol is represented by the equations.

$$Cr_2O_7^{2-} + 8H^+ \rightarrow 2Cr^{3+} + 4H_2O + 3O$$

$$CH_3.CH_2OH + 2O \rightarrow CH_3.COOH + H_2O$$

The small-scale preparation of ethanoic acid is carried out as described below.

Experiment

Put into a 50 cm³ pear-shaped flask 7 g of powdered sodium(I) dichromate(VI) and 10 cm³ of water. Add 7 cm³ of concentrated sulphuric(VI) acid in small portions, swirling the flask round until the orange-coloured salt has dissolved (considerable heat is evolved). Add to the flask a little powdered pumice, and then attach a reflux condenser to the flask (Fig. 11-1).

Prepare in a small beaker a mixture of 3 cm³ of industrial spirit and 12 cm³ of water. Add the mixture to the flask by pouring it down the condenser in portions of 1-2 cm³, regulating the additions so that the contents of the flask are kept gently boiling by the heat evolved in the oxidation (that is, without application of external heat). From time to time swirl the flask round to mix the contents thoroughly.

When all the alcohol-water mixture has been added boil the liquid over a small flame for five minutes to complete the reaction (using a Bunsen burner from which the chimney has been removed). Then by means of a still-head change over the condenser for direct distillation.

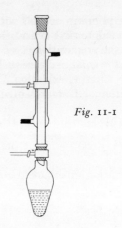

Fig. 11-1

Distil the contents of the flask until the liquid coming over is no longer acidic to litmus.

Purification. The distillate of crude ethanoic acid contains water, ethanal (test a small portion with Schiff's reagent), ethoxyethane (diethyl ether), and sulphuric(IV) acid (sulphurous acid). The ethanoic acid is purified by first converting it into its copper(II) salt or lead(II) salt, which is crystallized out. The crystals are then distilled with concentrated sulphuric(VI) acid, which liberates the ethanoic acid again.

(ii) By Hydrolysis of Esters. When an ester, such as ethyl ethanoate, or acetate, is hydrolysed by boiling aqueous sodium(I) hydroxide the alcohol is liberated, and the acid is left as the sodium(I) salt.

$$CH_3.COOC_2H_5 + NaOH \rightarrow C_2H_5OH + CH_3.COONa$$

The acid is then obtained from the sodium(I) salt by distillation with concentrated sulphuric(VI) acid. .

$$CH_3.COONa + H_2SO_4 \rightarrow CH_3.COOH + NaHSO_4$$

Details of this method of preparation are given in Chapter 13.

(iii) By Hydrolysis of Nitriles (Alkyl Cyanides). The most important property of the —CN group, when this is present in an organic molecule, is that it can be converted by hydrolysis into a —COOH group. Hydrolysis is brought about by boiling with concentrated hydrochloric acid or medium sulphuric(VI) acid (Chapter 14). Thus, if ethanonitrile, or acetonitrile, is boiled with concentrated hydrochloric acid, ethanoic, or acetic, acid and ammonium chloride are formed.

$$CH_3CN + 2H_2O + HCl \rightarrow CH_3.COOH + NH_4Cl$$

Manufacture and Uses of Alkanoic Acids

The chief alkanoic acids which are made on a large scale are methanoic, ethanoic, and propanoic acids. Of these ethanoic acid is the most important.

Methanoic acid is manufactured by heating carbon monoxide under pressure with sodium(I) hydroxide in an autoclave (an industrial form of pressure cooker). Sodium(I) methanoate is formed.

$$CO + NaOH \rightarrow HCOONa$$

The acid is obtained as a concentrated aqueous solution by distilling the sodium(I) methanoate with medium sulphuric(VI) acid.

Methanoic acid is chiefly used in the electroplating, tanning, and textile industries. It is also employed in disinfecting wine-casks (it is a strong antiseptic).

'Acetic' is derived from *acetum*, the Latin for vinegar. For many centuries vinegar, which is essentially a dilute solution of acetic acid, has been made by fermentation of wines and other alcoholic liquids by means of a mould (*Mycoderma aceti*). This method has no significance, however, as regards industrial acetic acid.

Until recently the main method of manufacturing ethanoic acid in Britain was by oxidation of ethanal. This is now being superseded by a process based on a light petroleum fraction consisting of C_5-C_7 alkanes. The latter are oxidized directly by air at a high pressure and elevated temperature, which is maintained by heat given out in the reaction. The products contain a large proportion of ethanoic acid, and both methanoic and propanoic acids are obtained as by-products.

Ethanoic acid is mainly employed as a solvent, and in making drugs (e.g. aspirin), synthetic resins, and esters (such as ethyl ethanoate and cellulose ethanoate). Propanoic acid is also used for the production of esters.

A few higher alkanoic acids, such as hexadecanoic acid and octadecanoic acid, are obtained by hydrolysis of naturally occurring esters present in vegetable oils and fats. The acids are used as 'plasticizers' (internal lubricants for plastics) and in making detergents.

Reactions of Alkanoic Acids

Alkanoic acids (like aldehydes and ketones) are electrophiles. They react with nucleophilic reagents such as water, OH^- ions, alcohols, and ammonia. Since the carboxyl (—COOH) group is composed of a carbonyl group and a hydroxyl group, alkanoic acids might be expected to give reactions typical of both. In practice most of the characteristic reactions of the carbonyl group, as shown by aldehydes and ketones, are absent from the acids. Thus the latter do not undergo the addition and addition-elimination reactions of the aldehydes and ketones. This is because the nucleophilic reagents concerned in most of these reactions are bases, and attack the acidic —OH group of the acids in preference to the carbonyl group.

Besides giving rise to reactions involving the —OH group, alkanoic acids undergo substitution in their alkyl radicals by halogens.

1. **Acidic Reactions.** The —OH group in carboxylic acids is more strongly acidic than the one in alcohols. In aqueous solution the lower alkanoic acids dissociate into carboxylate ions and oxonium ions.

$$R.COOH + H_2O \rightleftharpoons R.COO^- + H_3O^+$$

The solutions turn blue litmus red, and give off hydrogen with magnesium, iron, and zinc. They dissolve metal oxides, neutralize alkalies, and liberate carbon dioxide from carbonates and hydrogencarbonates (bicarbonates). In each case salts are formed. When the salts are treated with a strong mineral acid the organic acids are regenerated.

$$R.COONa + HCl \rightarrow R.COOH + NaCl$$

Three factors contribute to the acidic character of the —OH group in carboxylic acids:

(*i*) The —*I* effect of the oxygen atom causes an electron displacement away from the hydrogen atom.

(*ii*) The —*I* effect of the carbonyl oxygen atom is relayed to the O—H bond, increasing still further the electron displacement due to (*i*).

(*iii*) Mesomerism (resonance) occurs in the carboxyl group. As explained in Chapter 4, mesomerism in the carboxyl group results in a partial positive charge on the oxygen atom of the —OH group. This again increases the electron displacement towards the oxygen atom, the 'mesomeric effect', thus reinforcing the two inductive effects.

Alkanoic acids, however, are only weak acids in aqueous solution. Thus the degree of dissociation of ethanoic acid in molar solution is only 0·4 per cent, while for hydrochloric acid of the same concentration it is 80 per cent.

2. Esterification with Alcohols. Carboxylic acids react with alcohols in the presence of an acid catalyst to give esters. Thus, when ethanoic acid is heated with ethanol and concentrated sulphuric(VI) acid, ethyl ethanoate, or acetate, is formed.

$$CH_3.COOH + C_2H_5OH \underset{}{\overset{H^+}{\rightleftharpoons}} CH_3.COOC_2H_5 + H_2O$$

It will be noticed that an alcohol reacts differently from water with a carboxylic acid although both are nucleophiles and have similar structures. Alcohols are much poorer ionizing solvents than water, and are unable to ionize weak acids appreciably. Instead of removing the hydrogen atom as a proton from the —OH group, an alcohol molecule attacks the carbonyl carbon atom, which also has a partial positive charge. Alcohols, however, are only weak nucleo-

philes, and reaction at the carbon atom is very slow unless a mineral acid is added to supply H^+ ions. These convert carboxylic acid molecules into more reactive carbonium ions, in which the carbon atom has a full positive charge.

$$\begin{array}{ccc} \overset{\delta-}{O} & & OH \\ \| & & | \\ R\!-\!\overset{\delta+}{C} + H^+ & \rightleftharpoons & R\!-\!\overset{+}{C} \\ | & & | \\ OH & & OH \end{array}$$

The carbonium ion combines with an alcohol molecule as in the acid-catalysed addition of water or alcohol to an aldehyde (Chapter 10) The addition complex is however unstable It rapidly eliminates a water molecule, and a proton is removed by one of the bases present (H_2O or HSO_4^-) If the last two steps are combined, the further stages in the mechanism can be represented in a simplified form as follows:

$$\begin{array}{ccc} OH & & OH \\ | & & | \\ R\!-\!\overset{+}{C} \quad +:\!O\!-\!R' & \rightleftharpoons & R\!-\!\overset{+}{C}\!-\!\overset{+}{O}\!-\!R' \\ | \qquad | & & | \\ OH \quad\; H & & OHH \end{array}$$
$$\text{(unstable)}$$

$$\begin{array}{c} O \\ \| \\ \rightleftharpoons R\!-\!C\!-\!OR' + H_2O + H^+ \end{array}$$

3. Reduction. Oxidizing agents readily convert alcohols to aldehydes and then to carboxylic acids. The reversal of these changes with reducing agents is difficult. While aldehydes are easily reduced, the acids are affected only by the strongest reducing agents. Reduction to the corresponding primary alcohols can be brought about by means of lithium(I) tetrahydridoaluminate(III) (LiAlH$_4$). The reaction can be represented simply as follows:

$$R.COOH \xrightarrow{\text{2H}} R.CHO \xrightarrow{\text{2H}} R.CH_2OH$$

4. Reactions with Phosphorus Trichloride and Phosphorus Pentachloride. Both of these compounds react with alkanoic acids in the cold, the —OH group of the acid being replaced by a chlorine atom. Ethanoic acid yields a colourless liquid called *ethanoyl*, or *acetyl, chloride*.

$$3CH_3.COOH + PCl_3 \rightarrow 3CH_3.COCl + H_3PO_3$$
ethanoyl chloride

$$CH_3.COOH + PCl_5 \rightarrow CH_3.COCl + HCl + POCl_3$$

5. Substitution by Halogens. The hydrogen atoms attached to the carbon atom in the 2-position (that is, the carbon atom adjacent to the carboxyl group) are activated by the inductive effects of the oxygen atoms in the carbonyl group and the hydroxyl group. As a result they undergo substitution by chlorine and bromine more readily than the hydrogen atoms of an alkane. Substitution, however, is slow at ordinary temperatures and in the absence of a catalyst. If chlorine is passed into glacial ethanoic acid at 100°C in the presence of red phosphorus as a catalyst, substitution takes place as shown, giving *chloroethanoic*, or *chloroacetic*, *acid*.

$$CH_3.COOH + Cl_2 \rightarrow CH_2Cl.COOH + HCl$$
chloroethanoic acid

Catalysts like red phosphorus which assist the substitution of hydrogen by chlorine are called *chlorine carriers*.

By prolonging the chlorination the two remaining hydrogen atoms of the methyl radical can also be replaced, giving firstly dichloroethanoic acid (CHCl$_2$.COOH) and then trichloroethanoic acid (CCl$_3$.COOH). Bromine reacts in the same way as chlorine, but less readily. Iodine has no reaction.

As might be expected, chlorination of propanoic acid (or a higher acid) results mainly in substitution of the hydrogen atoms attached to the carbon atom in the 2-position, those joined to the carbon atom in the 3-position being replaced only to a minor extent. The products indicated are formed in turn.

$$CH_3.CHCl.COOH$$
2-chloropropanoic acid

$$CH_3.CCl_2.COOH$$
2,2-dichloropropanoic acid

With bromine substitution is confined to the hydrogen atoms linked to the carbon atom in the 2-position.

Chemical Tests for Alkanoic Acids

The following general tests can be used for recognizing the lower alkanoic acids:

(*i*) *Acidic Properties* Aqueous solutions of soluble alkanoic acids turn blue litmus paper red. The acids are soluble in sodium(I) carbonate solution, and effervescence occurs owing to liberation of carbon dioxide.

(ii) Ester Formation. If an alkanoic acid is warmed in a test-tube with ethanol and a few drops of concentrated sulphuric(VI) acid as a catalyst, an ethyl ester is produced. When the reaction mixture is poured into cold water the ester floats on the water and can be recognized by its pleasant fruity odour.

(iii) Iron(III) Chloride Test. This test depends on the fact that a deep-red *colour* is obtained when neutral iron(III) chloride solution is added to a neutral solution of a salt of an alkanoic acid. When the red solution is boiled a brown *precipitate* of the basic iron(III) salt is formed. If the acid itself is being tested, it must first be neutralized by adding ammonia solution until a piece of red litmus paper just turns blue. The slight excess of ammonia is then removed by boiling.

Iron(III) chloride solution usually contains free hydrochloric acid, which hinders the production of the red colour. The 'neutral' solution is made by adding one or two drops of sodium(I) hydroxide solution until a precipitate of iron(III) hydroxide just forms. The liquid is then filtered, and the filtrate is used for the test.

Structure of Ethanoic Acid

The empirical formula of ethanoic, or acetic, acid deduced from quantitative analysis is CH_2O. The molecular formula obtained from the relative density of the vapour is $C_2H_4O_2$. If the carbon, hydrogen, and oxygen atoms are given their usual valencies, several possible structures can be devised for the molecule, but only one of these agrees with the methods of preparation of the acid and its properties. This structure is represented by the following formula.

$$\begin{array}{ccc} H & & O \\ | & & \parallel \\ H-C-C & \\ | & \diagdown \\ H & & OH \end{array}$$

The validity of this structure is supported by the facts now given.

The two carbon atoms in the molecule are joined together. Ethanol, ethanal, and ethanoic acid all contain two carbon atoms in the molecule, and it is known that in ethanol the two carbon atoms are joined (Chapter 3). It is highly improbable that they become separated when ethanol is oxidized first to ethanal and then to ethanoic acid.

The molecule contains a hydroxyl group. This is shown by the reactions of glacial ethanoic acid with sodium to give hydrogen and with phosphorus pentachloride to give hydrogen chloride. Furthermore, when ethanoic acid is converted into sodium(I) ethanoate either by the action of sodium or by neutralization with sodium(I) hydroxide, only one of the four hydrogen atoms is removed from the molecule. It follows that only one of the four hydrogen atoms is present in a hydroxyl group; that is, the molecule contains one, and only one, hydroxyl group.

The molecule contains a CH_3- radical. Thus, in the preparation of the chloroethanoic acids three hydrogen atoms in the molecule are replaced in stages by chlorine atoms. Again, when a solution of sodium(I) ethanoate is electrolysed (Chapter 5) ethane is obtained, and the ethane molecule consists of two CH_3- radicals joined together.

There is little direct evidence that the ethanoic acid molecule contains the carbonyl group. The presence of the group is shown, however, by the infrared absorption spectrum of the acid, an absorption peak being obtained in the range of wavelengths characteristic of the group in other carbonyl compounds (see Chapter 3). Other absorption peaks confirm the presence of the CH_3- and $-OH$ groups.

METHANOIC ACID, HCOOH

Boiling point: 100·5°C   *Density:* 1·22 g cm^{-3}

Methanoic, or formic, acid is a colourless liquid with a choking smell, which resembles that of sulphur dioxide. The acid was called formic acid because of its occurrence in certain species of ants (Latin *formica* = ant). It was first prepared by distilling the liquid obtained by crushing red ants in water. The use of the acid by these insects is an illustration of chemical warfare in Nature. When red ants are attacked they squirt a jet of the poisonous liquid at their enemies.

The manufacture and uses of methanoic acid have been described earlier. In the laboratory the acid can be obtained by oxidizing methanol with sodium(I) dichromate(VI) and sulphuric(VI) acid. It can also be prepared by distilling ethanedioic acid (oxalic acid) crystals, $(COOH)_2.2H_2O$, with propane-1,2,3-triol (glycerol), which acts as a catalyst.

$$\begin{matrix} COOH \\ | \\ COOH \end{matrix} \rightarrow HCOOH + CO_2$$

Differences Between Methanoic Acid and Other Alkanoic Acids

1. Reaction with Concentrated Sulphuric(VI) Acid. If methanoic acid (or one of its salts) is warmed with concentrated sulphuric(VI) acid, the latter acts as a dehydrating agent and carbon monoxide is given off.

$$HCOOH - H_2O \rightarrow CO$$

Other alkanoic acids are unaffected by concentrated sulphuric(VI) acid.

2. Reducing Reactions. Methanoic acid is the only acid of the series to have strong reducing properties. It is oxidized to carbon dioxide and water by a warm acidified solution, or a cold alkaline solution, of potassium(I) manganate(VII) (potassium permanganate).

$$HCOOH + O \rightarrow H_2O + CO_2$$

A neutral solution of a methanoate, or formate, reduces an ammoniacal solution of silver(I) oxide on warming, giving either a silver 'mirror' or a black precipitate of silver.

When methanoic acid (or one of its salts) is boiled with mercury(II) chloride solution a white precipitate of mercury(I) chloride is obtained.

$$2HgCl_2 + HCOOH \rightarrow Hg_2Cl_2 \downarrow 2HCl + CO_2$$

This reaction distinguishes methanoic acid and its salts from other reducing carboxylic acids, e.g. ethanedioic acid (oxalic acid), and their salts.

The reducing actions of methanoic acid and methanoates call for some comment from the point of view of the structure of the acid. It will be noted that the molecule contains the aldehydic group.

$$H-C\underset{\textstyle OH}{\overset{\textstyle O}{\diagup}}$$

Methanoic acid is the only alkanoic acid which contains the aldehydic group, and it is also the only alkanoic acid with strong reducing properties. Therefore, since aldehydes are well known reducing agents, it has been customary to attribute the reducing powers of methanoic acid to the presence of the aldehyde group. It is doubtful whether this explanation is correct. While methanoic acid and methanoates, like aldehydes, reduce potassium(I) manganate(VII) and an ammoniacal solution of silver(I) oxide, they have no action on Fehling's solution, which is reduced by nearly all aldehydes. On the other hand, methanoic acid and methanoates reduce mercury (II) chloride solution—a reaction not given by aldehydes. Also, methanoic acid and its salts show none of the other typical reactions of an aldehyde. They do not impart a violet colour to Schiff's reagent, undergo reduction with zinc and hydrochloric acid, nor do they form a phenylhydrazone or an oxime.

3. Reactions with Phosphorus Halides. It is not possible to prepare methanoyl chloride, corresponding to ethanoyl chloride, by treating methanoic acid with phosphorus trichloride or pentachloride. This is because of the instability of methanoyl chloride, which decomposes spontaneously into carbon monoxide and hydrogen chloride.

$$HCOCl \rightarrow CO + HCl$$

4. Reaction with Chlorine. There is also no chloromethanoic acid corresponding to chloroethanoic acid. Chlorination of methanoic acid, using the same conditions as for ethanoic acid, produces carbon dioxide and hydrogen chloride.

$$HCOOH + Cl_2 \rightarrow CO_2 + 2HCl$$

5. Action of Heat on Sodium(I) Methanoate. The sodium(I) and potassium(I) salts of methanoic acid differ from the sodium(I) and potassium(I) salts of other alkanoic acids in yielding hydro-

gen when heated. The metal ethanedioate (oxalate) is also formed.

$$2HCOONa \rightarrow \begin{array}{c} COONa \\ | \\ COONa \end{array} + H_2$$

This reaction is used in the manufacture of ethanedioic acid (Chapter 15).

CHLOROETHANOIC ACIDS
Substitution by chlorine of the hydrogen atoms in the methyl radical of ethanoic acid yields chloroethanoic, or chloroacetic, acids. Fluorine, bromine, and iodine derivatives can also be obtained (the latter indirectly). The most important member of the group is monochloroethanoic acid, which is usually called chloroethanoic, or chloroacetic, acid.

Chloroethanoic Acid ($CH_2Cl.COOH$). This is prepared by passing chlorine into a known mass of glacial ethanoic acid contained in a flask partially immersed in a boiling water-bath. Red phosphorus, iodine, or phosphorus trichloride is used as a catalyst, or chlorine carrier. Chlorination is continued until the theoretical increase in mass (in accordance with the following equation) has taken place.

$$CH_3.COOH + Cl_2 \rightarrow CH_2Cl.COOH + HCl$$

The time (4–6 hours) required for the preparation can be shortened if direct sunlight is allowed to fall on the flask. Chloroethanoic acid is obtained from the remaining liquid by fractional distillation.

Chloroethanoic acid forms colourless crystals, which dissolve readily in water and ethanol. The crystals should not be handled as they blister the skin. The acid melts at 62°C, and boils at 189°C.

The chloroethanoic acid molecule is 'bifunctional.' It combines the reactions of a carboxylic acid with those of an alkyl halide. Like ethanoic acid, chloroethanoic acid forms salts with bases, esters with alcohols, and an acid chloride with phosphorus pentachloride.

$$CH_2Cl.COOH + PCl_5 \rightarrow CH_2Cl.COCl + POCl_3 + HCl$$
chloroethanoyl
chloride

On the other hand, the chlorine atom of chloroethanoic acid undergoes substitution by nucleophilic reagents in the same way as the halogen atom of an alkyl halide. Thus it gives rise to the reactions outlined below.

1. Hydrolysis. Chloroethanoic acid is hydrolysed by boiling water alone. Hydroxyethanoic acid (glycollic acid) is formed.

$$CH_2Cl.COOH + H_2O \rightarrow CH_2(OH).COOH + HCl$$
hydroxyethanoic
acid

Hydrolysis takes place more rapidly with boiling aqueous sodium(I) hydroxide. In this case the sodium(I) salt of hydroxy-ethanoic acid is left in solution.

2. Ammonolysis. If chloroethanoic acid is dissolved in excess of concentrated aqueous ammonia and the solution is allowed to stand for 24 hours, the chlorine atom is substituted by an amino group (—NH_2). The product is aminoethanoic acid (glycine).

$$CH_2Cl.COOH + 2NH_3 \rightarrow CH_2(NH_2).COOH + NH_4Cl$$
aminoethanoic
acid

3. Reaction with Potassium(I) Cyanide. When the potassium(I) salt of chloroethanoic acid is boiled with aqueous solution of potassium(I) cyanide the potassium(I) salt of cyanoethanoic acid is formed.

$$CH_2Cl.COOK + KCN \rightarrow CH_2(CN).COOK + KCl$$
potassium(I)
cyanoethanoate

On hydrolysis with a mineral acid potassium(I) cyanoethanoate yields the dicarboxylic acid propanedioic acid (malonic acid), $CH_2(COOH)_2$ (Chapter 16).

Dichloroethanoic acid ($CHCl_2.COOH$) and *trichloroethanoic acid* ($CCl_3.COOH$) can be obtained by prolonged chlorination of ethanoic acid, but are prepared more easily by indirect methods. The first is a colourless liquid, the second a crystalline solid. They are relatively unimportant compounds.

STRENGTHS OF CARBOXYLIC ACIDS

The relative strengths of alkanoic acids in aqueous solution depend on the position of equilibrium in the reaction shown.

$$R.COOH + H_2O \rightleftharpoons R.COO^- + H_3O^+$$

The stronger of two acids is the one which is ionized to the larger extent when the molecular concentrations are equal.

163

The relative strengths of weak acids can be expressed by their acid dissociation constants, K_a, but a more convenient scale is provided by their pK_a values, where $pK_a = -\log K_a$ (compare $pH = -\log[H^+]$. The relative strengths of some carboxylic-acids, including the first four alkanoic acids, are shown in Table 11-2. The stronger the acid the larger is the value of K_a, and the smaller the value of pK_a.

The strongest alkanoic acid is methanoic acid, the only member not containing an alkyl radical. Ethanoic acid is weaker than methanoic acid because the CH_3— radical has a $+I$ (electron-releasing) effect. This is relayed to the O—H bond as indicated, and by decreasing the electron displacement in the

$$CH_3 \rightarrow \overset{\overset{\textstyle O}{\|}}{C} \rightarrow O \rightarrow H$$

bond towards the oxygen atom it reduces the tendency of the hydrogen atom to ionize. At the same time the $+I$ effect increases the electron density on the oxygen atoms of the ethanoate ion (as explained in Chapter 4, a carboxylate ion is a resonance hybrid with a charge of $-\frac{1}{2}$ on each oxygen atom). Hence recombination of the ions occurs more readily. Thus, by hindering the forward change in the ionization reaction and promoting the backward one, the $+I$ effect causes ethanoic acid to ionize to a smaller extent than methanoic acid under similar conditions.

The C_2H_5— radical has a larger $+I$ effect than CH_3—, and therefore propanoic acid is weaker again than ethanoic acid. After C_2H_5—, however, there is practically no increase in the $+I$ effect of the alkyl radicals, so that higher alkanoic acids are about equal in strength.

Table 11-2. RELATIVE STRENGTHS OF SOME CARBOXYLIC ACIDS AT 25°C

Acid	Formula	$(K_a/\text{mol dm}^{-3}) \times 10^5$	pK_a
Methanoic	H—COOH	20·9	3·68
Ethanoic	CH_3—COOH	1·82	4·74
Propanoic	C_2H_5—COOH	1·41	4·85
Butanoic	$C_2H_5.CH_2$—COOH	1·51	4·82
Fluoroethanoic	F—CH_2.COOH	259	2·59
Chloroethanoic	Cl—CH_2.COOH	136	2·87
Bromoethanoic	Br—CH_2.COOH	125	2·94
Iodoethanoic	I—CH_2.COOH	67	3·17

If a chlorine atom, which has a $-I$ effect, is substituted for one of the hydrogen atoms of the CH_3— radical in ethanoic acid, it brings about the electron displacements shown. Similar displacements occur in the chloroethanoate ion. In this case the

$$Cl \leftarrow CH_2 \leftarrow \overset{\overset{\textstyle O}{\|}}{C} \leftarrow O \leftarrow H$$

inductive effect of the chlorine atom increases the partial positive charge on the hydrogen atom of the —OH group, while it decreases the partial negative charges on the oxygen atoms of the chloroethanoate ion. Ionization of the acid is thus made easier, and recombination of ions more difficult. Hence chloroethanoic acid is a stronger acid than ethanoic acid. Stronger still are dichloroethanoic acid and trichloroethanoic acid, in which the $-I$ effect of the halogen atoms progressively increases. Thus the degrees of dissociation of mono-, di-, and trichloroethanoic acids in molar solution at $25°C$ are 4 per cent, 20 per cent, and 65 per cent respectively.

The $-I$ effects of different halogens vary with the strength of their electronegative character. This decreases from fluorine to iodine. Hence fluoroethanoic acid is stronger than chloroethanoic acid, while bromoethanoic acid and iodoethanoic acid are weaker.

Determination of pK_a *(Half-neutralization Method).* The dissociation constant exponent, pK_a, of a weak acid is easily measured. *It is equal to the pH of an aqueous solution of the acid which has been half-neutralized by sodium(I) hydroxide.* This can be shown as follows:

In the case of a mixture of a weak acid and its sodium(I) salt in aqueous solution

$$\frac{[H^+][A^-]}{[HA]} = K_a$$

Therefore

$$[H^+] = K_a \times \frac{[HA]}{[A^-]}$$

Since the acid is only slightly dissociated and the salt highly dissociated, $[A^-]$ can be regarded as derived entirely from the salt. Then

$$[H^+] = K_a \times \frac{[acid]}{[salt]}$$

When the acid is half-neutralized $[acid] = [salt]$, and therefore

$$[H^+] = K_a$$

It follows that

$$pK_a = -\log K_a = -\log [H^+] = pH$$

In practice a known volume of the acid solution is titrated with sodium(I) hydroxide solution of suitable concentration. The actual concentrations of the acid and alkali need not be known. Phenolphthalein is used as indicator. The titration gives the amount of alkali solution required for complete neutralization of a certain volume of acid solution. Half the amount of alkali is then added to the same volume of acid solution, so that the acid is half-neutralized. Finally the pH of the liquid is measured by means of a pH meter or by the use of indicators.

RELATIVE MOLECULAR MASS OF A CARBOXYLIC ACID

General methods of finding the relative molecular masses of organic compounds have been given in Chapter 3. Sometimes these methods are unsuitable for carboxylic acids, or easier methods are available. The two methods now described are chemical ones. Both require a knowledge of the acid's basicity, which, in the case of a carboxylic acid, is equal to the number of —COOH groups in the molecule. The basicity can usually be found from the number of different salts which the acid forms with an alkali, or by other methods described in text-books of physical chemistry.

1. **Titration Method.** A known mass of the anhydrous acid is dissolved in water (or in ethanol if the acid is only sparingly soluble in water). The solution is made up to a known volume, and a portion is titrated with a standard solution of sodium(I) hydroxide, using phenolphthalein as indicator.

Let the formula of the acid be $X(COOH)_n$, where n is the basicity. Since one mole (40 grams) of NaOH neutralizes one mole of —COOH, $40n$ grams of NaOH are required to neutralize one mole of $X(COOH)_n$.

From the titration it is found that a mass m_1 of NaOH neutralizes a mass m_2 of $X(COOH)_n$. Hence $40n$ grams of NaOH neutralize $(m_2/m_1) \times 40n$ grams of $X(COOH)_n$.

The result is numerically equal to the relative molecular mass of the acid.

2. Silver(I) Salt Method. When the silver(I) salt of a carboxylic acid is heated ('ignited') it decomposes, leaving a residue of metallic silver. If the mass of silver obtained from a known mass of the salt is determined, the molar mass of the salt and then that of the acid can be deduced.

Let the formula of the silver salt be $X(COOAg)_n$, where n again is the basicity of the acid.

$$X(COOAg)_n \rightarrow n\ Ag$$

If the relative atomic mass of silver is taken to be 107·9,

$$\frac{\text{Mass of silver}}{\text{Mass of silver(I) salt}} = \frac{107 \cdot 9n \text{ grams}}{\text{molar mass of silver(I) salt}}$$

From the above relationship the molar mass of $X(COOAg)_n$ can be calculated. The molar mass of $X(COOH)_n$ is obtained by subtracting $107 \cdot 9n$ grams for silver and adding $(1 \times n)$ grams for hydrogen.

166

Example

0·664 g of the silver(I) salt of a dibasic organic acid left on ignition a residue of 0·431 g of silver. Calculate the relative molecular mass of the acid. (Ag = 107·9).

Let m be the molar mass of the silver(I) salt. Then from the equation given above

$$\frac{0 \cdot 431 \text{ g}}{0 \cdot 664 \text{ g}} = \frac{107 \cdot 9 \times 2 \text{ g}}{m}$$

From this $\qquad\qquad m = 332 \cdot 4 \text{ g}$

$\therefore$ molar mass of the acid $= 332 \cdot 4 - (107 \cdot 9 \times 2) + 2 \text{ g}$

$$= 118 \cdot 6 \text{ g}$$

Relative molecular mass of the acid = **118·6**

(*Note*. Exercises on the determination of relative molecular masses of carboxylic acids are given in Miscellaneous Problems at the end of the book.)

The silver(I) salts of organic acids are sparingly soluble and can usually be obtained readily by precipitation. The silver(I) salt is made from the acid by neutralizing the latter with ammonia solution, boiling off excess of ammonia, and adding the solution to silver(I) nitrate(V) solution. The silver(I) salt method cannot be applied in the case of acids which contain halogens or sulphur as the residue left on heating is a silver(I) halide or silver(I) sulphide. The method is also unsuitable for ethanedioic acid (oxalic acid) because the silver(I) salt of this acid undergoes sudden and violent decomposition.

1. In which of the following cases have the acids named been given correct formulæ: (a) methanoic acid, CH_2O_2; (b) ethanoic acid, $C_2H_4O_2$; (c) propanoic acid, $C_2H_5.COOH$; (d) butanoic acid, $C_2H_5.CH_2.CH_2.COOH$?

2. Show by means of formulæ connected by arrows the steps by which the following conversions could be brought about in the laboratory:

(a) Ethanol to propanoic acid.

(b) Ethene to ethanoic acid.

3. State which of the following contribute to the acidic character of the carboxyl group in alkanoic acids:

(a) The inductive effect of the hydroxyl oxygen atom.

(b) The inductive effect of the carbonyl oxygen atom.

(c) The inductive effect of an attached alkyl radical.

(d) Isomerism.

(e) Mesomerism.

4. Give the names and formulæ of the organic compounds formed in the reactions between propanoic acid and (a) ammonia, (b) phosphorus pentachloride, (c) methanol, (d) lithium(I) tetrahydrido-aluminate(III) (lithium aluminium hydride).

5. With which of the following reagents does methanoic acid react differently from other alkanoic acids: (a) sodium(I) hydrogencarbonate, (b) acidified potassium(I) manganate(VII) solution, (c) mercury(II) chloride solution, (d) concentrated sulphuric(VI) acid, (e) phosphorus pentachloride?

6. Name the organic products of the reactions between methanoic acid and the reagents given in question 5.

7. From a consideration of inductive effects arrange the following in order of their strengths as acids, putting the strongest one first:

(a) $CH_3.COOH$, (b) $C_2H_5.COOH$, (c) $CH_2Cl.COOH$,
(d) $CHCl_2.COOH$, (e) $CH_2Br.COOH$.

8. Give the structural formulæ of the following: (a) 2-methylpropanoic acid, (b) 2,2-dichloropropanoic acid, (c) chloroethanoyl chloride, (d) hydroxyethanoic acid, (e) methyl ethanoate.

12 Acid Derivatives—Acid Chlorides and Anhydrides

Alkanoic acids give rise to several important classes of derivatives besides salts and halogeno-acids (considered in the last chapter). This chapter and subsequent ones deal with acid chlorides, acid anhydrides, esters, acid amides, and nitriles (alkyl cyanides).

ACID (ACYL) CHLORIDES

Acid chlorides form an homologous series of general formula C_nH_{2n+1}. COCl, or R.COCl. They are also called *acyl* chlorides because they contain an acyl group of the type now shown.

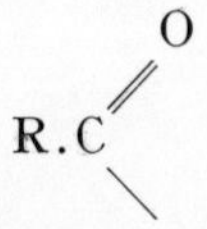

acyl group

Acyl fluorides, bromides, and iodides can also be prepared, but only the chlorides are used to any large extent and merit attention. The chlorides are derived from carboxylic acids by replacing the hydroxyl group of the latter by a chlorine atom; e.g.

HCOOH

methanoic, or formic, acid

(HCOCl)

(methanoyl, or formyl, chloride)

CH₃.COOH

ethanoic, or acetic acid

CH₃.COCl

ethanoyl, or acetyl, chloride

C₂H₅.COOH

propanoic acid

C₂H₅.COCl

propanoyl chloride

Methanoyl chloride is unstable, and has not been isolated. Ethanoyl chloride is outstandingly the most important member of the series. The lower acyl chlorides are colourless fuming liquids, the boiling points of which rise with increase in molecular mass. Higher members are solids.

Preparation of Acyl Chlorides

Acyl chlorides can be obtained by treating the anhydrous acid with phosphorus trichloride, phosphorus pentachloride, or sulphur dichloride oxide. In the case of ethanoic, or acetic acid the reactions are as follows:

$$3CH_3.COOH + PCl_3 \rightarrow 3CH_3.COCl + H_3PO_3$$

$$CH_3.COOH + PCl_5 \rightarrow CH_3.COCl + POCl_3 + HCl$$

$$CH_3.COOH + SOCl_2 \rightarrow CH_3.COCl + SO_2 + HCl$$

Sulphur dichloride oxide (thionyl chloride) is the acid chloride of sulphuric(IV) acid (sulphurous acid), $SO(OH)_2$. As seen in Chapter 9, it can also be used for substitution of an —OH group by a chlorine atom in the preparation of alkyl chlorides from alcohols.

For the laboratory preparation of ethanoyl chloride it is more convenient to use phosphorus trichloride or sulphur dichloride oxide, which are liquids, than phosphorus pentachloride, which is a solid. In industry sulphur dichloride oxide is preferred because it gives the best yield. Ethanoyl chloride is prepared from ethanoic acid and phosphorus trichloride as now described.

Experiment

The apparatus used should be dry because both phosphorus trichloride and ethanoyl chloride react with moisture and produce fumes of hydrogen chloride. Even so, hydrogen chloride is always evolved owing to the presence of a little water in the glacial ethanoic acid. The preparation should therefore be carried out in a fume-cupboard.

Set up the apparatus shown in Fig. 12-1. Use as receiver a small distilling-flask, to the side-arm of which is attached a calcium(II) chloride drying-tube to exclude atmospheric moisture. Cool the receiver in a trough of cold water. Put into the 50 cm³ pear-shaped flask 16 cm³ of glacial ethanoic acid. In the dropping-funnel place 8 cm³ of phosphorus trichloride (this is used in excess because some is wasted in side-reactions). Insert a plug of cotton wool into the top of the dropping-funnel to prevent fuming of the liquid.

Add the phosphorus trichloride to the flask a few drops at a time, shaking the flask during the addition to mix the liquids. When all the phosphorus trichloride has been added warm the water-bath to about 45°C, and keep it at this temperature for 15 minutes to drive off hydrogen chloride dissolved in the liquid. Then gradually heat the water to boiling point to distil off the ethanoyl chloride. Purify the crude product by redistillation, collecting the fraction obtained at 50°–54°C. The yield is about 11 g. Ethanoyl chloride boils at 52°C and has a density of 1·10 g cm⁻³.

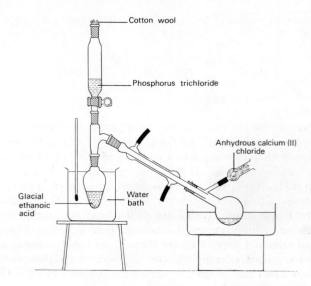

Fig. 12-1. Preparation of ethanoyl, or acetyl, chloride

Reactions of Acyl Chlorides

The functional group in these compounds is the *acyl chloride* group (—COCl). The latter has an exceptionally strong electrophilic character, so that acyl chlorides react vigorously with nucleophilic reagents, even when these are relatively weak nucleophiles, as in the case of water and alcohols. Most of the reactions are of the type shown, in which the chlorine atom

169

undergoes replacement by another atom or group.

$$R-\overset{\displaystyle O}{\underset{\displaystyle Cl}{C}} + HY \rightarrow R-\overset{\displaystyle O}{\underset{\displaystyle Y}{C}} + HCl$$

Alternatively the reaction may be regarded as substitution of the hydrogen atom of HY by the acyl group $R.CO—$. The introduction of an acyl group into a molecule in place of a hydrogen atom is called *acylation*. If the group is an ethanoyl, or acetyl, group ($CH_3.CO—$), the process is described as *ethanoylation*, or *acetylation*. Ethanoyl chloride is chiefly used as an ethanoylating agent. Of special importance are its reactions with compounds containing hydroxyl groups. In these reactions hydrogen chloride and an ethanoyl derivative are formed. As both products are easily recognized, ethanoyl chloride (like sodium and phosphorus pentachloride) can be used to detect the presence of an —OH group in a compound.

1. Reactions with Compounds Containing Hydroxyl Groups. Acyl chlorides undergo hydrolysis with water and alcoholysis with ethanol in the cold. The reactions are strongly exothermic. With water ethanoyl chloride yields ethanoic acid and hydrogen chloride, while with ethanol it gives ethyl ethanoate and hydrogen chloride.

$$CH_3.COCl + HOH \rightarrow CH_3.COOH + HCl$$

$$CH_3.COCl + C_2H_5OH \rightarrow CH_3.COOC_2H_5 + HCl$$

170

Experiment

Put 1 cm³ of water into a test-tube, and stand the tube in a rack. *Cautiously* add one or two drops of ethanoyl chloride from a dropping-tube. A violent reaction ocurs, the liquid in the test-tube becomes hot, and steamy fumes of hydrogen chloride are evolved.

Repeat the experiment, using ethanol in place of water. After addition of the ethanoyl chloride pour the contents of the tube into a beaker of water. The ethyl ethanoate floats on the surface, and its pleasant fruity odour is readily detected.

The OH⁻ ion is one of the strongest of nucleophilic reagents and, as might be expected, acyl chlorides react with aqueous alkalies even more violently than with water. The reactions with the lower acyl chlorides are too dangerous to be tried, but they can be carried out with the less reactive higher members.

$$R.COCl + 2OH^- \rightarrow R.COO^- + Cl^- + H_2O$$

2. Ammonolysis. When an acyl chloride is added to concentrated aqueous ammonia the liquid becomes hot, and a solution of an acid amide remains. Ethanoyl chloride gives *ethanamide*, or *acetamide*, and ammonium chloride.

$$CH_3.\overset{\displaystyle O}{\underset{\displaystyle Cl}{C}} + NH_3 \rightarrow CH_3.\overset{\displaystyle O}{\underset{\displaystyle NH_2}{C}} + HCl$$

ethanamide

$$NH_3 + HCl \rightarrow NH_4Cl$$

In a similar manner ethanoyl chloride ethanoylates amines, which are 'substituted ammonias.' With ethylamine it yields N-ethylethanamide, or N-ethylacetamide (the prefix N- indicates that the ethyl radical is attached to the nitrogen atom).

$$CH_3.C\!\!\begin{array}{c}O\\ \diagup\\ \diagdown\\ Cl\end{array} + C_2H_5NH_2 \rightarrow CH_3.C\!\!\begin{array}{c}O\\ \diagup\\ \diagdown\\ NHC_2H_5\end{array} + HCl$$

ethylamine N-ethylethanamide

3. Reduction. Acid chlorides are reduced by hydrogen and a palladium catalyst firstly to aldehydes and then to alcohols. The acid chloride is dissolved in an inert hydrocarbon solvent (usually xylene), and hydrogen is passed into the boiling solution in the presence of palladium. Reduction of ethanoyl chloride in this way provides a method of converting ethanoic acid, which is itself difficult to reduce, back to ethanol.

$$CH_3.COCl + H_2 \rightarrow CH_3.CHO + HCl$$

$$CH_3.CHO + H_2 \rightarrow CH_3.CH_2OH$$

Reaction Mechanism for Acyl Chlorides and Nucleophiles.

Acyl chlorides owe their strong electrophilic character to the combined $-I$ effects of the oxygen atom and the chlorine atom. These give the carbonyl carbon atom an unusually high partial positive charge (decreased slightly by the $+I$ effect of the alkyl radical).

Although most of the reactions of acyl chlorides with nucleophilic reagents result in substitution of the halogen atom, experimental evidence indicates that the mechanism involved differs from that for substitution in alkyl halides. As was seen in Chapter 9, the order of reactivity of alkyl halides with nucleophilic reagents is: iodides > bromides > chlorides > fluorides. For acyl halides the order of reactivity is the opposite, fluorides being the most reactive and iodides the least reactive.

In the case of acyl halides the mechanism probably consists of addition-elimination, as in some of the reactions of aldehydes and ketones with nucleophilic reagents. It is thought, for example, that in the reaction between water and ethanoyl chloride the first step consists of addition of a water molecule at the carbonyl carbon atom. The oxygen atom of the added molecule now has a positive charge, which attracts the electrons of the O—H bonds and weakens them. Another water molecule, acting as a base, is able to remove one of the hydrogen atoms as a proton. This leaves an unstable negative ion, from which a chlorine atom is eliminated as an anion.

$$CH_3\!-\!\overset{\overset{\delta-}{O}}{\underset{Cl}{C^{\delta+}}} + :OH_2 \underset{\longleftarrow}{\overset{slow}{\longrightarrow}} CH_3\!-\!\overset{\overset{\bar{O}}{\vert}}{\underset{Cl}{C}}\!\!-\!\overset{+}{O}H_2 \underset{\longleftarrow}{\overset{-H^+}{\longrightarrow}} CH_3\!-\!\overset{\overset{\bar{O}}{\vert}}{\underset{Cl}{C}}\!\!-\!OH$$

$$\underset{\longleftarrow}{\overset{-Cl^-}{\longrightarrow}} CH_3\!-\!\overset{\overset{O}{\vert\vert}}{C}\!\!-\!OH$$

The above mechanism is consistent with the order of reactivity of acyl halides. The slowest step in the mechanism is the first one. Since the —I effects of the halogens diminish from fluorine to iodine, the partial charge on the carbonyl carbon atom, and hence the rate of reaction, decreases in the same order. Further supporting evidence is the fact that acyl halides, like aldehydes and ketones, are able to form stable addition compounds with Grignard reagents.

Carbonyl Chloride, $COCl_2$. Carbonyl chloride, commonly called *phosgene*, is not an alkanoic acid chloride, but is included here for convenience. It is the acid chloride of carbonic acid.

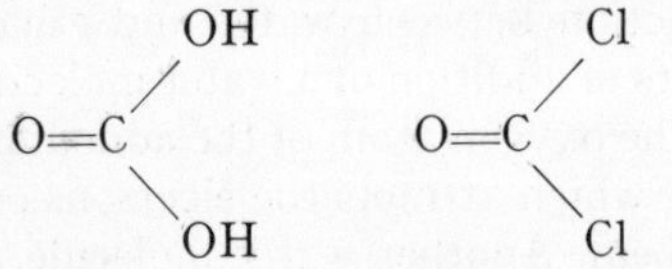

$$O=C\begin{smallmatrix} OH \\ \\ OH \end{smallmatrix} \qquad O=C\begin{smallmatrix} Cl \\ \\ Cl \end{smallmatrix}$$

carbonic acid   carbonyl chloride

Carbonyl chloride is the most important organic acid chloride in industry. It is chiefly used in the manufacture of synthetic resins and plastics, including polyurethane, an ingredient of high-gloss paints. It is made by passing a mixture of equal volumes of carbon monoxide and chlorine over a catalyst of carbon (activated charcoal) at $200°C$.

$$CO + Cl_2 \rightarrow COCl_2$$

Carbonyl chloride is a colourless gas with a smell of damp hay. As its boiling point ($8°C$) is only just below ordinary temperature it is easily liquefied by pressure, and it is usually handled in the form of the liquefied gas. It is highly toxic, and was used as a poison gas in the First World War.

Like other acyl chlorides carbonyl chloride is attacked by nucleophilic reagents, and yields corresponding products. The reactions of carbonyl chloride with water, ethanol, and ammonia are illustrated by the equations now given.

$$COCl_2 + H_2O \rightarrow CO_2 + 2HCl$$

$$COCl_2 + 2C_2H_5OH \rightarrow (C_2H_5)_2CO_3 + 2HCl$$
diethyl carbonate

$$COCl_2 + 4NH_3 \rightarrow CO(NH_2)_2 + 2NH_4Cl$$
urea

ACID ANHYDRIDES

The anhydrides of alkanoic acids are derived by elimination of one molecule of water from two molecules of the acid.

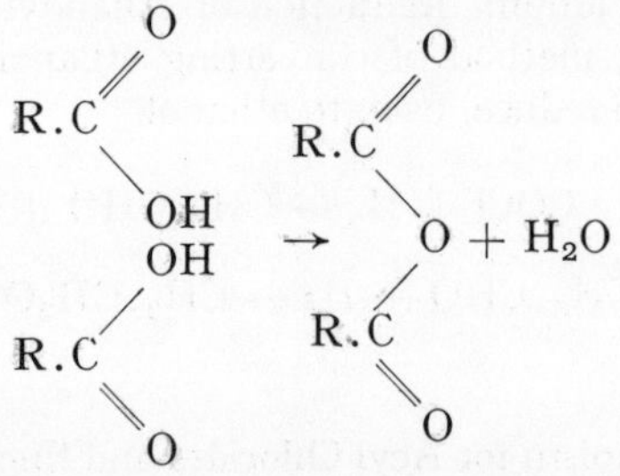

$$R.C\begin{smallmatrix}O\\ \\OH\end{smallmatrix} \quad R.C\begin{smallmatrix}O\\ \\OH\end{smallmatrix} \rightarrow \quad \begin{smallmatrix}R.C\diagdown\\ \quad O\\ R.C\diagup\end{smallmatrix} + H_2O$$

In practice it is difficult to dehydrate alkanoic acids directly to their anhydrides, and these are usually prepared indirectly. Methanoic acid does not form an anhydride of the type shown above. The only alkanoic acid anhydride of industrial importance is ethanoic, or acetic, anhydride.

Preparation of Ethanoic, or Acetic, Anhydride.

Ethanoic anhydride is prepared in the laboratory by adding ethanoyl chloride in small portions to anhydrous sodium(I) ethanoate. Considerable heat is given out in the reaction.

$$CH_3.COCl + CH_3.COONa \rightarrow (CH_3.CO)_2O + NaCl$$

The apparatus is similar to that used in the preparation of ethanoyl chloride (Fig. 12-1). When all the ethanoyl chloride has been added the ethanoic anhydride is distilled off.

On a large scale ethanoic anhydride is produced by catalytic oxidation of ethanal (acetaldehyde) by air at 60°C and a pressure of 5 atmospheres. The catalyst is a mixture of copper(II) ethanoate and cobalt(II) ethanoate.

$$2CH_3.CHO + O_2 \rightarrow (CH_3.CO)_2O + H_2O$$

The water is removed by distillation under reduced pressure before it has time to react with the ethanoic anhydride. The latter is then fractionally distilled to separate it from ethanoic acid, which is also produced in the oxidation.

Properties of Ethanoic Anhydride

Ethanoic anhydride is a colourless liquid with a pungent smell resembling that of ethanoic acid. It has a density of $1·08$ g cm^{-3}, and boils at 139°C. It does not fume in damp air, and, rather surprisingly, is only slightly soluble in cold water.

Ethanoic anhydride resembles ethanoyl chloride in reacting with nucleophilic reagents. It undergoes hydrolysis, alcoholysis, and ammonolysis in the same way, although it reacts much less vigorously. It is an ethanoylating agent, but when it is used for this purpose, only one of the two ethanoyl groups is used directly the other yielding ethanoic acid.

(*i*) Ethanoic anhydride is slowly hydrolysed by cold water to ethanoic acid. The reaction takes place rapidly on warming.

$$(CH_3.CO)_2O + H_2O \rightarrow 2CH_3.COOH$$

(*ii*) If ethanoic anhydride and ethanol are heated together, ethyl ethanoate and ethanoic acid are formed.

$$(CH_3.CO)_2O + C_2H_5OH \rightarrow CH_3.COOC_2H_5 + CH_3COOH$$

(*iii*) Heat is evolved when ethanoic anhydride is mixed with a cold concentrated solution of ammonia, and a solution containing ethanamide, or acetamide, and ammonium ethanoate is obtained.

$$(CH_3.CO)_2O + 2NH_3 \rightarrow CH_3.CONH_2 + CH_3.COONH_4$$

(*iv*) Ethanoic anhydride brings about ethanoylation of both aliphatic and aromatic amines. Ethylamine yields N-ethylethanamide and ethanoic acid.

$$(CH_3.CO)_2O + C_2H_5NH_2$$

$$\rightarrow CH_3.\overset{\overset{\displaystyle O}{\|}}{C}-NHC_2H_5 + CH_3.COOH$$

Ethanoic anhydride is used as a large-scale ethanoylating agent (usually in conjunction with glacial ethanoic acid). Three-quarters of the industrial output is absorbed in the manufacture of cellulose ethanoate fibre and plastic (Chapter 26). Another major use is in the preparation of drugs such as aspirin and diethanoyl morphine ('heroin').

REACTIVITIES OF ACID DERIVATIVES WITH NUCLEOPHILES

Earlier in this chapter the relative reactivities of different acyl halides with nucleophilic reagents were explained as due to the different inductive effects of the F, Cl, Br, and I atoms. The larger the $-I$ effect the greater is the partial positive charge on the carbonyl carbon atom, and the more reactive is the derivative. The relative reactivities of acid chlorides, anhydrides, esters, and amides with nucleophiles can be similarly explained. All of these can be represented by the following formula, in which Y is an electron-attracting atom or group ($-Cl$, $-O.OC.R$, $-OR$, or $-NH_2$):

$$\overset{\delta-}{\underset{\delta+}{\underset{\delta+}{R}-\overset{\overset{\delta-}{O}}{\underset{}{\overset{\|}{C}}}-\overset{\delta-}{Y}}}$$

The order of reactivity of the acid derivatives with nucleophiles (e.g. water) is in the order now shown.

Acid chlorides > acid anhydrides > esters > acid amides.

This order corresponds with that of the inductive effects of Y.

Another factor which may affect the reactivities of the above compounds is mesomerism. The latter arises because the atom attached to the carbonyl group has a lone pair of electrons; e.g.

$$R-\overset{\overset{\cdot\cdot}{O}:}{\underset{}{\overset{\|}{C}}}-\overset{\cdot\cdot}{\underset{\cdot\cdot}{Cl}}: \leftrightarrow R-\overset{\overset{\bar{\cdot\cdot}}{:O:}}{\underset{}{\overset{\|}{C}}}=\overset{+}{\underset{\cdot\cdot}{Cl}}:$$

Mesomerism has only a minor effect in the case of most acid derivatives. This is because the ionic structures have relatively high energies (owing to separation of charge), and make only small contributions to the actual structures of the resonance hybrids. Mesomerism, however, always tends to stabilize a molecule and make it less reactive. This is seen most clearly in carbonyl chloride, which is a resonance hybrid with *three* contributing structures. If the reactivity of this compound were determined solely by inductive effects, it should react with water even more rapidly than ethanoyl, or acetyl, chloride does. In practice carbonyl chloride is hydrolysed only slowly at ordinary temperatures.

EXERCISE 12

1. Name the following groups: (*a*) R.CO—; (*b*) $CH_3.CO$—, (*c*) $C_2H_5.CO$—.

2. Write equations for the reactions between propanoic acid and (*a*) phosphorus trichloride, (*b*) sulphur dichloride oxide.

3. Give the systematic names of the organic compounds formed when propanoyl chloride reacts with (*a*) water, (*b*) ethanol, (*c*) ammonia, (*d*) hydrogen (with a palladium catalyst), (*e*) sodium propanoate.

4. Write the structural formulæ of (*a*) butanoyl chloride, (*b*) ethanamide, (*c*) *N*-ethylethanamide. (*d*) ethanoic anhydride.

5. Represent the steps which are probably involved in the reaction between an aqueous alkali (OH^- ions) and an acyl chloride R.COCl.

6. Complete the following equations: (*a*) $(RCO)_2 O + C_2H_5OH \rightarrow$; *b*) $(RCO)_2 O + 2NH_3 \rightarrow$.

7. Which of the following statements are true?

(*a*) Methanoyl chloride normally exists as a colourless gas.

(*b*) Ethanoylation is the substitution of a hydrogen atom in a molecule by the $CH_3.CO$— group.

(*c*) Acyl chlorides are electrophiles.

(*d*) In acyl chlorides the partial positive charge on the carbonyl carbon atom is increased by the $+ I$ effect of the chlorine atom.

8. From a consideration of inductive effects arrange the following acyl halides according to the rates at which you would expect them to react with ethanol under similar conditions: (*a*) $CH_3.COCl$, (*b*) $CH_3.COBr$, (*c*) $CH_2Cl.COCl$, (*d*) $C_2H_5.COBr$.

13   Acid Derivatives *(continued)* —Esters

An ester *is the compound, other than water, formed by the combination of an alcohol with an acid (organic or inorganic).*

The reaction between an alcohol and an acid is reversible, and is represented in general by the equation

$$\text{ALCOHOL} + \text{ACID} \rightleftharpoons \text{ESTER} + \text{WATER}$$

Thus, when an acid is added to an alcohol, or an ester to water, an equilibrium is established in which all four substances are present. The forward reaction is described as *esterification* of the alcohol or acid, and the backward reaction as *hydrolysis* of the ester.

ESTERS OF ALKANOIC ACIDS

The general equation for the formation of an ester from a monohydric alcohol and an alkanoic acid can be written as follows:

$$\underset{\text{acid}}{R.\overset{\displaystyle O}{\overset{\|}{C}}-OH} + \underset{\text{alcohol}}{R'OH} \rightleftharpoons \underset{\text{ester}}{R.\overset{\displaystyle O}{\overset{\|}{C}}-OR'} + H_2O$$

It is important to remember that in writing the structural formula of an ester derived from a monohydric alcohol and an alkanoic acid the acid part of the molecule is put first. Thus the formula $CH_3.COOC_2H_5$ represents ethyl ethanoate, or acetate, while $C_2H_5.COOCH_3$ stands for methyl propanoate (methyl propionate). The structures of these esters are represented more fully as follows:

$$\underset{\text{ethyl ethanoate}}{CH_3-\overset{\displaystyle O}{\overset{\|}{C}}-O-C_2H_5} \qquad \underset{\text{methyl propanoate}}{C_2H_5-\overset{\displaystyle O}{\overset{\|}{C}}-O-CH_3}$$

Table 13-1. METHYL AND ETHYL ESTERS
(General formulæ: $C_nH_{2n}O_2$, $R.COOR'$)

Systematic name	Common name	Structural formula	Boiling point/°C
Methyl	*Methyl*		
Methanoate	Formate	$HCOOCH_3$	32
Ethanoate	Acetate	$CH_3.COOCH_3$	57
Propanoate	Propionate	$C_2H_5.COOCH_3$	80
Butanoate	*n*-Butyrate	$C_2H_5.CH_2.COOCH_3$	102
Ethyl	*Ethyl*		
Methanoate	Formate	$HCOOC_2H_5$	54
Ethanoate	Acetate	$CH_3.COOC_2H_5$	77
Propanoate	Propionate	$C_2H_5.COOC_2H_5$	99
Butanoate	*n*-Butyrate	$C_2H_5.CH_2.COOC_2H_5$	121

The methyl and ethyl esters of the lower alkanoic acids are shown in Table 13-1. The esters of the first two acids, like the acids themselves, may be named either by their systematic names or by their common names. For esters derived from higher acids systematic names are recommended.

Esters of the lower monohydric alcohols and lower alkanoic acids are colourless volatile liquids, which show a progressive rise in boiling point with increase in molecular mass. Higher esters are solids. The liquid members are lighter than water, and float on water owing to their small solubility. Higher esters are insoluble in water, but dissolve in ethanol or ethoxyethane.

Esters are well known for their pleasant fruity odours. The scents of flowers and the flavours of fruits are largely caused by mixtures of esters. Synthetic esters are used as perfumes and flavouring essences, some examples being ethyl ethanoate-apples, ethyl butanoate-pineapples, 3-methylbutyl ethanoate ('amyl' acetate)-pear-drops, octyl ethanoate-oranges.

Preparation of Esters

(i) *From the Alcohol and Acid*. A mixture of an alcohol and an alkanoic acid comes to equilibrium very slowly, even when boiled. The rate of esterification is increased, however, by hydrogen ions derived from a strong mineral acid, usually concentrated sulphuric(VI) acid. The latter not only acts as a catalyst, but also absorbs water formed in the reaction. In this way the position of equilibrium is displaced and the yield of ester increased. The mechanism of the reaction has been described in Chapter 11.

Ethyl ethanoate is prepared in the laboratory by refluxing together ethanol, glacial ethanoic acid, and concentrated sulphuric(VI) acid. If the reaction vessel is surrounded by a boiling water-bath, the temperature is kept down to 100°C, and this avoids the formation of ethoxyethane (diethyl ether) from the ethanol and sulphuric(VI) acid. Excess of ethanol is used as some is lost through charring by the sulphuric(VI) acid.

Experiment

Put into a 50 cm³ pear-shaped flask 12 cm³ of industrial spirit, 8 cm³ of glacial ethanoic acid, and a little powdered pumice. Add carefully in small portions 5 cm³ of concentrated sulphuric(VI) acid. Between each addition swirl the flask round to mix the liquids and cool it under the tap. When all the sulphuric(VI) acid has been added attach a reflux condenser to the flask, and partially immerse

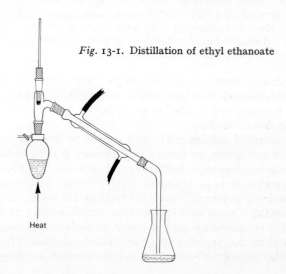

Fig. 13-1. Distillation of ethyl ethanoate

Heat

$$CH_3.COOH + C_2H_5OH \underset{}{\overset{H^+}{\rightleftharpoons}} CH_3.COOC_2H_5 + H_2O$$

the latter in a beaker of water standing on a tripod and gauze. Boil the water gently for 20 minutes.

Remove the water-bath, tripod, and gauze, and allow the flask to cool. Connect the flask through a still-head to the condenser, which has been arranged for direct distillation (Fig. 13-1). In the still-head insert a thermometer-pocket containing a few drops of mercury and a thermometer. Distil the contents of the flask directly over a small flame, collecting the liquid which comes over up to 85°C in a small conical flask.

Purification. The crude ester contains ethanol, ethanoic acid, water, and sulphur dioxide. Shake the liquid in a separating-funnel with 10 cm³ of fairly concentrated sodium(I) carbonate solution until the upper layer of ethyl ethanoate ceases to turn blue litmus paper red. During the shaking remove the stopper from time to time to allow carbon dioxide to escape. Discard the lower layer, and repeat the washing with 10 cm³ of water. Again separate the liquids, and dry the ethyl ethanoate with anhydrous calcium(II) chloride. Finally redistil the liquid, collecting the fraction which passes over at 74°–78°C. The yield is about 8 cm³ (7·2 g).

In industry the most important simple esters are *ethyl* ethanoate and *butyl ethanoate*, which are used as solvents in making lacquers. Their manufacture is similar in principle to the laboratory method described above.

Concentrated sulphuric(VI) acid cannot be used to prepare esters of methanoic, or formic, acid because it dehydrates the acid to carbon monoxide. Methanoic acid, however, is a sufficiently strong acid to serve as its own catalyst, although a longer period of refluxing is required. Thus ethyl methanoate can be obtained by refluxing ethanol with excess of methanoic acid for 24 hours.

Esters of secondary and tertiary alcohols also cannot be prepared in the presence of concentrated sulphuric(VI) acid. This is because of the ease with which the alcohols are converted to alkenes by sulphuric(VI) acid. In these cases hydrogen chloride is used as the catalyst. The gas is passed into the alcohol to give a fairly concentrated solution, which is then refluxed with the organic acid. This is called the *Fischer-Speier* method of esterification.

(*ii*) *By Transesterification.* If an ester of one alcohol is refluxed with another alcohol in the presence of a little alkali as a catalyst, the first alcohol is partially displaced by the second. The new ester can then be separated by fractional distillation. Transesterification is convenient when direct esterification employing sulphuric(VI) acid is unsuitable. It is often used for making esters of tertiary alcohols. For example, 1,1-dimethyl-ethyl ethanoate (*t*-butyl acetate) can be prepared in this way from ethyl ethanoate and 2-methylpropan-2-ol (*t*-butyl alcohol).

$$CH_3.COOC_2H_5 + (CH_3)_3COH \rightleftharpoons CH_3.COOC(CH_3)_3 + \\ + C_2H_5OH$$

(*iii*) *From an Alkyl Iodide and a Silver(I) Salt.* When an alkyl iodide is refluxed with the silver(I) salt of an alkanoic acid in ethanolic solution a yellow precipitate of silver(I) iodide is gradually formed, and an ester remains in solution. Again the ester is isolated by fractional distillation.

$$RI + R'.COOAg \rightarrow R'.COOR + AgI \downarrow$$

This method is also used for making esters which are difficult to obtain by direct esterification.

As noted in Chapter 12, esters are formed when alcohols are treated with acid chlorides or anhydrides.

Reactions of Esters

The key group in an ester molecule is the carbonyl group, although again direct evidence for the presence of the group is limited to infrared absorption spectra and the formation of stable addition compounds with Grignard reagents. The substances which react with esters are the usual nucleophilic reagents —water, OH^- ions, alcohols, ammonia, and reducing agents. These again attack the carbonyl carbon atom because of its low electron density. The reactions result in substitution of the alkoxide group, —OR, but, as in the case of acyl chlorides, the reactions probably take place by addition-elimination, and not by simple substitution. As noted in Chapter 12, an —OR group has a smaller $-I$ effect than a chlorine atom, and esters are weaker electrophiles than acyl chlorides. Hence they react less readily with nucleophilic reagents.

1. **Hydrolysis.** As water· is a weak nucleophile it hydrolyses esters only slowly, even on boiling. In practice hydrolysis is carried out by boiling the ester with either a mineral acid or an alkali. In the case of the acid the reaction is catalysed by hydrogen ions, and an equilibrium mixture is formed. Hydrolysis of ethyl ethanoate by boiling dilute sulphuric(VI) acid is represented by the equation

$$CH_3.COOC_2H_5 + H_2O \underset{}{\overset{H^+}{\rightleftharpoons}} CH_3.COOH + C_2H_5OH$$

The probable mechanism of the reaction is the converse of the mechanism given in Chapter 11 for the acid-catalysed esterification of the alcohol.

Hydrolysis takes place more rapidly with an alkali than with an acid, the OH^- ion being a strong nucleophile. In this case the reaction goes to completion.

$$CH_3.COOC_2H_5 + OH^- \rightarrow CH_3.COO^- + C_2H_5OH$$

Alkaline hydrolysis of an ester is sometimes described as *saponification* because it is used in the manufacture of soap, as described shortly. Ethanol and ethanoic acid are obtained from ethyl ethanoate as described below.

Experiment

Put 4 cm³ of ethyl ethanoate into a 50 cm³ pear-shaped flask. Dissolve 4 g of sodium(I) hydroxide pellets in 20 cm³ of water, and add the solution to the flask together with a little powdered pumice. Attach a reflux condenser, and boil the mixture gently until the upper layer of ethyl ethanoate has disappeared (30 minutes). Change over the condenser for direct distillation and distil the liquid, collecting the first 8 cm³ of distillate. This is a dilute solution of ethanol (test for the latter by the tri-iodomethane (iodoform) reaction). Absolute ethanol can be obtained from the dilute solution, but this requires the experiment to be done on a larger scale.

The residue in the flask contains sodium(I) ethanoate and excess of alkali. Transfer the liquid to an evaporating-dish, and neutralize the alkali by adding dilute sulphuric(VI) acid in small portions until a piece of blue litmus paper just turns red. Evaporate the neutralized liquid to dryness over a tripod and gauze. Scrape the remaining mixture of sodium(I) ethanoate and sodium(I) sulphate(VI) into the pear-shaped flask, and add 5 cm³ of concentrated sulphuric(VI) acid. Fit the flask with a still-head and thermometer, and attach a condenser and receiver. Distil the liquid, collecting the fraction which passes over at 115°–120°C. This is glacial ethanoic acid. Test the acid with 'neutral' iron(III) chloride solution as described in Chapter 11.

The rate of hydrolysis is proportional to the concentrations of both ester and hydroxyl ions. This agrees with the slow, or rate-controlling, step in the reaction mechanism being addition of the OH^- ion to the carbonyl carbon atom. The addition complex is unstable, and splits off the alkoxide group as an anion. The latter then forms the alcohol by capturing a proton from a water molecule or an acid molecule.

$$R-\underset{\underset{OR'}{|}}{\overset{\overset{O}{||}}{C}} + :OH^- \rightleftharpoons R-\underset{\underset{OR'}{|}}{\overset{\overset{\overline{O}}{|}}{C}}-OH \rightleftharpoons R'O^- + R-\overset{\overset{O}{||}}{C}-OH$$

(unstable)

$$R'O^- + HOH \rightleftharpoons R'OH + OH^-$$

2. Alcoholysis. Ethanol, like water, is too weak a nucleophile to react appreciably with esters by itself. However, a reaction occurs if ethanol is refluxed with an ester (other than an ethyl ester) in the presence of a little alkali as a catalyst. The latter converts some of the ethanol into ethoxide ions, which are strongly nucleophilic. Under these conditions ethanol partially replaces the alcohol already present in the ester, and an ethyl ester is formed.

$$R.COOR' + C_2H_5O^- \rightleftharpoons R.COOC_2H_5 + R'O^-$$

$$R'O^- + H_2O \rightleftharpoons R'OH + OH^-$$

As seen earlier, this type of reaction is used in the transesterification method of preparing esters.

3. Ammonolysis. The alkoxide group of an ester can be replaced by an amino ($-NH_2$) group by shaking the ester with concentrated aqueous ammonia at ordinary temperatures. Ethyl ethanoate and ammonia yield ethanamide (acetamide) and ethanol.

$$CH_3.COOC_2H_5 + NH_3 \rightleftharpoons CH_3.CONH_2 + C_2H_5OH$$

ethanamide

4. Reduction. Under suitable conditions esters can be reduced to alcohols by sodium and ethanol or by lithium(I) tetrahydridoaluminate(III). Usually a given ester forms *two* alcohols. Ethyl ethanoate, however, produces only one alcohol, ethanol (in the equation given below R is CH_3 and R' is C_2H_5).

$$R.COOR' + 4H \rightarrow R.CH_2OH + R'OH$$

Isomerism of Esters

The general formula of esters ($C_nH_{2n}O_2$) is the same as that of alkanoic acids. Esters are therefore isomeric with alkanoic acids which contain the same number of carbon atoms in the molecule. Thus methyl methanoate ($HCOOCH_3$) is isomeric with ethanoic acid ($CH_3.COOH$). In addition esters can be isomeric with other esters, so that the number of isomers increases rapidly as the number of carbon atoms in the molecule becomes larger. There are three compounds with the molecular formula $C_3H_6O_2$. These are:

$$HC\overset{\displaystyle O}{\underset{\displaystyle OC_2H_5}{\big<}} \qquad CH_3.C\overset{\displaystyle O}{\underset{\displaystyle OCH_3}{\big<}} \qquad C_2H_5.C\overset{\displaystyle O}{\underset{\displaystyle OH}{\big<}}$$

ethyl methanoate   methyl ethanoate   propanoic acid

When n is 4—$C_4H_8O_2$— there are seven isomeric compounds. These include four esters, two acids, and 3-hydroxybutanal (aldol), the dimer of ethanal, C_2H_4O (Chapter 10). The student is recommended to write the structural formulæ of the esters and acids and to identify them.

Identification of Esters

In industrial laboratories, which have the necessary apparatus, an ester is quickly identified by means of its infrared absorption spectrum. Chemical identification involves the splitting of the ester by hydrolysis into the alcohol and acid, which are then identified separately by suitable tests. To obtain the alcohol and acid the ester is hydrolysed by alkali as described earlier. In the final stage, however, the sodium(I) salt of the acid is distilled with *dilute* sulphuric(VI) acid. This is necessary because some aliphatic acids (e.g. methanoic acid) are decomposed by concentrated sulphuric(VI) acid. In the case of certain esters, such as ethanedioates (oxalates), the acid liberated is non-volatile and is not obtained as a distillate in the final stage. If no acidic distillate is produced, the liquid left in the distilling-flask is tested for the acid.

OILS AND FATS

The most important esters commercially are those of propane-1,2,3-triol (glycerol). These esters, which are commonly called 'glycerides,' occur naturally in oils and fats. In this connection the term 'oil' includes vegetable and animal oils, but not mineral oils like petroleum, which consist of hydrocarbons.

Some examples of ester-containing oils and fats are olive-oil, coconut-oil, palm-oil, whale-oil, lard, and tallow (beef or mutton fat). These oils and fats consist chiefly of the esters formed by propane-1,2,3-triol with octadecanoic acid ($C_{17}H_{35}.COOH$),

hexadecanoic acid ($C_{15}H_{31}.COOH$), and octadecenoic acid ($C_{17}H_{33}.COOH$). The first two are alkanoic acids commonly called stearic acid and palmitic acid, while the third, known as oleic acid, belongs to a class of compounds known as 'unsaturated acids.' Since propane-1,2,3-triol is a trihydric alcohol, its normal esters contain one molecule of the alcohol combined with three molecules of a monobasic acid. The common names of the above esters are thus glyceryl tristearate, glyceryl tripalmitate, and glyceryl trioleate, these names being often abbreviated to *tristearin*, *tripalmitin*, and *triolein*. The systematic name of tristearin, which may be taken as representative, is propane-1,2,3-triyl trioctadecanoate. Its formula is shown in the next section.

Soap

Ordinary soap is essentially a mixture of the sodium(I) salts of the three acids mentioned in the last section, although other substances are often added for special purposes. It is made by boiling oils and fats with sodium(I) hydroxide solution. Hydrolysis occurs, resulting in the formation of the sodium(I) salts of the three acids and the liberation of propane-1,2,3-triol.

$$\begin{array}{l} CH_2O.OC.C_{17}H_{35} \\ | \\ CHO.OC.C_{17}H_{35} \\ | \\ CH_2O.OC.C_{17}H_{35} \\ \quad\text{tristearin} \end{array} + 3NaOH \rightarrow \begin{array}{l} CH_2OH \\ | \\ CHOH \\ | \\ CH_2OH \\ \text{propane-1,2,3-triol} \end{array} + 3C_{17}H_{35}.COONa \\ \qquad\qquad\qquad\qquad\text{sodium(I)} \\ \qquad\qquad\qquad\qquad\text{octadecanoate}$$

In the laboratory soap can be prepared from beef-fat, mutton-fat, or olive-oil as now described.

Experiment

Put about 10 g of beef dripping (which must be free from salt) or 7 cm³ of olive-oil into a beaker with 40 cm³ of 20 per cent sodium(I) hydroxide solution and 10 cm³ of industrial spirit. The alcohol hastens the reaction by acting as a common solvent for the reactants. Add a few fragments of porous pot to promote steady boiling. Cover the beaker with a clock-glass, and heat it over a gauze. When the mixture begins to boil continue the heating with a very small flame (which should be removed temporarily if considerable frothing occurs).

After twenty minutes' boiling allow the beaker to cool, and skim off any fat, or pour off any oil, which remains on the surface. Add two teaspoonfuls of salt, and dissolve it by stirring. Soap is insoluble in salt solution, and is precipitated on the surface of the liquid. Run off the liquid from the beaker, keeping back the soap by means of a spoon. Swill the soap in the beaker with two or three lots of cold water to free it from alkali and salt. Put a small piece of the soap into a test-tube with some warm water, and shake the tube. An abundant lather is produced.

Butter and Margarine

Butter is an emulsion of water in fat (about 90 per cent) together with small amounts of casein, milk-sugar, and salts. Butter fat consists chiefly of the esters of propane-1,2,3-triol with hexadecanoic acid (palmitic acid) and octadecenoic acid (oleic acid). It differs from other fats in containing a relatively high percentage of the ester of propane-1,2,3-triol with butanoic acid.

Margarine is an emulsion made from soured skimmed milk, oils, and fats. The oils and fats may be of animal origin (e.g. whale-oil and tallow) or of vegetable origin (e.g. groundnut-oil, palm-oil, etc.). Salt, colouring-matter, and vitamins A and D are also incorporated. The desired flavour and aroma are produced by treating the skimmed milk with lactic acid bacteria.

An important part of margarine manufacture is the *hardening* of oils, such as whale-oil and groundnut-oil, to convert them into solid fats. This is accomplished by hydrogenation, the unsaturated carbon atoms of triolein being made to combine with hydrogen to yield tristearin. The fundamental reaction can be represented as follows:

$$\underset{\substack{| \\ }}{-C} \!\!=\!\! \underset{\substack{| \\ }}{C}- \; + \; H_2 \xrightarrow{\text{Ni}} \; -\underset{\substack{| \\ H}}{\overset{\substack{H \\ |}}{C}} - \underset{\substack{| \\ H}}{\overset{\substack{H \\ |}}{C}} -$$

Hydrogenation is carried out catalytically by the method of Sabatier and Senderens. Finely divided nickel is added to the oil, the latter is heated to 180°C, and hydrogen is blown in under a pressure of 2–5 atmospheres.

ESTERS OF INORGANIC ACIDS

The esters formed by monohydric alcohols with hydrochloric acid, hydrobromic acid, and hydriodic acid are alkyl halides, and have been described in Chapter 9.

Esters of Sulphuric(VI) Acid

Since sulphuric(VI) acid is a dibasic acid it forms both acid esters and neutral esters. As seen in Chapter 8, when a primary alcohol (such as ethanol) is refluxed with concentrated sulphuric(VI) acid at about 80°C the alkyl hydrogen sulphate(VI) is formed.

$$C_2H_5OH + H_2SO_4 \rightleftharpoons C_2H_5HSO_4 + H_2O$$

Secondary and tertiary alcohols are mostly dehydrated to alkenes on heating with concentrated sulphuric(VI) acid.

The lower alkyl hydrogen sulphates(VI) are syrupy liquids, which are difficult to isolate. They are chiefly important as intermediates in the manufacture of alcohols from alkenes (Chapter 8).

The neutral esters *dimethyl sulphate(VI)*, $(CH_3)_2SO_4$, and *diethyl sulphate(VI)*, $(C_2H_5)_2SO_4$, are obtained by distilling the alkyl hydrogen sulphates(VI) under reduced pressure. Under these conditions the acid esters decompose.

$$2C_2H_5HSO_4 \rightarrow (C_2H_5)_2SO_4 + H_2SO_4$$

Dimethyl sulphate(VI) and diethyl sulphate(VI) are heavy colourless liquids. Their vapours are extremely poisonous, and great care is necessary in handling the compounds. They are used as methylating and ethylating agents.

Detergents. A detergent is a substance which, when dissolved in water, helps to remove dirt from an object. The chief factor in the soiling of, say, a fabric is a film of oil or grease, which attaches the dirt particles to the fibres. To remove the dirt efficiently a detergent must have the following properties:

(*i*) It must have a fairly large solubility in water.

(*ii*) It must lower the surface tension of water considerably, so that the water, or detergent solution, can 'wet' the surface of the fabric, that is, spread over the surface and insert itself between, and under, the greasy dirt particles.

(*iii*) It must be an emulsifying agent, so that the dirt particles, when removed from the surface, are kept in suspension ready to be rinsed away.

The traditional detergent, soap, fulfils all the above requirements, but suffers from two disadvantages. Its solubility in water is relatively small, and it forms a scum, or precipitate, with the calcium and magnesium ions present in hard water. When soap dissolves in water it dissociates into Na^+ ions and anions such as octadecanoate ions $(C_{17}H_{35}.COO^-)$. The detergent action of soap depends on the anions, the sodium(I) ions playing no part at all. Solubility and wetting power are due to the hydrophilic $-COO^-$ group, while emulsifying power is associated with the hydrophobic hydrocarbon chain $C_{17}H_{35}-$. Now the $-SO_4^-$ group is more strongly hydrophilic than the $-COO^-$ group, and therefore the sodium(I) salts of the acid sulphates(VI) derived from $C_{12}-C_{18}$ alcohols have a larger solubility and higher wetting power than the sodium(I) salts of the corresponding carboxylic acids. At the same time they possess good emulsifying power owing to their long hydrocarbon chains. Furthermore, the calcium(II) and magnesium(II) salts are soluble, so that no precipitate is formed with hard water.

Two detergents in common use are sodium(I) dodecyl sulphate(VI) $(C_{12}H_{25}SO_4Na)$, which is used in shampoos, and sodium(I) hexadecyl sulphate(VI) $(C_{16}H_{33}SO_4Na)$, an industrial detergent. Many detergents are now manufactured from alkenes of the $C_{12}-C_{18}$ group produced by 'cracking' oils and recombining the molecular fragments (Chapter 5). These unsaturated hydrocarbons combine directly with concentrated sulphuric(VI) acid to yield alkyl hydrogen sulphates(VI); e.g.

$$C_9H_{19}.CH{=}CH.CH_3 + H_2SO_4 \rightarrow C_9H_{19}.CH_2{-}\underset{\underset{\displaystyle SO_4H}{|}}{\overset{\overset{\displaystyle CH_3}{|}}{CH}}$$

The acid esters are converted into the sodium(I) salts by neutralization. The sodium(I) salt of the product represented in the equation is called sodium(I) 1-methylundecyl sulphate(VI), and is contained in the well-known detergent 'Teepol.'

Esters of Nitric(III) Acid (Nitrous Acid)

The only important esters of nitric(III) acid are ethyl nitrate(III) (ethyl nitrite), $C_2H_5NO_2$, and 3-methylbutyl nitrate(III) (isoamyl nitrite), $C_5H_{11}NO_2$. Both compounds are used in medicine.

Ethyl nitrate(III) is a colourless liquid of a low boiling point (17°C). It is prepared by slowly adding a well-cooled mixture of medium sulphuric(VI) acid and ethanol to a solution of sodium(I) nitrate(III) at 0°C.

$$C_2H_5OH + HNO_2 \rightleftharpoons C_2H_5NO_2 + H_2O$$

When the reaction mixture is allowed to stand ethyl nitrate(III) separates out as an upper oily layer.

Another method of obtaining ethyl nitrate(III) is to reflux an ethanolic solution of iodoethane and silver(I) nitrate(III) for about two hours on a water-bath. In this case ethyl nitrate(III) and its isomer nitroethane are produced in the approximate proportion of 1:2. A precipitate of silver(I) iodide is also formed.

$$C_2H_5I + AgNO_2 \rightarrow C_2H_5NO_2 + AgI \downarrow$$

During the refluxing it is necessary to pass iced water through the condenser to prevent the escape of the very volatile ethyl nitrate(III). The boiling point (114°C) of nitroethane is much higher than that of ethyl nitrate(III), and the two liquids can be easily separated by fractional distillation.

Nitroethane is not an ester, but a *nitroalkane*. The preparation of nitroalkanes from alkanes has been described in Chapter 5. Nitric(III) acid, ethyl nitrate(III), and nitroethane have the structures shown.

$$H-O-N{=}O$$
nitric(III) acid

$$C_2H_5-O-N{=}O$$
ethyl nitrate(III)

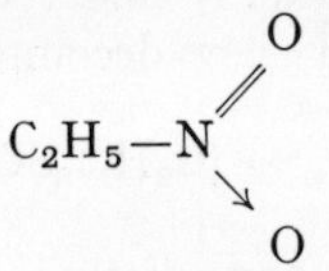

$$C_2H_5-N{\nwarrow}^{O}_{O}$$
nitroethane

The arrow in the formula of nitroethane represents a co-ordinate covalent bond.

An alkyl nitrate(III) can be distinguished from a nitroalkane by its different behaviour on reduction. Thus with zinc and hydrochloric acid ethyl nitrate(III) is reduced to ethanol and ammonia, which forms ammonium chloride with the acid present. If the liquid left is warmed with excess of sodium(I) hydroxide solution, ammonia is evolved and can be recognized by its smell.

$$C_2H_5-O-N{=}O + 6H \rightarrow C_2H_5OH + NH_3 + H_2O$$

Under similar conditions nitroethane is reduced to ethylamine. This can be tested for by the isocyanobenzene reaction described in Chapter 9.

$$C_2H_5NO_2 + 6H \rightarrow C_2H_5NH_2 + 2H_2O$$
ethylamine

Esters of Nitric(V) Acid

The simple esters of nitric(V) acid are of little importance. They can be obtained by refluxing together the alkyl iodide and silver(I) nitrate(V) in ethanolic solution, as illustrated by the following equation:

$$C_2H_5I + AgNO_3 \rightarrow C_2H_5NO_3 + AgI \downarrow$$
$$\text{ethyl nitrate(V)}$$

Ethyl nitrate(V) is a colourless liquid (b.p. 20°C) with a pleasant smell. It is liable to explode when heated.

Nitroglycerine. The two most important esters of nitric(V) acid are those formed with propane-1,2,3-triol (glycerol) and cellulose. The esters are propane-1,2,3-triyl nitrate(V) and cellulose nitrate(V), which are commonly, but incorrectly, called 'nitroglycerine' and 'nitrocellulose' (neither compound contains the nitro group). Cellulose nitrate(V) is described in Chapter 18.

Nitroglycerine, $C_3H_5(NO_3)_3$, is the basic material in the manufacture of several kinds of high explosive. It is produced by adding propane-1,2,3-triol at a carefully controlled rate to a mixture of concentrated nitric(V) acid and concentrated sulphuric(VI) acid. The latter fulfils its usual dual rôle in esterification as a catalyst and dehydrating agent.

$$\begin{array}{l} \text{CH}_2\text{OH} \\ | \\ \text{CHOH} \\ | \\ \text{CH}_2\text{OH} \end{array} + 3\text{HNO}_3 \rightleftharpoons \begin{array}{l} \text{CH}_2\text{NO}_3 \\ | \\ \text{CHNO}_3 \\ | \\ \text{CH}_2\text{NO}_3 \end{array} + 3\text{H}_2\text{O}$$

The resulting mixture is run into water, and the nitroglycerine separates as an oily liquid.

If a small amount of nitroglycerine is ignited it burns quietly. When it is heated quickly, given a mechanical shock, or detonated it explodes with great violence. It can be handled safely if it is absorbed in kieselguhr. This mixture is known as *dynamite*. Other high explosives incorporating nitroglycerine are *gelignite* and *cordite*. Gelignite contains 60 per cent of nitroglycerine, about 25 per cent of potassium(I) nitrate(V), and small amounts of other ingredients. Cordite is described Chapter 18.

EXERCISE 13

1. Which of the following combinations can be used to prepare esters of alkanoic acids: (a) the acid and an alcohol, (b) another ester and an alcohol, (c) the sodium(I) salt of the acid and an alcohol, (d) the acid and an alkyl iodide, (e) the silver(I) salt of the acid and an alkyl iodide?

2. Give the systematic names of the following esters: (a) $CH_3.COOCH_3$, (b) $C_2H_5COOC_2H_5$, (c) $CH_3.COOCH_2.C_2H_5$, (d) $HCOOCH(CH_3)_2$.

3. Represent the steps involved in the probable mechanism of hydrolysis of methyl methanoate by OH^- ions.

4. From a consideration of inductive effects arrange the following esters according to the rates at which you would expect them to be hydrolysed by aqueous sodium(I) hydroxide under similar conditions: (a) $HCOOCH_3$, (b) $CH_3.COOCH_3$, (c) $CH_2Cl.COOCH_3$, (d) $C_2H_5.COOCH_3$.

5. Name the organic compounds formed when methyl propanoate is treated under suitable conditions with (a) dilute sulphuric(VI) acid, (b) aqueous sodium(I) hydroxide, (c) butan-1-ol, (d) aqueous ammonia, (e) sodium and ethanol.

6. Give the systematic names and structural formulæ of the esters which have the molecular formula $C_4H_8O_2$.

7. Which is the odd one out of the following: olive-oil, beef dripping, soap, butter, margarine?

8. Sodium(I) dodecyl sulphate(VI) is a detergent used in shampoos.

(a) What is its chemical formula?

(b) Which group of atoms is responsible for its high solubility in water?

(c) Which group of atoms is responsible for its emulsifying power?

(d) What advantage over soap has it for use with hard water?

14 Acid Derivatives *(continued)* —Acid Amides and Nitriles

An acid amide is derived from a monocarboxylic acid by replacing the hydroxyl group of the latter by an amino ($-NH_2$) group. Thus ethanamide, or acetamide, is obtained from ethanoic, or acetic, acid.

ethanoic acid	ethanamide

The functional group in acid amides is the $-CONH_2$ group, which is called the *amido* group.

Alternatively acid amides may be regarded as derived from ammonia by replacing a hydrogen atom of the ammonia molecule by an acyl group.

ammonia	ethanamide

Strictly speaking, ethanamide is a *primary* acid amide. Secondary and tertiary acid amides such as diethanamide, $(CH_3.CO)_2NH$, and triethanamide, $(CH_3.CO)_3N$, can be prepared. They are of little importance, however, and can be ignored.

Acid amides are named according to the acids from which they are derived. Table 14-1 shows the first four 'straight-chain'

Table 14-1. ALKANOIC ACID AMIDES
(General formula: $R.CONH_2$)

Systematic name and Structural formula	Common name	Melting point/°C	Boiling point/°C
Methanamide $HCONH_2$	Formamide	2	193
Ethanamide $CH_3.CONH_2$	Acetamide	82	222
Propanamide $C_2H_5.CONH_2$	Propionamide	77	213
Butanamide $C_2H_5.CH_2.CONH_2$	*n*-Butyramide	116	216

members of the series derived from alkanoic acids. With the exception of methanamide, or formamide, which is a liquid at ordinary temperatures, the acid amides are crystalline solids. Their melting points and boiling points are much higher than would be expected from their relative molecular masses. This is explained by hydrogen-bonding between the $-NH_2$ group in one molecule and the carbonyl oxygen atom in another. The irregular increase in melting point and boiling point of the lower amides is probably associated with variations in the extent and strength of hydrogen-bonding.

The lower solid acid amides are deliquescent, and dissolve freely in water owing to hydrogen-bonding with water molecules. As usual the solubility decreases with increase in length of the hydrophobic hydrocarbon chain, and higher members are insoluble in water. Amides dissolve readily in ethanol.

In the pure state acid amides are odourless. Commercial ethanamide usually smells of mice, but the cause of this is uncertain. It has been attributed to the presence of traces of diethanamide.

The simple alkanoic acid amides have no industrial applications, but dimethylmethanamide, $HCON(CH_3)_2$, is an important solvent used in the manufacture of synthetic fibres.

Laboratory Preparation of Acid Amides

Acid amides can be obtained by treating the acid chloride, acid anhydride, or an ester with concentrated aqueous ammonia; e.g.

$$CH_3.COCl + 2NH_3 \rightarrow CH_3.CONH_2 + NH_4Cl$$

$$(CH_3.CO)_2O + 2NH_3 \rightarrow CH_3.CONH_2 + CH_3.COONH_4$$

$$CH_3.COOC_2H_5 + NH_3 \rightarrow CH_3.CONH_2 + C_2H_5OH$$

Starting from the acid, the simplest method of preparing the amide is to make the ammonium salt, which is then dehydrated by heating. Thus when ammonium ethanoate is heated it forms ethanamide.

$$CH_3.COONH_4 \rightleftharpoons CH_3.CONH_2 + H_2O$$

If the ammonium salt is heated by itself, the yield of amide is low because a high proportion of the salt dissociates into ammonia and the free acid.

$$CH_3.COONH_4 \rightleftharpoons CH_3.COOH + NH_3$$

Dissociation can be reduced by heating the salt with the anhydrous acid, which causes the equilibrium point in the reaction to be displaced to the left. The bulk of the salt then undergoes dehydration to the amide. The preparation of ethanamide is carried out as now described.

Experiment

Put 15 g of ammonium ethanoate and 15 cm³ of glacial ethanoic acid into a 50 cm³ pear-shaped flask, and add a little powdered pumice. Attach to the flask a reflux air-condenser, that is, an ordinary condenser without water passing through. (A water-condenser would condense ethanamide vapour to solid, which might block the condenser). Heat the flask on a sand-tray over a tripod (without gauze), using an ordinary Bunsen flame, so that the contents are kept gently boiling (Fig. 14-1a). After a time oily streaks of molten ethanamide will be observed running down the condenser. Continue the heating for about 45 minutes.

Allow the remaining liquid to cool partially. Remove the sand-tray and tripod, and then by means of a still-head fitted with a 250°C thermometer change over the condenser for direct distillation (Fig. 14-1b). Again use the condenser without water. Distil the liquid with a small direct flame, using a Bunsen burner without the chimney. At first water and ethanoic acid distil over, and then the temperature rises rapidly. Change the receiver when the thermometer registers 180°C, and collect the ethanamide in a dry beaker. The amide solidifies as it cools.

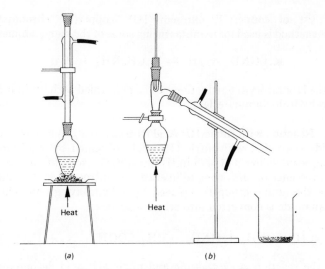

Fig. 14-1. Preparation of ethanamide

The crude product, which has a strong smell of mice, is purified by recrystallization from propanone or, preferably, a mixture of benzene (three parts) and dry ethyl ethanoate (one part). After recrystallization the ethanamide should be odourless.

Reactions of Acid Amides

In an acid amide ($RCONH_2$) the carbonyl carbon atom again has a partial positive charge owing to the $-I$ effect of the oxygen atom. The partial charge is decreased by the $+I$ effect of the alkyl radical, but increased by the $-I$ effect of the amino group. Owing to the relatively small $-I$ effect of the $-NH_2$ group the partial charge on the carbon atom is smaller than in the acid derivatives already encountered. Hence acid amides are less strongly electrophilic. Thus they react extremely slowly with the weaker nucleophilic reagents such as water and ethanol. Some of the reactions of acid amides are complex, and the manner in which they occur is still obscure.

The $-NH_2$ group is normally a basic group, but in aqueous solution acid amides are neutral to litmus. This may be attributed to the basic character of the amino group being roughly balanced by the acidic character of the acyl group $RCO-$. Under suitable conditions, however, acid amides can act either as feeble bases or as feeble acids.

I. Hydrolysis. As stated above, acid amides react very slowly with water. Even boiling with water for several hours produces only slight hydrolysis. Ethanamide and water yield traces of ammonium ethanoate.

$$CH_3.CONH_2 + H_2O \rightleftharpoons CH_3.COOH + NH_3$$
$$\rightleftharpoons CH_3.COONH_4$$

As in the case of esters, both mineral acids and alkalies serve as catalysts for the hydrolysis. Thus when ethanamide is refluxed for about half an hour with medium hydrochloric acid it forms ethanoic acid and ammonium chloride.

$$CH_3.CONH_2 + H_2O + HCl \rightarrow CH_3.COOH + NH_4Cl$$

Hydrolysis takes place more rapidly with alkalies than with acids. If ethanamide is boiled with dilute sodium(I) hydroxide solution, ammonia is rapidly evolved and sodium(I) ethanoate remains in solution.

$$CH_3.CONH_2 + NaOH \rightarrow CH_3.COONa + NH_3$$

The rapid hydrolysis of amides by alkalies is again due to the strong nucleophilic character of the hydroxyl ion. The probable mechanism of the reaction (which should be compared with the mechanism given in Chapter 13 for hydrolysis of esters) is as follows:

$$R-\underset{\underset{NH_2}{|}}{\overset{\overset{O}{\|}}{C}} + \ ^-:\!OH \ \rightleftharpoons \ R-\underset{\underset{NH_2}{|}}{\overset{\overset{\bar{O}}{|}}{C}}-OH \ \rightleftharpoons \ R-\overset{\overset{O}{\|}}{C}-OH + NH_2^-$$

amide ion

(unstable)

$$H_2O + NH_2^- \rightleftharpoons OH^- + NH_3$$

2. Alcoholysis. The reaction by which an acid amide is prepared from an ester and ammonia is reversible. The reverse reaction can be brought about by refluxing the amide with ethanol, but takes place very slowly unless an alkali is added as a catalyst. The alkali partially converts ethanol into ethoxide $(C_2H_5O^-)$ ions, which are more strongly nucleophilic. The overall reaction in the case of ethanamide is represented by the following equation:

$$CH_3.CONH_2 + C_2H_5OH \overset{OH^-}{\rightleftharpoons} CH_3.COOC_2H_5 + NH_3$$

Alcoholysis of acid amides has no practical importance.

3. Reduction. Acid amides, like alkanoic acids themselves, are difficult to reduce. Reduction to the corresponding amine can be brought about by hydrogenation assisted by heat, pressure, and a catalyst of copper(II) chromate(III) (copper(II) chromite). This method is used for manufacturing some of the higher amines.

$$R.CONH_2 + 2H_2 \rightarrow R.CH_2NH_2 + H_2O$$

In the laboratory the reduction can be performed with lithium(I) tetrahydridoaluminate(III).

4. Reaction with Nitric(III) Acid (Nitrous Acid). When an acid amide is treated with nitric(III) acid—obtained by acidifying an aqueous solution of sodium(I) nitrate(III) (sodium nitrite)—effervescence occurs owing to liberation of nitrogen. At the same time the amino group is replaced by a hydroxyl group. Thus ethanamide is converted into ethanoic acid.

$$CH_3.CONH_2 + HNO_2 \rightarrow CH_3.COOH + N_2 + H_2O$$

As the reaction is exothermic and nitric(III) acid decomposes rapidly even at ordinary temperatures (giving brown fumes of nitrogen dioxide), the reaction is carried out as now described.

Experiment

Dissolve 2 g of ethanamide in 10 cm³ of water in a boiling-tube, and add an equal volume of dilute hydrochloric acid. Cool the tube (preferably in iced water). Clamp a dropping-funnel so that its stem reaches nearly to the bottom of the mixture in the boiling-tube (Fig. 14-2). In the funnel place a solution of about 5 g of sodium(I) nitrate(III) in 15 cm³ of water. Run the solution in small portions into the boiling-tube, cooling the latter from time to time. There is a vigorous effervescence owing to liberation of nitrogen. Ethanoic acid can be obtained from the remaining liquid by fractional distillation.

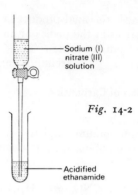

Sodium (I)
nitrate (III)
solution

Fig. 14-2

Acidified
ethanamide

5. Dehydration. When an alkanoic acid amide is distilled with phosphorus(V) oxide (phosphorus pentoxide), which is a vigorous dehydrating agent, an acid nitrile (alkyl cyanide) distils over. The reaction is used for the laboratory preparation of ethanonitrile (methyl cyanide) from ethanamide.

$$CH_3.CONH_2 - H_2O \rightarrow CH_3CN$$

6. Reaction with Bromine and Sodium(I) Hydroxide Solution. If an acid amide is treated with bromine and aqueous sodium(I) hydroxide the carbonyl group is eliminated and an amine is formed. Ethanamide yields methylamine (CH_3NH_2). This important change, which is known as *Hofmann's reaction*, is described in the next chapter.

7. Acidic Character and Basic Character. Weak acidic character arises in acid amides from the attraction of the nitrogen atom

for the electrons of the N—H bonds. This is reinforced by the —*I* effect of the acyl group (RCO—). Electron displacement towards the nitrogen atom weakens the bonds, or activates the hydrogen atoms, so that one of the latter can be removed by a sufficiently strong electron donor. Thus, if ethanamide is warmed with metallic sodium, hydrogen is evolved and a salt, sodium(I) ethanamide, is left.

$$CH_3.CONH_2 + Na \rightarrow [CH_3.CONH]^- Na^+ + \tfrac{1}{2}H_2$$
$$\text{sodium(I) ethanamide}$$

Aqueous acetamide also forms salts with certain metal oxides (e.g. HgO).

Feeble basic character in acid amides is shown by their reactions with strong acids to form salts. Thus ethanamide combines with dry hydrogen chloride to give ethanoylammonium chloride.

$$CH_3.CONH_2 + HCl \rightarrow [CH_3.CONH_3]^+ Cl^-$$

The salts are unstable in aqueous solution. When they are added to water they are almost completely hydrolysed, giving the free amide again.

Chemical Tests for Acid Amides

The most suitable reactions to use for the identification of fatty acid amides are the following:

(*i*) The evolution of ammonia when the amide is heated with sodium(I) hydroxide solution.

(*ii*) The liberation of the vapour of an alkanoic acid (with characteristic smell) when the amide is boiled with dilute sulphuric(VI) acid.

(iii) The occurrence of effervescence, due to evolution of nitrogen, when the amide solution is treated with nitric(III) acid (nitrous acid) in the cold.

(iv) The production of an amine, which has a characteristic fishy smell, by Hofmann's reaction (Chapter 15).

The first two tests, but not the second two, are also given by ammonium salts of alkanoic acids. A further method of distinguishing between an acid amide and an ammonium salt of an alkanoic acid is by means of 'neutral' iron(III) chloride solution, which gives a blood-red coloration with a solution of an ammonium salt of an alkanoic acid.

CARBAMIDE, $CO(NH_2)_2$

Melting point: 132°C   *Boiling point:* (decomposes)

'Carbamide' is an abbreviation of 'carbonamide.' Carbamide is the acid amide of carbonic acid (H_2CO_3). Equally 'recommended' is the name *urea*.

carbonic acid   carbamic acid   carbamide

Carbamic acid exists only in the form of its salts, such as ammonium carbamate, which is produced as an intermediate compound in the manufacture of carbamide from ammonia and carbon dioxide, as described shortly.

Carbamide, or urea, is the final product in the breakdown of waste nitrogenous matter in the bodies of many animals. Thus, about two per cent of human urine consists of carbamide, and an adult excretes about 28 g of carbamide daily.

Laboratory Preparation of Carbamide

Carbamide is prepared by evaporating to dryness an aqueous solution containing ammonium sulphate(VI) and potassium(I) cyanate in the proportion of one mole to two moles. Ammonium cyanate is first formed.

$$NH_4^+ + CNO^- \rightleftharpoons NH_4CNO$$

The ammonium cyanate then undergoes isomeric change into carbamide. The change is reversible, and an equilibrium exists between the two isomers in solution.

$$NH_4CNO \rightleftharpoons CO(NH_2)_2$$

At 100°C about 95 per cent of the ammonium cyanate is converted into carbamide. The latter, however, has the smaller solubility in water, and when the solution is evaporated the equilibrium is disturbed by the crystallizing out of carbamide. The dry residue, therefore, consists only of carbamide and potassium(I) sulphate(VI). The carbamide is separated by extraction with hot absolute ethanol, in which potassium(I) sulphate(VI) is insoluble. Colourless crystals of carbamide are deposited when the ethanolic solution is concentrated and cooled.

Manufacture and Uses of Carbamide

Carbamide is made on a large scale by heating carbon dioxide with excess of ammonia at 150°–200°C and a pressure of 200 atmospheres. The carbon dioxide is a by-product from the manufacture of synthesis gas (Chapter 5). The first product of the reaction is ammonium *carbamate*, but in the presence of excess of ammonia this is largely converted into carbamide.

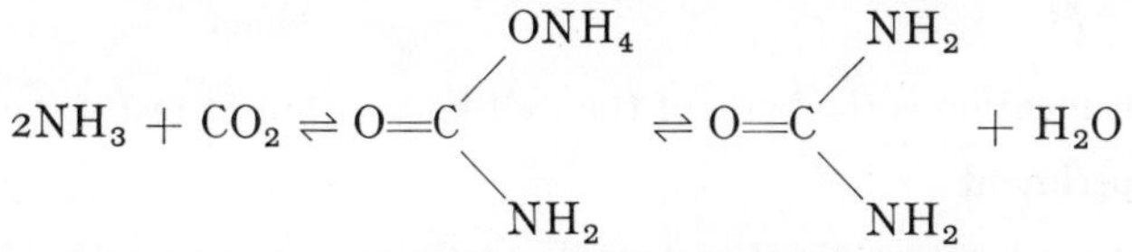

$$2NH_3 + CO_2 \rightleftharpoons O=C\begin{array}{c} ONH_4 \\ \\ NH_2 \end{array} \rightleftharpoons O=C\begin{array}{c} NH_2 \\ \\ NH_2 \end{array} + H_2O$$

ammonium carbamate   carbamide

Carbamide, or urea, is chiefly used in the manufacture of synthetic resins and plastics, such as carbamide-methanal, or urea-formaldehyde (Chapter 26). It is also employed on a large scale (particularly in the U.S.A.) as a fertilizer. A minor application is in making drugs (e.g. phenobarbitone).

Reactions of Carbamide

Some of the reactions of carbamide correspond with those of alkanoic acid amides, but others are different. Carbamide might be expected to be more strongly electrophilic than an alkanoic acid amide because the electron-repelling alkyl radical present in the latter is replaced in carbamide by an electron-attracting amino group. This should result in a larger partial positive charge on the carbonyl carbon atom. In practice there is little difference between the reactivities of carbamide and ethanamide with the common nucleophilic reagents. The low reactivity of carbamide compared with that expected from inductive effects is explained by mesomerism, or resonance. Carbamide is a resonance hybrid with the three contributing structures shown below.

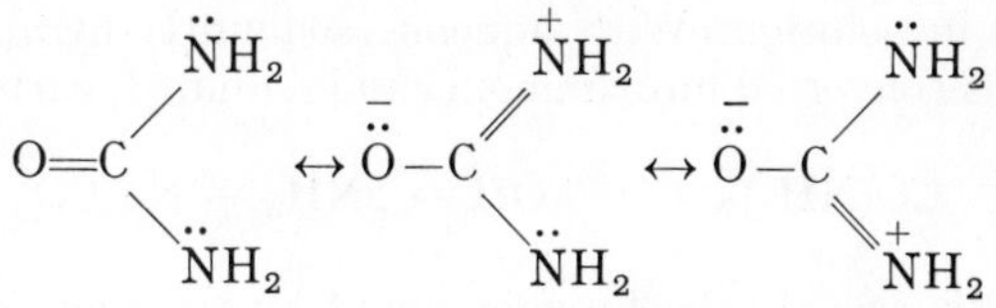

As seen earlier, mesomerism stabilizes a molecule and the larger the number of contributing structures the less reactive is the molecule. Mesomerism also occurs in alkanoic acid amides, but in this case there are only two contributing structures.

1. Basic Character. Although an aqueous solution of carbamide is neutral to litmus, the compound behaves as a monoacidic base, and forms stable salts with both inorganic and organic acids. The two best known salts are carbamide, or urea, nitrate(V), $CO(NH_2)_2 . HNO_3$, and carbamide ethanedioate-2-water, or urea oxalate, $2CO(NH_2)_2 . (COOH)_2 . 2H_2O$. Both salts are sparingly soluble in water, and are readily obtained by precipitation. The structures of the salts are uncertain.

Experiment

To a concentrated aqueous solution of carbamide add a few drops of concentrated nitric(V) acid or a concentrated solution of ethanedioic acid (oxalic acid). A white crystalline precipitate of the corresponding carbamide, or urea, salt is formed.

There is some evidence that when carbamide forms salts the proton from the acid is added to the oxygen atom of the base. This would explain why carbamide acts as a monoacidic base.

2. Hydrolysis. Like other acid amides carbamide is hydrolysed when boiled with either a mineral acid or an alkali. With hydrochloric acid carbon dioxide is evolved, and ammonium chloride remains in solution. With aqueous sodium(I) hydroxide carbamide is converted into ammonia and sodium(I) carbonate.

$$CO(NH_2)_2 + 2NaOH \rightarrow 2NH_3 + Na_2CO_3$$

Hydrolysis can also be brought about by treating an aqueous solution of carbamide with *urease*, an enzyme obtained from soya beans. The enzyme changes the amide quantitatively into ammonium carbonate.

$$CO(NH_2)_2 + 2H_2O \rightarrow (NH_4)_2CO_3$$

This reaction is used for estimating the concentration of carbamide in an aqueous solution, the ammonium carbonate produced being estimated volumetrically.

3. Reaction with Nitric(III) Acid (Nitrous Acid). Carbamide also reacts with nitric(III) acid (acidified sodium(I) nitrate(III) solution) in the same way as other acid amides. In this case both amino groups are replaced by hydroxyl groups. The resulting acid, carbonic acid, decomposes into carbon dioxide and water.

$$CO(NH_2)_2 + 2HNO_2 \rightarrow CO_2 + 3H_2O + 2N_2$$

The reaction is carried out as described earlier for ethanamide.

4. Action of Heat (*Biuret Test for Carbamide*). When carbamide is heated to a temperature just above its melting point it loses ammonia and biuret is formed. This is an example of condensation

in which ammonia is eliminated instead of water.

$$NH_2-CO-NH\underline{[H + NH_2]}-CO-NH_2 \rightarrow$$

$$NH_3 + NH_2-CO-NH-CO-NH_2$$
biuret

This reaction is the basis of the well-known 'biuret test' for urea

Experiment

Warm a few crystals of carbamide gently in a dry test-tube. When the crystals melt continue warming until the liquid solidifies owing to the much higher melting point of biuret. The ammonia given off during the heating can be readily detected by its smell. Allow the tube to cool, and dissolve the remaining solid by warming it with a little dilute sodium(I) hydroxide solution. Now add to the tube one drop of very dilute copper(II) sulphate(IV) solution. The liquid turns pink or violet.

The biuret test is given by compounds containing the —NH—CO— group, which is known as the *peptide group*, or *peptide linkage*. The second name refers to the fact that the group is present in proteins (Chapter 16), where it links together different parts of the long molecular chains.

NITRILES

Aliphatic nitriles are alkyl cyanides. They can be regarded as derived from alcohols by replacing the —OH group by the *cyano* (—CN) group. They are also derivatives of carboxylic acids because they are the final products of dehydration of ammonium salts of the acids. Acid amides are an intermediate stage.

$$\underset{\text{ammonium salt}}{R-\overset{\overset{\textstyle O}{\|}}{C}-ONH_4} \rightleftharpoons \underset{\text{acid amide}}{R-\overset{\overset{\textstyle O}{\|}}{C}-NH_2} \rightleftharpoons \underset{\text{nitrile}}{R-C{\equiv}N}$$

The changes are reversible. When a nitrile is hydrolysed it is first converted into the acid amide and then into the ammonium salt of the acid (or, more usually, into the acid itself, since a strong mineral acid is used for the hydrolysis). Nitriles are named according to the acids which they yield on hydrolysis. Thus ethanonitrile, or acetonitrile (CH_3CN), gives ethanoic, or acetic, acid ($CH_3.COOH$). The first three members of the homologous series are shown in Table 14-2.

Table 14-2. ALIPHATIC NITRILES

(General formula: $C_nH_{2n+1}CN$, or RCN)

Systematic name	Trivial name	Structural formula	Boiling point/°C
Ethanonitrile	Acetonitrile	CH_3CN	82
Propanonitrile	Propionitrile	C_2H_5CN	98
Butanonitrile	*n*-Butyronitrile	$C_2H_5.CH_2CN$	119

The lower nitriles are colourless neutral liquids, which are highly poisonous. As usual melting points and boiling points rise with increase in size of the molecules. The lower members are moderately soluble in water, but higher members are only slightly soluble. They dissolve readily, however, in organic solvents.

Nitriles have only a limited application in industry. Their chief importance is as intermediates in the synthesis of other compounds.

Preparation of Nitriles

(*i*) *By Dehydration of Acid Amides.* Nitriles are usually prepared by warming the acid amides with phosphorus(V) oxide, which combines with the water removed to form polytrioxophosphoric(V) acid (metaphosphoric acid). The nitrile is subsequently obtained by distillation.

$$CH_3.CONH_2 + P_2O_5 \rightarrow CH_3CN + 2HPO_3$$

(*ii*) *From an Alkyl Halide and Potassium(I) Cyanide.* Nitriles can also be prepared by dissolving an alkyl halide (preferably the iodide) in ethanol and refluxing the solution with an aqueous solution of potassium(I) cyanide. Thus iodoethane gives propanonitrile.

$$C_2H_5I + CN^- \rightarrow C_2H_5CN + I^-$$

The mixture of solvents is required to keep both reactants in solution. Refluxing is continued on a water-bath for about 8 hours, after which the nitrile is separated by fractional distillation.

Reactions of Nitriles

The triple bond in the cyano group ($-C{\equiv}N$) resembles the one in ethyne ($HC{\equiv}CH$) in being composed of a σ bond and two π bonds (Chapter 7). In ethyne, however, the triple bond is between two like atoms, whereas in the cyano group it joins two different atoms. Since nitrogen is more electronegative than carbon, there is an electron displacement in the bond towards the former, as shown.

$$\overset{\delta+}{R}-\overset{}{C}\equiv\overset{\delta-}{N}$$

Also, polarization of the multiple bond $C\equiv N$ is larger than that of a single bond $C-N$ because of the greater mobility of the π electrons (compare polarization of the $C=O$ bond described in Chapter 10). Thus, although the partial charges portrayed are reduced to some extent by the $+I$ effect of the alkyl radical R, nitriles have fairly high electric dipole moments.

Nitriles are unsaturated, but owing to polarization of the triple bond they resemble aldehydes and ketones rather than alkynes. The latter form addition compounds with electrophilic reagents, while nitriles, like aldehydes and ketones, react additively with nucleophilic reagents, which can attack the carbon atom with its partial positive charge. Thus, under suitable conditions, nitriles combine with water, ethanol, ammonia, hydrogen, and Grignard reagents. In most cases further reactions occur, and the overall reactions may be complex. The two most important reactions, are hydrolysis and reduction.

1. **Hydrolysis.** As explained earlier, hydrolysis of a nitrile takes place in two stages.

(i)

$$R-C\equiv N + H_2O \rightleftharpoons R-C\overset{\displaystyle O}{\underset{\displaystyle NH_2}{\big<}}$$

(ii) $R.CONH_2 + HOH \rightleftharpoons R.COOH + NH_3 \rightleftharpoons R.COONH_4$

Strictly speaking, only the second stage consists of hydrolysis. In the first stage a water molecule is added to the nitrile molecule to give the acid amide.

In practice conversion of the nitrile to a carboxylic acid requires the use of a strong mineral acid or an alkali as a catalyst. The acid-catalysed reaction is carried out by refluxing the nitrile for about three hours with concentrated hydrochloric acid or 50 per cent sulphuric(VI) acid. The reaction can be stopped at the acid amide stage by employing phosphoric(V) acid as the catalyst.

$$CH_3CN + 2H_2O + HCl \rightarrow CH_3.COOH + NH_4Cl$$

In alkaline hydrolysis the nitrile is refluxed with concentrated aqueous sodium(I) hydroxide for about one hour. Ammonia is evolved and the sodium(I) salt of the acid is left in solution.

$$CH_3CN + H_2O + NaOH \rightarrow CH_3.COONa + NH_3$$

2. **Reduction.** Nitriles are reduced to primary amines when treated with sodium and ethanol. With ethanonitrile, or acetonitrile, the reaction takes place slowly at ordinary temperatures, but rapidly on warming.

$$CH_3CN + 4H \rightarrow CH_3.CH_2.NH_2$$
$$\text{ethylamine}$$

An alternative method of reduction is to pass a mixture of hydrogen and the vapour of the nitrile over a nickel catalyst at 180°C.

Tests for Nitriles

Nitriles can be recognized by their infrared absorption spectra, in which the absorption band for the —CN group occurs at a characteristic wave number. The following chemical tests can also be used:

(*i*) Ammonia is liberated when a nitrile is boiled with a concentrated solution of sodium(I) hydroxide. The ammonia can be recognized by its smell and by turning wet red litmus paper blue.

(*ii*) The reduction of a nitrile to a primary amine by sodium and ethanol can be established by means of the isocyano reaction (Chapter 9). To carry out the test dissolve one drop of ethanonitrile in 5 cm³ of ethanol, and add a small piece of sodium. After the reaction add 1 cm³ of trichloromethane and warm the mixture (sodium(I) hydroxide is already present owing to interaction of some of the sodium with water in the ethanol). The characteristic nauseating odour of an isocyano-compound shows that the original compound was a nitrile.

The first test, but not the second, is also given by acid amides and ammonium salts of organic acids.

Ascent of an Homologous Series

By means of the general reactions given in this and previous chapters it is possible to pass from one member of an homologous series to the next one higher in the series. The key reaction in this synthesis is clearly the one by which an extra carbon atom is added to the molecule. The easiest way of introducing an additional carbon atom is to treat an alkyl halide with potassium(I) cyanide, and this is utilized in the synthesis of, say, ethanol from methanol. The stages involved in this transformation are as follows:

$$CH_3OH \xrightarrow{P \text{ and } I_2} CH_3I \xrightarrow{KCN} CH_3CN \xrightarrow[(H^+)]{H_2O} CH_3.COOH \xrightarrow{LiAlH_4}$$

methanol   iodomethane   ethanonitrile   ethanoic acid

$$CH_3.CH_2OH$$

ethanol

HYDROGEN CYANIDE, HCN

Boiling point: 26°C   *Density:* 0·70 g cm⁻³

Hydrogen cyanide is also known as hydrocyanic acid. Although, strictly speaking, it is not a nitrile (it does not contain an alkyl radical) it behaves as if it were the nitrile of methanoic, or formic, acid. It can therefore be regarded as methanonitrile, or formonitrile. It can be made by distilling ammonium methanoate, or formate, with phosphorus(V) oxide. Methanamide, or formamide, is an intermediate stage in the conversion.

$$HCOONH_4 \xrightarrow{-H_2O} HCONH_2 \xrightarrow{-H_2O} HCN$$

A simpler laboratory method of preparation is to heat potassium(I) cyanide with dilute sulphuric(VI) acid.

$$KCN + H_2SO_4 \rightarrow HCN + KHSO_4$$

Hydrogen cyanide is a colourless volatile liquid with an almond-like odour. It is one of the most poisonous substances known, and *its preparation should be attempted only by experienced chemists* who are familiar with the necessary precautions. The liquid is moderately soluble in water. The aqueous solution is an extremely weak acid, being weaker even than carbonic acid. The vapour burns in air with a violet flame. forming water, carbon dioxide, and nitrogen.

Hydrogen cyanide undergoes hydrolysis in the manner typical of nitriles. When its aqueous solution is allowed to stand in a sealed tube it is converted first into methanamide and then into ammonium methanoate. These changes occur more rapidly than the corresponding changes with ethanonitrile.

$$HCN \xrightarrow{+H_2O} HCONH_2 \xrightarrow{+H_2O} HCOONH_4$$

Hydrolysis takes place more quickly in the presence of hydrochloric acid, and in this case methanoic acid and ammonium chloride are obtained. Sodium(I) hydroxide solution combines with hydrogen cyanide to give the salt sodium(I) cyanide, but an aqueous solution of the latter slowly changes on standing into one of sodium methanoate and ammonia.

$$CN^- + 2H_2O \rightarrow HCOO^- + NH_3$$

Like ethanonitrile, hydrogen cyanide can be reduced to a primary amine. Thus, if an ethanolic solution of the compound is treated with sodium, methylamine is formed.

$$HCN + 4H \rightarrow CH_3NH_2$$

In spite of its poisonous character hydrogen cyanide is used on a considerable scale in industry. It is manufactured by passing a mixture of methane, ammonia, and air over a platinum catalyst at 1000°C.

$$CH_4 + NH_3 + 1\tfrac{1}{2}O_2 \rightarrow HCN + 3H_2O$$

Large amounts of hydrogen cyanide are used to make sodium(I) cyanide, which is employed in electroplating and in the extraction of gold. Sodium(I) cyanide and potassium(I) cyanide are obtained by absorbing hydrogen cyanide in aqueous sodium(I) hydroxide and potassium(I) hydroxide. Other applications of hydrogen cyanide are in the manufacture of high polymers (e.g. nylon) and in extermination of vermin.

EXERCISE 14

1. State which of the following pairs of substances can be used to obtain propanamide:
   (a) Sodium propanoate and aqueous ammonia.
   (b) Ammonium propanoate and propanoic acid.
   (c) Propanoyl chloride and aqueous ammonia.
   (d) Propanoic anhydride and aqueous ammonia.
   (e) Ethyl propanoate and aqueous ammonia.
2. Write equations for the reactions of an acid amide $R.CONH_2$ with (a) aqueous sodium(I) hydroxide, (b) nitric(III) acid (nitrous acid), (c) phosphorus(V) oxide.
3. When a compound X of formula C_3H_7NO was refluxed with medium hydrochloric acid the remaining solution was found to contain ammonium chloride, and on distillation it gave an alkanoic acid of formula $C_3H_6O_2$. Identify X and write the equation for the reaction.
4. State which of the following properties apply to ethanamide only, to carbamide only, to both, or to neither:
   (a) Its aqueous solution turns red litmus blue.
   (b) It gives off ammonia when heated.
   (c) It gives off ammonia when heated with aqueous sodium(I) hydroxide.
   (d) Its concentrated aqueous solution yields a white precipitate with concentrated nitric(V) acid:
   (e) It gives off nitrogen when treated with nitric(III) acid (nitrous acid).

5. Which of the following statements apply only to the —C≡C— group in alkynes, only to the —C≡N group in nitriles, to both, or to neither?

(a) The triple bond is composed of a π bond and two σ bonds.

(b) The triple bond is highly polarized.

(c) The group undergoes addition with electrophilic reagents.

(d) The group adds on hydrogen under suitable conditions.

(e) The group reduces Baeyer's reagent.

6. Name three classes of organic compounds which liberate ammonia when heated with aqueous sodium(I) hydroxide.

7. Indicate the stages by which the following conversions can be brought about (using not more than two intermediate stages): (a) methanol to ethylamine, (b) ethanol to propanoic acid.

8. Name the following groups: (a) —NH₂, (b) —CONH₂, (c) —CN, (d) —NH—CO—.

15 Amines and Quaternary Ammonium Compounds

AMINES

An aliphatic amine *is a derivative of ammonia in which one or more hydrogen atoms have been replaced by the corresponding number of alkyl groups.*

According to the number of hydrogen atoms in the molecule of ammonia which have undergone replacement *primary, secondary,* and *tertiary* amines are distinguished.

$$N \begin{cases} CH_3 \\ H \\ H \end{cases} \qquad N \begin{cases} CH_3 \\ CH_3 \\ H \end{cases} \qquad N \begin{cases} CH_3 \\ CH_3 \\ CH_3 \end{cases}$$

methylamine (primary amine)   dimethylamine (secondary amine)   trimethylamine (tertiary amine)

A *primary* amine can also be regarded as derived theoretically from an alkane by substituting a hydrogen atom of the latter by an amino $(-NH_2)$ group. For example, methylamine (CH_3NH_2) may be considered as derived in this way from methane (CH_4).

Secondary and tertiary amines are called *simple* amines if their alkyl radicals are similar. If they are different, the amine is said to be a *mixed* one. In this case the names of the different alkyl radicals are given alphabetically (irrespective of their number) in the name of the amine. This is illustrated by the examples of mixed amines given below.

$$CH_3NHC_2H_5 \qquad\qquad (CH_3)_2N(C_2H_5)$$
ethylmethylamine   ethyldimethylamine

Some of the lower primary, secondary, and tertiary amines are listed in Table 15-1. Methylamine, dimethylamine, and trimethylamine are gases at ordinary temperatures, but most of the lower and intermediate amines are volatile liquids. They are well known for their characteristic fishy odours. Many of the amines occur naturally as products of protein decay. Thus all three methylamines are found in herring-brine. The higher amines are solids and have no smell.

Table 15-1. LOWER ALIPHATIC AMINES
(General formula: $C_nH_{2n+1}NH_2$, or RNH_2)

Name	Structural formula	Boiling point/°C	Normal physical state
(Ammonia)	(NH_3)	$(-33\cdot5)$	(Gas)
Primary Amines			
Methylamine	$CH_3.NH_2$	-7	Gas
Ethylamine	$C_2H_5.NH_2$	19	Liquid
Propylamine	$C_2H_5.CH_2.NH_2$	48	Liquid
Butylamine	$C_2H_5.CH_2.CH_2NH_2$	77	Liquid
Secondary Amines			
Dimethylamine	$(CH_3)_2NH$	7	Gas
Diethylamine	$(C_2H_5)_2NH$	55	Liquid
Tertiary Amines			
Trimethylamine	$(CH_3)_3N$	$3\cdot5$	Gas
Triethylamine	$(C_2H_5)_3N$	89	Liquid

The —NH$_2$ group is hydrophilic and the lower amines, like ammonia, are very soluble in water. This is explained by hydrogen-bonding between water molecules and the nitrogen atoms of the amino groups. The solubility decreases, however, as the hydrophobic alkyl radicals increase in size, and higher amines are insoluble in water.

Amines also resemble ammonia in being bases. Their aqueous solutions turn red litmus blue, and they combine with acids to form salts. One difference from ammonia is that they burn readily in air. Methylamine on combustion yields carbon dioxide, water, and nitrogen.

$$4CH_3NH_2 + 9O_2 \rightarrow 4CO_2 + 10H_2O + 2N_2$$

PRIMARY AMINES

Laboratory Preparation

(*i*) *From Acid Amides*. The lower primary amines are usually prepared by treating an acid amide with bromine and aqueous sodium(I) hydroxide. This brings about elimination of the carbonyl group.

$$RCONH_2 \rightarrow RNH_2$$

The mechanism of the reaction, which is known as *Hofmann's reaction*, is complex, but there are three main stages. In the case of ethanamide, or acetamide, these are as follows:

(*a*) Ethanamide is converted by substitution with bromine into *N*-bromoethanamide (*N*- indicating attachment of the bromine atom to the nitrogen atom). This compound has a pale-straw colour. Addition of dilute alkali removes excess of bromine.

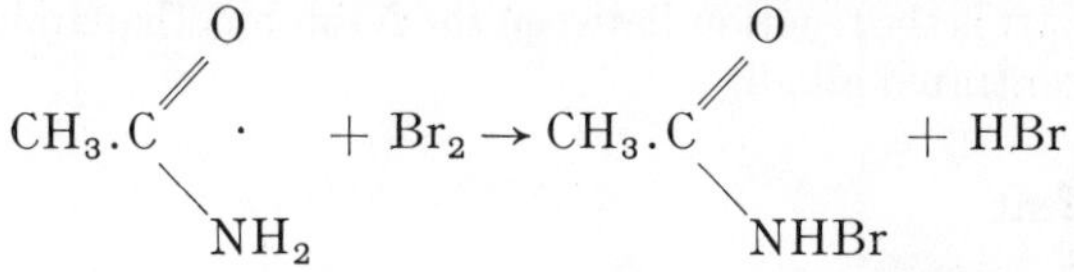

(*b*) When the *N*-bromoethanamide is heated with concentrated alkali hydrogen bromide is eliminated, and this combines with the alkali. The residue then undergoes a molecular rearrangement in which the nitrogen atom takes up a position between the two carbon atoms, giving *methyl isocyanate*.

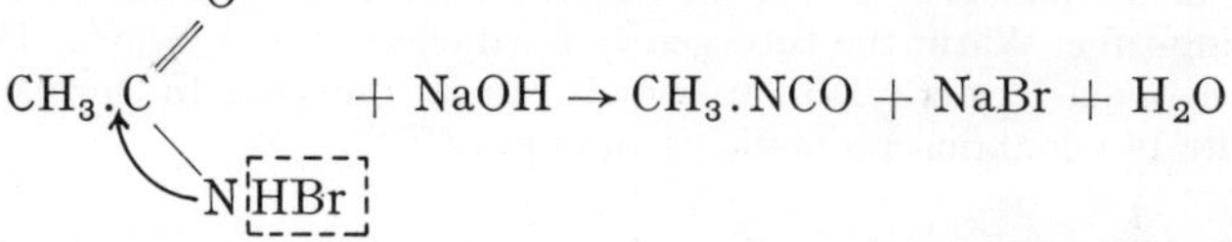

(*c*) The methyl isocyanate reacts with excess of the hot concentrated alkali to form *methylamine* and sodium(I) carbonate. It is at this stage that a carbon atom is eliminated from the molecule.

$$CH_3.NCO + 2NaOH \rightarrow CH_3NH_2 + Na_2CO_3$$

By means of a similar series of reactions ethylamine ($C_2H_5NH_2$) can be obtained from propanamide ($C_2H_5.CONH_2$). The importance of Hofmann's reaction lies not only in its use for preparing primary amines, but also in the fact that it provides a method of eliminating a carbon atom from a molecule.

In practice Hofmann's reaction is carried out in two parts. The first consists of preparation of the N-bromoethanamide, the second part is the reaction between the N-bromoethanamide and the concentrated alkali.

Experiment

Put into a boiling-tube about 2 g of ethanamide and then (in the fume-cupboard) about 2 cm³ of liquid bromine (this is very corrosive and must be handled with great care). Shake the two substances gently together for half a minute, and remove excess of bromine by adding 'bench'sodium(I) hydroxide solution until the colour turns to pale-yellow. The liquid left contains N-bromoethanamide.

Make a concentrated solution of sodium(I) hydroxide by dissolving 10 g of the pellets in 20 cm³ of water, and add this solution to the boiling-tube. Warm the tube gently until effervescence begins. The gas evolved is methylamine (which can be detected by its fishy smell), but contains ammonia as an impurity.

Methylamine can be collected as a gas by downward delivery (it is heavier than air). Usually, however, methylamine and ethylamine are prepared as aqueous solutions by absorption in water. Alternatively, they can be absorbed in a dilute acid. Evaporation of the remaining solution then leaves the amine in the form of a substituted ammonium salt, as illustrated below.

$$CH_3NH_2 + HCl \rightarrow [CH_3NH_3]^+Cl^-$$

methylammonium chloride

Compare

$$NH_3 + HCl \rightarrow [NH_4]^+Cl^-$$

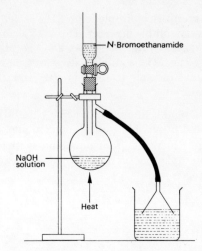

Fig. 15-1. Preparation of an aqueous solution of methylamine or one of its salts

The apparatus used for making an aqueous solution of methylamine or one of its salts is shown in Fig. 15-1. (Apparatus with ground-glass joints should not be used because the concentrated alkali tends to cement the parts together.) The solutions of N-bromoethanamide and concentrated alkali are prepared as described previously. The former is put into the dropping-funnel and the latter into the 100 cm³ distilling-flask, to which some porous pot is also added. The alkali is heated to 60°–70°C, and the N-bromoethanamide solution is slowly run into the flask. The gas evolved is passed into water or dilute acid through an inverted funnel to prevent 'sucking back' of the liquid owing to the high solubility of the gas.

202

Methylamine prepared as described always contains ammonia as an impurity. The pure amine can be obtained by preparing methylammonium chloride crystals and recrystallizing these from absolute ethanol. The ammonium chloride impurity in the crystals is almost insoluble in ethanol. The purified crystals are then warmed with aqueous sodium(I) hydroxide to regenerate methylamine.

(*ii*) *By Ammonolysis of an Alkyl Halide*. It was seen in Chapter 9 that when an alkyl halide (preferably an iodo-compound) is heated with excess of an ethanolic solution of ammonia in a sealed vessel an amine is produced. This is obtained chiefly as a salt, but also to some extent as the free amine. Iodoethane yields ethylammonium iodide and ethylamine.

$$C_2H_5I \xrightleftharpoons{NH_3} [C_2H_5NH_3]^+I^- \xrightleftharpoons{NH_3} C_2H_5NH_2 + [NH_4]^+I^-$$

In this reaction a little diethylamine and triethylamine are also produced.

(*iii*) *By Reduction of a Nitrile*. Primary amines can be obtained by reduction of nitriles with sodium and ethanol. Reduction of ethanonitrile by this method gives ethylamine. The reaction is carried out as described in Chapter 14.

$$CH_3CN + 4H \rightarrow CH_3.CH_2.NH_2$$

(*iv*) *By Reduction of a Nitroalkane*. Nitroalkanes can be converted to primary amines by reducing them with zinc (or tin) and hydrochloric acid. Nitromethane yields methylamine, which is left as a salt.

$$CH_3NO_2 + 6H \rightarrow CH_3NH_2 + H_2O$$

This method is seldom used for preparing aliphatic amines, but is important for making aromatic amines (Chapter 23).

Reactions of Primary Amines

Primary amines are closely related to ammonia, and, like ammonia, are nucleophiles. The key atom in their molecules is the nitrogen atom, which possesses a partial negative charge (due

$$\overset{\delta+}{\underset{\underset{\overset{|}{H}}{\delta+}}{\overset{H}{\overset{|}{\underset{R}{\delta+}{-}}\overset{\cdot\cdot}{N}{:}^{\delta-}}}}$$

to its $-I$ effect) and a pair of unshared electrons. As in the case of ammonia, the lone pair of electrons is used to form a co-ordinate covalent bond with an atom of low electron density in an electrophilic molecule. Thus amines react with electrophilic reagents like acids, alkyl halides, and the various classes of compounds (aldehydes, ketones, etc.) which contain the carbonyl group.

1. **Reactions as Bases.** When an amine is dissolved in water the nitrogen atom shares its lone pair of electrons with a proton from a water molecule, the latter acting as an acid. Methylamine and water form methylammonium ions and hydroxyl ions.

$$CH_3 \overset{H}{\underset{\underset{H}{|}}{\overset{|}{N}}}: + H-O-H \rightleftharpoons \left[CH_3 \overset{H}{\underset{\underset{H}{|}}{\overset{|}{N}}}-H \right]^+ + OH^-$$

Like ammonia solution, the methylamine solution turns red litmus blue, and precipitates metal hydroxides, such as iron(III) hydroxide, from solutions of the metal salts.

Amines are only weak bases in aqueous solution. Their strengths can be compared by their dissociation constants K_b or the dissociation constant exponents pK_b, where $pK_b = -\log K_b$. The values of these at 25°C for ammonia, methylamine, and ethylamine are shown below. K_b increases, and pK_b decreases, as the base becomes stronger.

	$(K_b/\text{mol dm}^{-3})$ $\times 10^5$	pK_b
Ammonia (NH_3)	1·79	4·75
Methylamine (CH_3NH_2)	43·8	3·36
Ethylamine ($C_2H_5NH_2$)	56·0	3·25

Methylamine is a stronger base than ammonia because the $+I$ effect of the methyl radical increases the electron density on the nitrogen atom. Methylamine, however, is a weaker base than ethylamine because the C_2H_5— radical has a larger $+I$ effect than the CH_3— radical.

As mentioned earlier, the salts formed by amines are substituted ammonium salts, as exemplified by methylammonium chloride ($CH_3NH_3{}^+Cl^-$). Sulphuric(VI) acid, being dibasic, forms both an acid salt and a normal salt with amines. The names and formulæ of the salts obtained from methylamine and sulphuric(VI) acid are shown below.

$$CH_3NH_3{}^+HSO_4{}^-$$
methylammonium
hydrogensulphate(VI)

$$(CH_3NH_3{}^+)_2SO_4{}^{2-}$$
methylammonium
sulphate(VI)

The salts of amines are colourless crystalline compounds, which dissolve readily in water. Like ammonium salts they liberate the free bases when warmed with alkalies.

$$CH_3NH_3{}^+ + OH^- \rightarrow CH_3NH_2 + H_2O$$

2. Alkylation. It has been seen earlier that when an alkyl iodide is heated with an ethanolic solution of ammonia in a sealed vessel one of the hydrogen atoms of the ammonia molecule is substituted by an alkyl radical and a primary amine is formed. The reaction, however, does not stop at this stage. Further substitution produces secondary and tertiary amines (chiefly in the form of their salts) and eventually a *quaternary ammonium compound*. In the case of methyl iodide the series of reactions may be represented simply as follows:

$$CH_3I + NH_3 \rightarrow CH_3NH_2 + HI$$
methylamine

$$CH_3I + CH_3NH_2 \rightarrow (CH_3)_2NH + HI$$
dimethylamine

$$CH_3I + (CH_3)_2NH \rightarrow (CH_3)_3N + HI$$
trimethylamine

$$CH_3I + (CH_3)_3N \rightarrow N(CH_3)_4I$$
tetramethylammonium
iodide

The last reaction is analogous to the combination of hydrogen chloride with ammonia to form ammonium chloride.

The reaction between an alkyl iodide and an ethanolic solution of ammonia or an amine invariably yields a mixture of products, the relative amounts of each depending on the proportions of the reactants, the temperature, and the duration of the reaction.

3. Ethanoylation, or Acetylation.

When a concentrated aqueous solution of a primary amine is mixed with ethanoyl chloride or ethanoic anhydride one of the hydrogen atoms of the amino group is substituted by the ethanoyl, or acetyl, group. Ethylamine yields *N*-ethylethanamide, or *N*-ethylacetamide; e.g.

$$C_2H_5NH_2 + CH_3COCl \rightarrow C_2H_5NH \cdot COCH_3 + HCl$$

N-ethylethanamide

Secondary amines behave in the same way as primary amines. Tertiary amines, however, fail to react because they do not contain a hydrogen atom which can be replaced by an acyl group.

4. The Isocyano Reaction.

When a primary amine is warmed with trichloromethane and an ethanolic solution of potassium(I) hydroxide an isocyano-compound is produced. This has a characteristic nauseating smell (and is poisonous). Ethylamine (used in the form of one of its salts) gives isocyanoethane.

$$C_2H_5NH_2 + CHCl_3 + 3KOH \rightarrow C_2H_5NC + 3KCl + 3H_2O$$

isocyanoethane

This reaction, which is specific for primary amines, has been described previously in connection with trichloromethane (Chapter 9).

5. Reaction with Nitric(III) Acid (Nitrous Acid).

Although both primary amines and acid amides contain the $-NH_2$ group they have few reactions in common. One reagent with which they react similarly is nitric(III) acid (acidified sodium(I) nitrate(III) solution). In both cases nitrogen is evolved, and the amino group is replaced by a hydroxyl group. Ethylamine forms ethanol.

$$C_2H_5NH_2 + HNO_2 \rightarrow C_2H_5OH + N_2 + H_2O$$

The reaction is carried out as described for ethanamide (see Fig. 14-2). Again ethylamine is used in the form of a salt. The remaining solution is heated to decompose any nitric(III) acid. It is then neutralized with ammonia and tested for the presence of ethanol by means of the tri-iodomethane reaction (Chapter 8).

The reactions of primary amines with nitric(III) acid are more complicated than indicated by the simple equation given. Although nitrogen is always evolved, a variety of other products (alcohols, ethers, and alkenes) may be obtained. The nature of these and their proportions depend on the primary amine and the conditions used. Even under the best conditions methylamine gives only low yield of methanol (the main product is methoxymethane). Ethylamine produces a yield of ethanol which is about 60 per cent of the theoretical yield, and propylamine a mixture of propan-1-ol and propan-2-ol totalling about 40 per cent of the theoretical amount.

Isomerism of Amines

All the primary amines except methylamine show isomerism, and the same molecular formula may represent primary, secondary, and tertiary amines. Thus there are two amines of formula

C_2H_7N and four amines of formula C_3H_9N. The isomers are given below.

Molecular Formula: C_2H_7N

$C_2H_5NH_2$
ethylamine

$CH_3.NH.CH_3$
dimethylamine

Molecular Formula: C_3H_9N

$CH_3.CH_2.CH_2.NH_2$
propylamine

$(CH_3)_2CH.NH_2$
(1-methylethyl)amine

$CH_3.NH.C_2H_5$
ethylmethylamine

$(CH_3)_3N$
trimethylamine

The number of isomers increases rapidly as the number of carbon atoms in the molecule becomes larger. There are eight amines (four primary, three secondary, and one tertiary) corresponding to the molecular formula $C_4H_{11}N$. The student should try to identify these compounds.

Individual amines can be identified by the characteristic melting points of certain sparingly soluble derivatives. Amongst these are the salts of hexachloroplatinic(IV) acid, which is made by dissolving platinum in aqua regia. The salts are precipitated when a solution of the acid is added to a solution of the alkyl-ammonium chloride. For example, with methylammonium chloride solution a yellow precipitate is obtained.

$$2CH_3NH_3^+ + PtCl_6^{2-} \rightarrow (CH_3NH_3^+)_2PtCl_6^{2-} \downarrow$$

methylammonium
hexachloroplatinate(IV)

SECONDARY AND TERTIARY AMINES

Dimethylamine and trimethylamine are gases, and like monomethylamine are very soluble in water. Diethylamine and triethylamine are colourless liquids which are miscible with water in all proportions. All these compounds have a fishy odour and burn readily in air. Like the primary amines they are bases and form salts with acids.

Secondary and tertiary amines can be prepared in the laboratory by heating an ethanolic solution of the primary amine with the alkyl iodide in a sealed vessel, a larger proportion of the alkyl iodide being used to obtain the tertiary compounds. (The equations have already been given under 'Reactions of Primary Amines'). A mixture of products, however, results. The individual amines can be separated either by fractional distillation or by utilizing the different solubilities of their salts in a suitable solvent (fractional crystallization).

Distinction between Primary, Secondary, and Tertiary Amines

Primary amines are easily recognized by the isocyano reaction, which is not given by secondary and tertiary amines. The usual method of distinguishing between the three classes of amines is by means of their different behaviour with nitric(III) acid (nitrous acid).

(*i*) *Primary Amines.* Although, as seen earlier, primary amines give rise to more than one organic product with nitric(III) acid, they all produce nitrogen. Primary amines can therefore be recognized by *effervescence* and evolution of a colourless gas when they are treated in the cold with acidified sodium(I) nitrate(III) solution.

$$CH_3NH_2 + HNO_2 \rightarrow CH_3OH + N_2 + H_2O$$

(*ii*) *Secondary Amines*. If a secondary amine is treated in the cold with nitric(III) acid no effervescence occurs, but a *yellow oil* separates. This contains the nitroso (—NO) group, and is called a nitrosoamine. The yellow oil obtained from diethylamine is *N*-nitrosodiethylamine, in which the nitroso group is attached to a nitrogen atom.

$$(C_2H_5)_2NH + HONO \rightarrow (C_2H_5)_2N.NO + H_2O$$
$$\text{\textit{N}-nitrosodiethylamine}$$

(*iii*) *Tertiary Amines*. Tertiary amines do not give any apparent reaction with nitric(III) acid in the cold. Actually the organic base combines with protons from the acid, and ions of a salt, e.g. $(CH_3)_3NH^+ NO_2{}^-$, remain in solution.

Manufacture and Uses of Amines

The most important simple amines are the methylamines and ethylamines. The former are manufactured by passing a mixture of methanol vapour and ammonia under pressure over a heated catalyst (aluminium(III) oxide).

$$CH_3OH + NH_3 \xrightarrow{\text{Al}_2\text{O}_3} CH_3NH_2 + H_2O$$

Some dimethylamine and trimethylamine are also formed, the proportions of the products depending on the proportions of the reactants. The different amines are separated (as liquids under pressure) in a fractionating tower.

The three ethylamines are made on a large scale by passing a mixture of ethanal vapour, ammonia, and hydrogen under pressure over a heated nickel catalyst.

$$CH_3.CHO + H_2 + NH_3 \xrightarrow{\text{Ni}} CH_3.CH_2NH_2 + H_2O$$

The aldehyde is probably reduced by hydrogen to ethanol, which then reacts with ammonia. As before, all three amines are obtained simultaneously.

The chief uses of the methylamines and ethylamines are as intermediates in the manufacture of other compounds, including quaternary ammonium compounds. Amines themselves are used to some extent as solvents, anti-oxidants (e.g. for rubber latex), and rocket propellants.

QUATERNARY AMMONIUM COMPOUNDS

A quaternary ammonium compound *is a compound in which all four hydrogen atoms of the ammonium ion have been substituted by alkyl or aryl radicals.* (Aryl radicals are radicals derived from aromatic hydrocarbons.) A typical quaternary ammonium compound is tetraethylammonium iodide, $[N(C_2H_5)_4]^+I^-$. Other compounds of this type include chlorides, bromides, nitrates(V), sulphates(VI), and hydroxides, all of which have an ionic composition.

Quaternary ammonium halides can be prepared by heating an ethanolic solution of a tertiary amine with excess of an alkyl halide in a sealed vessel. Tetraethylammonium iodide is obtained by combination of triethylamine and iodoethane.

$$(C_2H_5)_3N + C_2H_5I \rightarrow [N(C_2H_5)_4]^+I^-$$

Compounds containing mixed alkyl radicals can be prepared by choosing reactants with the appropriate radicals.

Quaternary ammonium salts resemble ammonium salts and salts of amines in being colourless crystalline compounds, which dissolve readily in water. The quaternary ammonium salts differ, however, from ammonium salts and salts of amines in not reacting with warm aqueous sodium(I) hydroxide. With this reagent ammonium salts liberate ammonia, and amine salts give the free amine. In both these cases the cation possesses an acidic hydrogen atom, that is, a hydrogen atom which can be removed as a proton by an OH^- ion; e.g.

$$CH_3NH_3^+ + OH^- \rightarrow CH_3NH_2 + H_2O$$

Cations of quaternary ammonium salts do not contain an acidic hydrogen atom, and hence cannot undergo the above type of reaction. Thus they do *not* liberate the tertiary amine when heated with sodium hydroxide solution.

If tetraethylammonium iodide is warmed with an aqueous suspension of silver(I) oxide (which furnishes silver(I) ions and hydroxyl ions) silver(I) iodide is precipitated. The remaining solution, when filtered and evaporated, yields tetraethylammonium hydroxide.

$$[N(C_2H_5)_4]^+ + I^- + Ag^+ + OH^-$$
$$\rightarrow AgI \downarrow + [N(C_2H_5)_4]^+OH^-$$

The quaternary ammonium hydroxides form colourless crystals. Their aqueous solutions are strong alkalies, comparable in strength with sodium(I) hydroxide. The solutions absorb carbon dioxide from air, and liberate ammonia from ammonium salts. It is interesting to note that, while ammonium hydroxide has never been isolated, organic analogues in the form of quaternary ammonium hydroxides are well known compounds.

Certain quaternary ammonium compounds are used commercially. Hexadecyltrimethylammonium chloride is employed as a detergent of the 'cationic' type, in which the detergent action depends on the cation instead of the anion. This compound is manufactured from hexadecylamine and chloromethane.

$$C_{16}H_{33}NH_2 + 3CH_3Cl \rightarrow [C_{16}H_{33}(CH_3)_3N]^+Cl^- + 2HCl$$

The corresponding bromide is used in dusting-powders for babies (most quaternary ammonium compounds are antiseptics). Other compounds of this class are employed as froth-flotation agents and in making ion-exchange resins.

EXERCISE 15

1. Write the structural formulæ of (a) ethylpropylamine, (b) 1-methylethylamine, (c) diethylmethylamine, (d) 2-methylpropylamine.

2. Classify the amines in question 1 as primary, secondary, or tertiary.

3. (a) Write equations for the three stages in the conversion of propanamide into ethylamine by means of bromine and aqueous sodium(I) hydroxide; (b) name the two intermediate compounds.

4. With which (if any) of the following reagents do primary aliphatic amines *not* react: (a) water, (b) ethanol, (c) aqueous sodium(I) hydroxide, (d) sulphuric(VI) acid, (e) bromoethane.

5. Name the organic products of the reactions between ethylamine and (a) hydrobromic acid, (b) iodomethane, (c) ethanoyl chloride, (d) trichloromethane and ethanolic potassium(I) hydroxide.

6. Give the structural formulæ and names of the secondary and tertiary amines which have the molecular formula $C_4H_{11}N$.

7. Which of the following statements are true?

(a) Acid amides behave chiefly as electrophilic reagents, amines as nucleophilic reagents.

(b) The inductive effect of the alkyl radical R in an amine RNH_2 increases the basic character of the —NH_2 group.

(c) The isocyano reaction can be used to distinguish between a secondary amine and a tertiary amine.

(d) Nitrogen is evolved when a primary amine, but not a secondary or tertiary one, is treated with nitric(III) acid (nitrous acid).

(e) A quaternary ammonium salt liberates an amine when it is heated with aqueous sodium(I) hydroxide.

16 Dicarboxylic Acids, Hydroxycarboxylic Acids, Amino Acids and Proteins

DICARBOXYLIC ACIDS

Ethanedioic acid (oxalic acid) is the first member of a series of acids (called the ethanedioic acid series) which contain two carboxyl groups in the molecule and are therefore described as *bifunctional*. These acids have the general formula $HOOC.(CH_2)_n.COOH$. They are formed from alkanes by substituting the two CH_3- radicals at the ends of the main carbon chain by two carboxyl groups. They are named systematically by adding the suffix *dioic acid* to the name of the alkane.

Alkane	*Dicarboxylic Acid*
Ethane $CH_3.CH_3$	Ethanedioic acid (oxalic acid) $HOOC.COOH$
Propane $CH_3.CH_2.CH_3$	Propanedioic acid (malonic acid) $HOOC.CH_2.COOH$
Butane $CH_3.CH_2.CH_2.CH_3$	Butanedioic acid (succinic acid) $HOOC.CH_2.CH_2.COOH$ etc.

The dicarboxylic acids are solid crystalline substances of high melting point. The lower members dissolve readily in water, and,

owing to the presence of two carboxyl groups in the molecule, they are stronger acids than the alkanoic acids. The most important member of the series is ethanedioic acid.

ETHANEDIOIC ACID-2-WATER
$$(COOH)_2.2H_2O$$

Melting point: $101°C$ (hydrate)   *Boiling point:* (decomposes)

Ethanedioic acid occurs naturally in beet leaves, while the acid salt potassium(I) hydrogenethanedioate is found in rhubarb leaves, dock leaves, and wood sorrel. Both the acid and its salts are highly poisonous.

Preparation of Ethanedioic Acid

Ethanedioic acid can be prepared both in the laboratory and on a large scale from sodium(I) methanoate (sodium formate). When the latter is heated hydrogen is given off and sodium(I) ethanedioate remains.

$$2HCOONa \rightarrow H_2 + (COONa)_2$$

The sodium(I) ethanedioate is dissolved in water and boiled with solid calcium(II) hydroxide, which forms a precipitate of calcium(II) ethanedioate.

$$(COONa)_2 + Ca(OH)_2 \rightarrow (COO)_2Ca \downarrow + 2NaOH$$

The precipitate is filtered and treated with dilute sulphuric(VI) acid. This liberates ethanedioic acid in solution, and gives a precipitate of calcium(II) sulphate(VI), which is separated by filtration. The remaining solution of ethanedioic acid is concentrated and crystallized.

$$(COO)_2Ca + H_2SO_4 \rightarrow (COOH)_2 + CaSO_4 \downarrow$$

Ethanedioic acid can also be obtained by oxidizing ethane-1,2-diol with concentrated nitric(V) acid. The dihydric alcohol contains two primary alcohol groups in the molecule, and on oxidation these are converted to carboxyl groups.

$$\begin{matrix} CH_2OH \\ | \\ CH_2OH \end{matrix} + 4O \rightarrow \begin{matrix} COOH \\ | \\ COOH \end{matrix} + 2H_2O$$

Properties of Ethanedioic Acid

Ethanedioic acid forms colourless crystals containing two molecules of water of crystallization. The crystals are moderately soluble in water and ethanol. When the crystals are heated they melt at 101°C, and give off their water of crystallization. At 180°–200°C the anhydrous acid partly melts and partly decomposes.

1. Acidic Properties. Although ethanedioic acid is only a weak acid, it is appreciably stronger than any of the alkanoic acids. The acid is dibasic and forms both normal and acid salts. *Three* salts can be obtained from the acid and potassium(I)hydroxide according to the amounts of acid and alkali combined together. These salts have the following names and formulæ:

Potassium(I) ethanedioate--1-water: $\quad (COOK)_2 . H_2O$

Potassium(I) hydrogenethanedioate: $\quad \begin{matrix} COOK \\ | \\ COOH \end{matrix}$

Potassium(I) trihydrogen diethanedioate-2-water: $\quad \left(\begin{matrix} COOK \\ | \\ COOH \end{matrix}\right) . \left(\begin{matrix} COOH \\ | \\ COOH \end{matrix}\right) . 2H_2O$

The trivial names of the three salts are normal potassium oxalate, potassium binoxalate, and potassium quadroxalate respectively.

The esters dimethyl ethanedioate and diethyl ethanedioate are made by refluxing anhydrous ethanedioic acid with excess of the appropriate alcohol for two hours and distilling off the product. Dimethyl ethanedioate forms colourless crystals (m.p. 54°C) at ordinary temperatures. With phosphorus pentachloride ethanedioic acid gives an acyl chloride called *ethanedioyl dichloride* (oxalyl chloride).

$$(COOH)_2 + 2PCl_5 \rightarrow (COCl)_2 + 2POCl_3 + 2HCl$$

Ethanedioic acid does not form an anhydride. Ethanediamide, the acid amide, is described shortly.

2. Dehydration by Concentrated Sulphuric(VI) Acid. When ethanedioic acid is warmed with concentrated sulphuric(VI) acid it undergoes dehydration, yielding a mixture of carbon dioxide and carbon monoxide.

$$\begin{matrix} COOH \\ | \\ COOH \end{matrix} -H_2O \rightarrow CO_2 + CO$$

3. Oxidation. A solution of ethanedioic acid (or one of its salts) quickly decolorizes a warm acidified solution of potassium(I) manganate(VII) (potassium permanganate). This reaction is utilized in volumetric analysis for the estimation of ethanedioic acid and its salts.

$$\begin{matrix} COOH \\ | \\ COOH \end{matrix} \quad +O \rightarrow 2CO_2 + H_2O$$

Ethanedioic acid differs from methanoic acid (the only reducing alkanoic acid) in not reducing a warm *alkaline* solution of potassium(I) manganate(VII) (except very slowly). Also it does not reduce boiling mercury(II) chloride solution to give a white precipitate of mercury(I) chloride.

Chemical Tests for Ethanedioic Acid

(*i*) *Reducing Action*. Warm a solution of ethanedioic with a few drops of: (*a*) acidified potassium(I) manganate(VII) solution; (*b*) a solution of this reagent to which sodium(I) carbonate solution has been added instead of acid. The colour of the potassium(I) manganate(VII) is destroyed in the first case, but not in the second.

(*ii*) *Concentrated Sulphuric(VI) Acid*. Warm a few crystals of ethanedioic acid (or one of its salts) with a few drops of concentrated sulphuric(VI) acid. Carbon monoxide is given off and can be ignited at the mouth of the tube. If the gases evolved are passed into calcium(II) hydroxide solution, the latter is turned milky by the carbon dioxide present. Ethanedioic acid is not charred by concentrated sulphuric(VI) acid.

(*iii*) *Calcium(II) Chloride Solution*. To a solution of ethanedioic acid add 10 per cent calcium(II) chloride solution. A white precipitate of calcium(II) oxalate is formed. The precipitate is insoluble in ethanoic, or acetic, acid, but dissolves in dilute hydrochloric acid.

Ethanediamide

Ethanediamide (oxamide) is a sparingly soluble crystalline solid. It can be made by heating ammonium ethanedioate, but is usually prepared by shaking diethyl ethanedioate with ammonia. Ethanediamide is precipitated.

$$\begin{matrix} COOC_2H_5 \\ | \\ COOC_2H_5 \end{matrix} + 2NH_3 \rightarrow \begin{matrix} CONH_2 \\ | \\ CONH_2 \end{matrix} + 2C_2H_5OH$$

Many of the chemical reactions of ethanediamide are parallel with those of ethanamide (acetamide). If ethanediamide is boiled with sodium(I) hydroxide solution, it is hydrolysed and ammonia is liberated.

$$(CONH_2)_2 + 2NaOH \rightarrow (COONa)_2 + 2NH_3$$

Nitric(III) acid (nitrous acid), obtained by acidifying sodium(I) nitrate(III) solution, converts ethanediamide into ethanedioic acid with evolution of nitrogen.

$$(CONH_2)_2 + 2HNO_2 \rightarrow (COOH)_2 + 2N_2 + 2H_2O$$

When ethanediamide is warmed with phosphorus(V) oxide the amide is dehydrated to the gas cyanogen.

$$(CONH_2)_2 - 2H_2O \rightarrow (CN)_2$$

PROPANEDIOIC ACID, $CH_2\begin{smallmatrix}COOH\\COOH\end{smallmatrix}$

Melting point: 136°C   *Boiling point:* (decomposes)

Propanedioic acid (*malonic acid*) is a colourless crystalline compound, which occurs in beetroot as the calcium(II) salt. It is soluble in water and ethanol. It is prepared by synthesis from the potassium(I) salt of chloroethanoic acid. The following are stages in the synthesis:

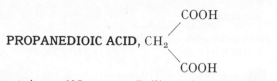

potassium(I) chloroethanoate	potassium(I) cyanoethanoate	propanedioic acid

If propanedioic acid is heated to 140–150°C, it loses carbon dioxide and forms ethanoic acid.

$$CH_2\begin{smallmatrix}COOH\\COOH\end{smallmatrix} \rightarrow CO_2 + CH_3.COOH$$

Decarboxylation in this way is a general reaction of all acids containing two carboxyl groups attached to the same carbon atom.

Propanedioic acid is chiefly of importance because of its esters, e.g. diethyl propanedioate, $CH_2(COOC_2H_5)_2$, which are much used in organic synthesis.

BUTANEDIOIC ACID, $\begin{smallmatrix}CH_2.COOH\\ \\CH_2.COOH\end{smallmatrix}$

Melting point: 185°C   *Boiling point:* 235°C

Butanedioic acid (*succinic acid*) forms colourless crystals which are moderately soluble in cold water and ethanol. The acid can be prepared synthetically from ethene in the following stages:

$$\underset{\text{ethene}}{\begin{smallmatrix}CH_2\\ \|\\ CH_2\end{smallmatrix}} \xrightarrow{Br_2} \underset{\substack{\text{1,2-dibromo-}\\\text{ethane}}}{\begin{smallmatrix}CH_2Br\\ |\\ CH_2Br\end{smallmatrix}} \xrightarrow{KCN} \underset{\substack{\text{butane-}\\\text{dionitrile}}}{\begin{smallmatrix}CH_2CN\\ |\\ CH_2CN\end{smallmatrix}} \xrightarrow[H^+]{H_2O} \underset{\substack{\text{butane-}\\\text{dioic acid}}}{\begin{smallmatrix}CH_2.COOH\\ |\\ CH_2.COOH\end{smallmatrix}}$$

Like other dicarboxylic acids butanedioic acid forms both normal and acid salts and esters. Most of the salts are soluble in water, but those of barium(II) calcium(II), and silver(I) are only sparingly soluble. If a solution of potassium(I) butanedioate is electrolysed, ethene and carbon dioxide are given off at the anode (Chapter 6).

When butanedioic acid is heated partial dehydration occurs, some of the acid being converted into *butanedioic anhydride*.

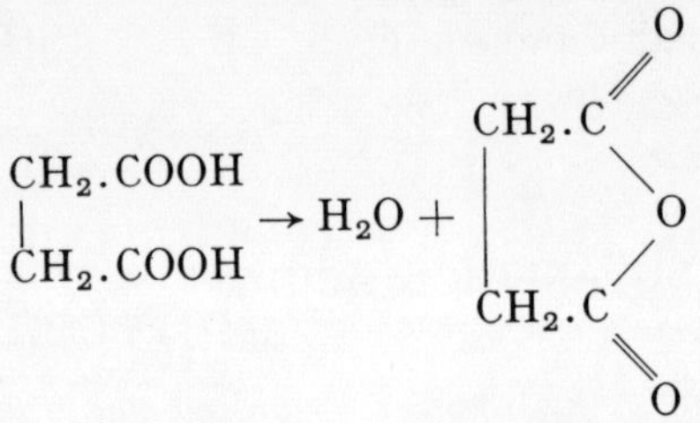

$$CH_2.COOH \atop CH_2.COOH \rightarrow H_2O +$$

$$\underset{\substack{\text{butanedioic} \\ \text{anhydride}}}{}$$

The anhydride is usually prepared by refluxing the acid for about an hour with ethanoic anhydride, which combines with the water produced. Butanedioic anhydride is deposited when the remaining liquid is cooled. The anhydride is a white crystalline compound (m.p. 120°C).

HYDROXYCARBOXYLIC ACIDS

I Hydroxymonocarboxylic Acids

The hydroxy*mono*carboxylic acids contain one hydroxyl group and one carboxyl group in the molecule, and form an homologous series of compounds of general formula $C_nH_{2n}(OH).COOH$. They can be regarded as derived from alkanoic acids by replacing a hydrogen atom of the alkyl group in the latter by a hydroxyl group.

Alkanoic Acid	*Hydroxycarboxylic Acid*
Methanoic acid HCOOH	Hydroxymethanoic acid (carbonic acid) (HO)COOH
Ethanoic acid $CH_3.COOH$	Hydroxyethanoic acid (glycollic acid) $CH_2(OH).COOH$
Propanoic acid $CH_3.CH_2.COOH$	2-Hydroxypropanoic acid (lactic acid) $CH_3.CH(OH).COOH$
	3-Hydroxypropanoic acid (hydracrylic acid) $CH_2(OH).CH_2.COOH$

HYDROXYETHANOIC ACID, $CH_2(OH).COOH$

Melting point: 80°C   *Boiling point:* (decomposes)

Hydroxyethanoic acid is a colourless crystalline solid, which dissolves readily in water and organic solvents. It occurs in unripe grapes, but is usually prepared from chloroethanoic acid. The acid is first neutralized with potassium(I) carbonate solution, and the solution of the potassium(I) salt is then refluxed with water for sixteen hours.

$$CH_2Cl.COOK + H_2O \rightarrow CH_2(OH).COOH + KCl$$

The remaining solution is evaporated to dryness, and the hydroxyethanoic acid is extracted with propanone, in which potassium(I) chloride is insoluble.

Hydroxyethanoic acid combines the chemical properties of a primary alcohol (due to its —CH_2OH group) with those of a carboxylic acid. The *primary alcohol group* only is affected when hydroxyethanoic acid is heated with concentrated nitric(V) acid. This group is oxidized to a carboxyl group and ethanedioic acid is formed. The *carboxyl group* only is concerned in the reactions with alkalies and alcohols. The former yield salts of hydroxyethanoic acid, and the latter esters. If an ethereal solution of hydroxyethanoic acid is treated with sodium the hydrogen atoms from both hydroxyl groups are displaced. Similarly phosphorus pentachloride brings about the substitution of both hydroxyl groups by chlorine atoms with the formation of chloroethanoyl chloride ($CH_2Cl.COCl$).

2-HYDROXYPROPANOIC ACID, $CH_3.CH(OH).COOH$

Melting point: 18°C   *Boiling point:* (decomposes)

2-Hydroxypropanoic acid (*lactic acid*) is produced in the souring of milk as a result of bacterial action. The acid also occurs in the muscle tissues of animals and can be prepared from meat extract. The acid obtained from this source is often called *sarcolactic acid* (Greek *sarkos* = flesh) to distinguish it from the acid made by fermentation or chemical methods. Chemically the two are identical, but they differ in their effects on polarized light. The 'optical isomerism' of the acids is discussed in the next chapter.

2-Hydroxypropanoic acid can be prepared by hydrolysing the addition compound formed by ethanal with hydrogen cyanide (Chapter 10).

$$CH_3.CH\overset{\displaystyle OH}{\underset{\displaystyle CN}{\diagup\diagdown}} + 2H_2O + HCl$$

$$\rightarrow CH_3.CH\overset{\displaystyle OH}{\underset{\displaystyle COOH}{\diagup\diagdown}} + NH_4Cl$$

2-Hydroxypropanoic acid can also be obtained from 2-chloropropanoic acid in the same way that hydroxyethanoic acid is prepared from chloroethanoic acid. The acid is neutralized with potassium(I) carbonate solution, and the potassium(I) salt is hydrolysed with boiling water.

$$CH_3.CHCl.COOK + H_2O \rightarrow CH_3.CH(OH).COOH + KCl$$

It is difficult to crystallize 2-hydroxypropanoic acid. The acid is usually obtained as a colourless, syrupy, sour-smelling liquid, which is miscible with water and ethanol in all proportions.

2-Hydroxypropanoic acid combines the properties of a secondary alcohol with those of a carboxylic acid. When the acid is oxidized with alkaline potassium(I) manganate(VII) the secondary alcohol group is converted into a carbonyl group, and 2-oxopropanoic acid (pyruvic acid) is formed.

$$CH_3.CH(OH).COOH + O \rightarrow CH_3.CO.COOH + H_2O$$
2-oxopropanoic acid

3-*Hydroxypropanoic acid* (hydracrylic acid), $CH_2OH.CH_2.COOH$, is mainly of interest because of its isomerism with 2-hydroxypropanoic acid. It is a syrupy liquid, which can be prepared by methods analogous to those used for its isomer. It has no practical importance.

II Hydroxy-Derivatives of Dibasic Acids

2,3-DIHYDROXYBUTANEDIOIC ACID,

$$CH(OH).COOH$$
$$|$$
$$CH(OH).COOH$$

Melting point: 167°C (approx.)   *Boiling point:* (decomposes)

Common names used for this acid are dihydroxysuccinic acid and *tartaric acid*. It is one of the most important organic acids. Large amounts are used, along with sodium(I) hydrogencarbonate, in many kinds of health salts and effervescent drinks. The acid occurs in various fruits, notably grapes, either as the free acid or as the acid potassium(I) salt. The latter is the chief constituent of *argol*, a brown crusty solid deposited on the sides of casks in which grape-juice is fermented to wine. The acid is obtained commercially from this source.

Properties of 2,3-Dihydroxybutanedioic acid

The acid is a colourless crystalline solid, which is very soluble in water and moderately soluble in ethanol. When the acid is heated it melts, but, since decomposition occurs at the same time, the melting point is indefinite.

2,3-Dihydroxybutanedioic acid and its salts are reducing agents, and decolorize both acidified and alkaline solutions of potassium(I) manganate (VII). A neutral solution of a salt of the acid gives a silver mirror with Tollens's reagent. The oxidation products of the acid vary with the nature of the oxidizing agent, but usually the acid is broken down into simpler substances. Strong oxidizing agents convert the acid to ethanedioic acid (oxalic acid), or carbon dioxide, and water.

2,3-Dihydroxybutanedioic acid forms two potassium(I) salts, a normal salt and an acid salt. These have the following formulæ:

$$\left(\begin{matrix} CH(OH).COOK \\ | \\ CH(OH).COOK \end{matrix} \right) \tfrac{1}{2}H_2O \qquad \begin{matrix} CH(OH).COOK \\ | \\ CH(OH).COOH \end{matrix}$$

potassium(I) 2,3-dihydroxy-butanedioate-½-water (normal potassium tartrate)   potassium(I) hydrogen-2,3-dihydroxybutanedioate (potassium hydrogen tartrate)

The acid salt, which is only slightly soluble in water, is much more important than the normal salt, which is readily soluble in water. The acid salt is one of the few potassium salts which can be precipitated, and it is therefore used as a means of identifying potassium in qualitative analysis. The acid salt is known commercially as 'cream of tartar.' It is used, together with sodium(I) hydrogencarbonate, in baking-powder.

Potassium(I) sodium(I) 2,3-dihydroxybutanedioate-4-water has the formula $KNaC_4H_4O_6.4H_2O$. This salt is called 'Rochelle salt' from its association with the port of La Rochelle in south-west France. It is made by neutralizing a warm solution of cream of tartar with sodium(I) carbonate and evaporating the solution. It is used on a large scale for silvering mirrors (by reduction of an ammoniacal solution of silver(I) oxide). In the laboratory Rochelle salt is employed in the preparation of Fehling's solution.

Chemical Tests for 2,3-Dihydroxybutanedioic Acid

(*i*) *Decomposition by Heat.* Heat a little of the acid in an ignition-tube. The acid melts, turns brown, and finally leaves a black residue of carbon. During the heating a strong smell of 'burnt sugar' is given off.

(*ii*) *Concentrated Sulphuric(VI) Acid.* Warm a little 2,3-dihydroxybutanedioic acid with a few drops of concentrated sulphuric(VI) acid. Carbon monoxide (which burns with a blue flame) and sulphur dioxide (which has a characteristic smell) are evolved, and at the same time heavy charring takes place.

(*iii*) *Tollens's Reagent.* This test is carried out as described in Chapter 10. A silver mirror is obtained.

(*iv*) *Fenton's Reagent.* To 5 cm³ of a solution of 2,3-dihydroxy-butanedioic acid (or one of its salts) add one drop of iron(II) sulphate(VI) solution and two drops of 10 per cent hydrogen peroxide solution. A deep yellow solution is formed, which turns violet when excess of dilute sodium(I) hydroxide solution is added.

(*v*) *Calcium(II) Chloride Solution.* Add 10 per cent calcium(II) chloride solution to a concentrated neutral solution of a 2,3-dihydroxybutanedioate. A white precipitate of the calcium(II) salt is slowly formed. (The precipitate is obtained more quickly if the inside of the test-tube is 'scratched' with a glass rod). The precipitate is soluble in ethanoic acid (distinction from ethanedioate).

III Hydroxy-derivatives of Tribasic Acids

2-HYDROXYPROPANE-1,2,3-TRICARBOXYLIC ACID-1-WATER

$$(C_6H_8O_7 . H_2O)$$

Melting point: 100°C (hydrate)   *Boiling point:* (decomposes)

This acid is better known under its trivial name of *citric acid*, and for convenience we shall use this name. The constitution of the anhydrous acid is represented by the formula now given.

$$CH_2.COOH$$
$$|$$
$$C(OH).COOH$$
$$|$$
$$CH_2.COOH$$

The acid occurs naturally in many fruits, including gooseberries, raspberries, and the citrus fruits. Lemon-juice contains 5–10 per cent of citric acid, and is still used to some extent as a commercial source of the acid. The latter is chiefly manufactured nowadays, however, by fermentation of a solution of glucose or beet molasses by means of a mould (*Aspergillus niger*).

Citric acid is the only common tribasic aliphatic acid. It forms three salts with sodium(I) hydroxide according to the proportions of acid and alkali which are made to react together. It is less easily oxidized than 2,3-dihydroxybutanedioic acid. Thus, although it decolorizes a warm acidified solution of potassium(I) manganate(VII), it does not give a silver mirror with Tollens's reagent.

Citric acid is used as a mordant in dyeing and in the preparation of pharmaceutical products. It is an ingredient of artificial lemonade powder and many kinds of effervescing health-salts. The addition of sodium(I) citrate to blood prevents the latter from clotting, a property utilized in preserving blood given by blood donors.

Chemical Tests for Citric Acid

(*i*) *Decomposition by Heat*. Heat a little citric acid in an ignition-tube. The acid melts, turns brown, and finally leaves a black residue of carbon. During the decomposition a faint smell of 'burnt sugar' can be detected, but this is largely masked by a copious evolution of white fumes which are extremely irritating to the nose.

(*ii*) *Concentrated Sulphuric(VI) Acid*. Warm a little citric acid, or one of its salts, with a few drops of concentrated sulphuric(VI) acid. Carbon monoxide (which burns with a blue flame) and carbon dioxide are evolved. The contents of the tube turn yellow or brown after a time, but there is no separation of free carbon.

(*iii*) *Calcium(II) Chloride Solution*. Add 10 per cent calcium(II) chloride solution to a cold concentrated neutral solution of a citrate. No precipitate is formed. Now boil the mixture. A precipitate of calcium(II) citrate forms, because the latter is less soluble in hot water than in cold water.

AMINO ACIDS AND PROTEINS

Amino Acids

Amino acids are compounds containing at least one amino group and one carboxyl group in the molecule. They are of great biological importance because their molecules are the chemical units from which the complex molecules of proteins are built up.

When a protein (such as egg albumen) is refluxed with a mineral acid (usually hydrochloric acid) it slowly breaks down into its constituent amino acids. These can be separated (e.g. by paper partition chromatography) and identified. In this way some twenty different amino acids can be obtained from naturally occurring proteins, most of them being present in any one protein. They are all 2-amino acids; that is, they have an —NH_2 group attached to the carbon atom adjacent to the —COOH group. They are represented by the following general formula.

$$\overset{\displaystyle NH_2}{\underset{\displaystyle \ }{X-CH-COOH}}$$

X in this formula may be a hydrogen atom or one of a wide variety of groups.

Three of the simplest 2-amino acids are aminoethanoic, or aminoacetic, acid (glycine), 2-aminopropanoic acid (alanine), and 2-amino-3-hydroxypropanoic acid (serine). (In the context of biochemistry the trivial names given in brackets are generally used.)

Acid	Structural formula
Glycine	$H-CH(NH_2).COOH$
Alanine	$CH_3-CH(NH_2).COOH$
Serine	$CH_2(OH)-CH(NH_2).COOH$

Although the majority of 2-amino acids obtained from proteins contain only one amino group and one carboxyl group in the molecule, this is not true for all of them. In a few cases two amino groups or two carboxyl groups are present.

Amino acids containing the —NH_2 group attached to a carbon atom in the 3- or 4-position are also known, but they are unimportant and can be ignored

Preparation of 2-amino Acids from Halogeno Acids

A simple method of preparing a 2-amino acid is to treat a halogeno acid such as chloroethanoic acid with ammonia.

Glycine is formed when a concentrated aqueous solution of chloroethanoic acid is mixed with a large excess of concentrated ammonia and the mixture is allowed to stand in a stoppered flask for 24 hours.

$$CH_2Cl.COOH + 2NH_3 \rightarrow CH_2(NH_2).COOH + NH_4Cl$$

The remaining liquid is boiled to drive off excess of ammonia, and while it is still hot excess of copper(II) carbonate is added to form the copper(II) salt of glycine. The blue solution is filtered, and hydrogen sulphide is passed through the filtrate to precipitate the copper as copper(II) sulphide, which is removed by filtration. Glycine is obtained by concentrating and crystallizing the solution.

$$(CH_2(NH_2).COO)_2Cu + H_2S \rightarrow CuS + 2CH_2(NH_2).COOH$$

Alanine, $CH_3.CH(NH_2).COOH$, can similarly be prepared from 2-chloropropanoic acid ($CH_3.CHCl.COOH$).

Properties of 2-amino Acids

The 2-amino acids are of great interest because of their peculiar constitution and the manner in which this is reflected in their properties. These properties are illustrated by glycine.

Glycine has a surprisingly high melting point in comparison with those of related acids. Thus ethanoic acid melts at 17°C, chloroethanoic acid at 62°C, and glycine at 235°C. Glycine is readily soluble in water, but almost insoluble in ethanol, ethoxyethane, and hydrocarbon solvents. The aqueous solution, unlike similar solutions of ethanoic acid and chloroethanoic acid, is neutral to litmus. These properties are more characteristic of an ionic compound like sodium(I) chloride than of a weak organic acid. It is now known that glycine and similar amino acids can exist in two forms and that in aqueous solution these are in equilibrium, as shown in the following equation:

$$H_2N.CH_2.COOH \rightleftharpoons \overset{+}{H_3N}.CH_2.\overset{-}{COO}$$
$$\text{(form A)} \qquad\qquad \text{(form B)}$$

The type of isomerism in which two isomers exist together in equilibrium is called *tautomerism*, or *dynamic isomerism*.

In form B a hydrogen ion, or proton, has apparently 'migrated' from the carboxyl group to the amino group. This form, in which different parts of the same molecule are oppositely charged, is described as a *zwitterion*, or an *inner salt*. In aqueous solution the proportion of form B to form A is about 250 000 to one. In the solid state glycine probably consists of zwitterions only. Since there is an equilibrium in aqueous solution between forms A and B, glycine can react as either according to the nature of the substance with which it is treated. If it reacts as A the equilibrium is disturbed and B changes into A. If it reacts as B the converse applies. Glycine behaves as the zwitterion in its reactions with acids and bases; in its other reactions it can be regarded simply as having the covalent structure.

The zwitterion form of glycine is an amphoteric electrolyte; that is, it behaves as a base towards acids and as an acid towards bases. If an aqueous solution of glycine is treated with hydrochloric acid and the solution is evaporated, a salt of formula $[\overset{+}{H_3}N.CH_2.COOH]Cl^-$ is obtained. This is commonly called glycine hydrochloride. With sodium(I) hydroxide solution the salt sodium(I) aminoethanoate $[H_2N.CH_2.COO^-]\,Na^+$ is obtained.

Although an aqueous solution of glycine is neutral to litmus, the solution reacts with metal carbonates and hydrogencarbonates to give carbon dioxide. The reactions take place more slowly, however, than with ethanoic acid.

Amino acids show the characteristic behaviour of an amino group in liberating nitrogen with nitric(III) acid (nitrous acid), the —NH_2 group being replaced by an —OH group. Glycine forms hydroxyethanoic acid.

$$CH_2(NH_2).COOH + HNO_2$$
$$\rightarrow CH_2(OH).COOH + N_2 + H_2O$$

Chemical Tests on Glycine

(i) *Aqueous Solution*. Prepare a fairly concentrated solution of glycine in water. Test the solution with litmus paper, and verify that the solution is neutral. Add ethanol to the solution and shake the tube. Glycine is precipitated.

(ii) *Sodium(I) Hydrogencarbonate*. Add a little of this reagent to an aqueous solution of glycine. Carbon dioxide is slowly evolved, and a drop of calcium(II) hydroxide solution introduced into the tube on the end of a glass rod turns milky.

(iii) *Copper(II) Sulphate(VI) Solution*. Add two drops of the solution to a solution of glycine. A deep-blue solution is produced by formation of the copper(II) salt of glycine.

(iv) *Iron(III) Chloride Solution*. Add two drops of 'neutral' iron(III) chloride solution to a solution of glycine. A deep-red colour is obtained. This is similar to that given by ethanoic acid after neutralization.

(v) *Isocyano Test*. Since glycine contains the primary amine group —CH_2NH_2, it gives the isocyano reaction with trichloromethane and an ethanolic solution of potassium(I) hydroxide. Carry out the test as described in Chapter 9.

Polypeptides

Two molecules of a 2-amino acid can be joined together by converting the acid into the more reactive acyl chloride, which is then made to react with more of the acid. The overall change can be regarded as condensation of two molecules of the acid with elimination of one molecule of water.

$$H_2N.CH_2.CO\,\boxed{OH + H}\,NH.CH_2.COOH$$
$$\text{glycine}$$

$$\rightarrow H_2N.CH_2CO-NH.CH_2.COOH + H_2O$$
$$\text{glycylglycine}$$

The product, glycylglycine, is described as a *dipeptide* because the molecule consists of two amino acid residues joined together. As shown in the last chapter, the linkage —CO—NH— is known as the 'peptide linkage' (or 'peptide bond').

Since glycylglycine still has an amino group at one end of its molecule and a carboxyl group at the other end, condensation can be continued at either end with further molecules of glycine.

In this way it is possible to build up a long molecular chain, which is composed of amino acid residues and which always has an amino group and a carboxyl group at the two ends. The process need not be restricted to the same amino acid. Long molecular chains can also be made in which different amino acid residues are joined together. In practice the synthesis becomes more difficult as the length of the molecular chain increases, and with the acyl chloride technique a limit is reached when the molecule contains about eighteen residues. Compounds consisting of a number of 2-amino acid residues joined by peptide linkages are called *polypeptides*. Proteins are polypeptides.

PROTEINS

Proteins occur in both plants and animals. In the latter they make up over half of the dry weight of the body. Plants are able to synthesize their own proteins from simple materials like carbon dioxide, water and nitrates but animals cannot do this. They must consume ready-made proteins, which are present in lean meat, eggs, cheese, etc. In the body the proteins are broken down into 2-amino acids, which are then recombined to give the proteins required by the organism. In the latter process some dozens or, possibly, hundreds of molecules condense together to give long polypeptide chains.

Proteins are divided into two main classes—*fibrous* and *globular* proteins. Members of the first class have fibre-like molecules, and serve chiefly as structural materials. They include myosin (in muscle tissue), collagen (in tendons and cartilage), and keratin (in skin, hair and nails). They are insoluble in water, and are chemically inactive. Globular proteins have more or less spherical molecules, are soluble in water, and are chemically reactive. Examples are egg albumen, hæmoglobin,

enzymes, and some hormones (e.g. insulin).

Molecular Structure of Proteins

An aqueous solution of a protein gives the *biuret test* (Chapter 14), which shows the presence of the peptide linkage —CO—NH— in the molecules. When a protein is boiled with hydrochloric acid it breaks down into individual amino acids through hydrolysis of the peptide bonds. This may be illustrated in a simple way by the hydrolysis of the dipeptide glycylglycine, which yields simply glycine.

$$H_2N.CH_2CO{-}NH.CH_2COOH \xrightarrow{+H_2O} 2H_2N.CH_2COOH$$

If hydrolysis of a protein is carried out under milder conditions (e.g. with a suitable enzyme), hydrolysis is only partial, and shorter polypeptide chains of varying length are obtained. (Some proteins, known as 'conjugated' proteins, occur combined with non-protein substances such as sugars or phosphoric(V) acid, and these additional substances are liberated when the proteins are hydrolysed).

From the above evidence one can deduce that protein molecules have the following type of molecular structure:

$$\underbrace{-NH-\overset{\overset{\textstyle X}{|}}{C}H-CO}_{\substack{\text{2-amino acid} \\ \text{residue}}}-\underbrace{NH-\overset{\overset{\textstyle Y}{|}}{C}H-CO}_{\substack{\text{2-amino acid} \\ \text{residue}}}-\underbrace{NH-\overset{\overset{\textstyle Z}{|}}{C}H-CO}_{\substack{\text{2-amino acid} \\ \text{residue}}}-$$

In this formula X, Y, and Z stand for various side-groups which are attached to the main polypeptide chain. It is to be understood that at any part of the chain there may be a succession of amino acid residues containing the same side-group.

An essential preliminary step in finding the molecular structure of a protein is to obtain the latter in a pure state and in sufficient quantity for the investigations required. At present this can be done only with globular proteins, and then only with considerable difficulty. One reason is that the tissue extracts from which proteins are prepared often contain a score or more of these substances. Another is that proteins are extremely sensitive to heat, organic solvents, and most chemical reagents. These bring about a change called *denaturation* in which the proteins become insoluble and lose their biological activity. (An example of this change is the coagulation of egg-white by boiling water.) By careful control of the conditions, however, it is possible to obtain pure globular proteins by selective precipitation followed by recrystallization.

The lengths of the polypeptide chains in globular proteins vary with the nature of the proteins as shown by their relative molecular masses. Special methods have to be used to determine the relative molecular masses because of their high values. One method is to measure the osmotic pressure of an aqueous solution of known concentration in a special form of osmometer. Another depends on measuring the rate of sedimentation of the protein when an aqueous solution is whirled round in an ultracentrifuge. The following are some average values obtained by these methods: insulin—12 000, egg albumen—42 000, hæmoglobin—67 000.

To find which amino acid residues are present in a particular protein the latter is first completely hydrolysed with hydrochloric acid. The amino acids obtained are then separated and identified by *two-way paper-partition chromatography*. This

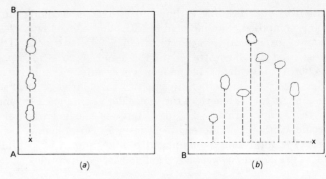

Fig. 16-1. Separation of amino acids by two-way paper-partition chromatography

process is carried out in two stages. In the first stage a drop of the mixture is placed at one corner of a square sheet of filter-paper and allowed to dry. The bottom edge of the paper is immersed in a suitable solvent in a closed vessel, and as the solvent rises up the paper it separates the amino acids partially (Fig. 16-1a). The paper is removed and dried when the solvent front has nearly reached the top of it.

In the second stage the paper is turned at right angles, and the procedure is repeated with a second solvent. Amino acids not separated by the first solvent are separated by the second one. The position reached by the solvent front is marked, and the paper is again dried. Since the amino acids are colourless, their positions cannot be seen until the paper is 'developed.' This is done by spraying with ninhydrin solution, which gives intense colours with amino acids. The positions of the individual acids are shown by small coloured patches of irregular shape (Fig. 16-1b).

The acids are identified by their retention factors, R_f. This is the ratio of the distance moved by the acid in a given time to the distance moved by the solvent front. The various R_f values are compared with those obtained by carrying out a parallel experiment with known amino acids.

When the amino acid residues present in a protein have been identified further investigations are made to establish the following:

(*i*) The number of residues of each acid contained in one molecule of the protein. Some residues may occur only once or twice, while others may occur 20–30 times.

(*ii*) The sequence of the amino acid residues in the polypeptide chain (or chains).

(*iii*) The number of polypeptide chains which make up the molecule. Some protein molecules consist of a single polypeptide chain, others of two or more chains joined together sideways. The insulin molecule contains two polypeptide chains, which have 31 and 20 amino acid residues respectively. The chains are connected at two points by disulphide linkages (—S—S—), as represented diagrammatically in Fig. 16-2.

Fig. 16-2

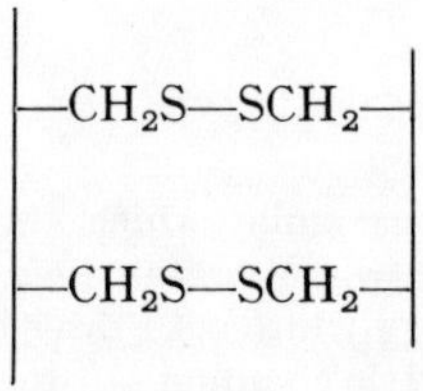

$$|\quad\quad\quad\quad\quad\quad|$$
$$-\mathrm{CH_2S-SCH_2}-$$

$$-\mathrm{CH_2S-SCH_2}-$$
$$|\quad\quad\quad\quad\quad\quad|$$

Space Structure of Protein Molecules

Information about the three-dimensional structure of proteins is derived almost entirely from X-ray analysis of protein crystals. This shows that in globular proteins the polypeptide chains may have two forms. In one form the chain resembles a coil wound round a cylinder in a clockwise direction. This form, which is illustrated in Fig. 16-3a and b, is described as an α-helix. It occurs when the side-groups (X,Y,Z, etc.) are the same or closely similar, so that they can be accommodated in the helix without the atoms interfering with each other. The dimensions of the helix are constant and such that the maximum number of hydrogen bonds can be formed between an —NH— group in one turn and a —CO— group in the next turn. The hydrogen bonds serve to stabilize the helix. When the side-groups differ appreciably, and cannot be incorporated in the helix without strain, the polypeptide chain opens out and follows a zigzag pattern. The proportions of α-helix to open chain vary with the nature of the protein.

In globular proteins the polypeptide chains are folded into a compact volume by twisting of the non-helical portions of the chain (Fig. 16-3c). The shape of the complete molecule is thus roughly spherical. The rigidity of the structures is maintained partly by the hydrogen bonds mentioned above and partly by cross-linkages between some of the side-groups which face each other in parallel portions of the chain. The cross-linkages may consist of disulphide 'bridges' (—S—S—) or they may be ionic bonds formed by interaction of basic —NH_2 groups and acidic —COOH groups.

The manner in which the polypeptide chain is folded in space has great biological significance. For example, a globular protein

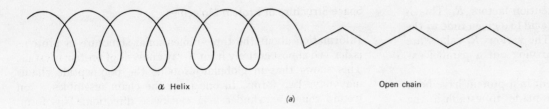

α Helix

Open chain

(a)

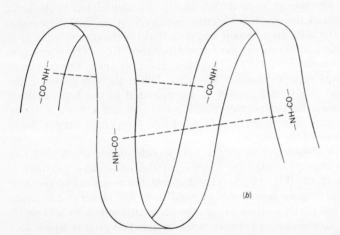

—CO-NH— —CO-NH—

—NH-CO— —NH-CO—

(b)

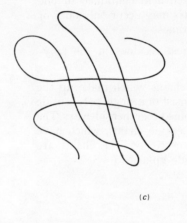

(c)

Fig. 16-3. Space structure of a globular protein molecule. (*a*) α-Helix and open-chain forms of the polypeptide chain; (*b*) part of an α-helix, showing hydrogen bonding; (*c*) schematic representation of folding of the polypeptide chain (actual folding is in three dimensions)

owes its solubility in water to most of the hydrophilic groups being situated on the outside of the structure. The catalytic action of enzymes also depends on the location of particular groups which can react with outside molecules to form intermediate compounds. These groups thus form centres of catalytic activity.

Tests on Proteins

For the general tests use milk (which contains caseinogen) or a solution of white of egg (which contains egg albumen). The solution is prepared by mixing egg-white with three or four times its volume of water and adding a little common salt to facilitate dissolving.

(i) *Denaturing.* Show that coagulation of the protein takes place when milk or egg-white solution is (a) boiled, (b) mixed with ethanol, (c) mixed with dilute sulphuric(VI) acid. The curd obtained by coagulation of milk protein is called *casein.*

(ii) *Lassaigne Test for Nitrogen and Sulphur.* Carry out these tests with a fragment of cheese as described in Chapter 3.

(iii) *Biuret Test.* To 2 cm³ of milk or egg-white solution add an equal volume of dilute sodium(I) hydroxide solution and then two drops of copper(II) sulphate(VI) solution. A violet colour is obtained.

(iv) *Nitric(V) Acid Test.* To 2 cm³ of milk or egg-white solution add three or four drops of concentrated nitric(V) acid. An intense yellow colour is produced. The same reaction occurs if concentrated nitric(V) acid comes into contact with the skin.

EXERCISE 16

Dicarboxylic Acids and Hydroxycarboxylic Acids

1. State which of the following apply to ethanedioic acid only, to methanoic acid only, to both, or to neither:

(a) Its aqueous solution gives a white precipitate with calcium(II) chloride solution.

(b) It decolorizes a warm acidified solution of potassium(I) manganate(VII).

(c) Its aqueous solution gives a white precipitate when boiled with mercury(II) chloride solution.

(d) It evolves carbon monoxide when heated with concentrated sulphuric(VI) acid.

2. Name the organic compounds formed by heating (a) sodium(I) methanoate, (b) ammonium ethanedioate, (c) propanedioic acid, (d) butanedioic acid.

3. Give the chemical formulæ of (a) ethanedioic acid crystals, (b) ethanediamide, (c) ethanedioyl chloride, (d) dimethyl propanedioate.

4. Indicate the stages by which the following conversions can be brought about (using not more than two intermediate stages): (a) ethanol to 2-hydroxypropanoic acid, (b) ethene to 3-hydroxy-propanoic acid.

Amino Acids and Proteins

5. State which of the following are 2-amino acids:

(a) $H_2N.CH_2.COOH$,   (b) $CH_3.CH_2.CONH_2$,
       (c) $CH_3.CH(NH_2).COOH$,
       (d) $HOOC.CH(NH_2).CH_2.COOH$

6. Name the chief organic compounds formed when glycine reacts with (a) copper(II) carbonate, (b) aqueous sodium(I) hydroxide, (c) nitric(III) acid, (d) ethanol (in the presence of a mineral acid).

7. Give the terms used to describe the following:

(a) The existence of two isomers together in equilibrium.

(b) A molecule with a full positive charge and a full negative charge on different parts of the molecule.

(c) A compound formed by condensing together of a number of 2-amino acid molecules.

(d) The class of proteins to which keratin belongs.

(e) The change which occurs when a solution of a protein is heated.

8. Is it true that enzymes (a) are globular proteins, (b) are soluble in ethanol, (c) contain the peptide linkage, (d) are more efficient as catalysts at higher temperatures.

9. Name three types of bonds which may contribute to the rigidity of the structure of a globular protein molecule.

17 Stereoisomerism (Space Isomerism)

Usually the isomerism of two compounds can be explained by the fact that, although the compounds have the same molecular formula, their structural formulæ are different. Some examples of isomerism, however, cannot be explained so simply. The cases which arise fall mainly under two headings: (*i*) *optical isomerism*, in which the chief difference between the isomers is their ability to rotate the plane of polarized light; (*ii*) *geometrical isomerism*, in which the isomers do not possess optical activity (except in special cases). Both types of isomers are called *stereoisomers* because their existence can be explained only by reference to the arrangement of the atoms in space.

Light is regarded as consisting of transverse vibrations (that is, the direction of vibration is at right angles to the direction of propagation). Just as a circle has an infinite number of diameters, so there are an infinite number of planes of vibration. If the light falls on a Nicol prism (a crystal of Iceland spar cut in a special way) only those vibrations in one plane pass through. The light is then said to be 'plane polarized.' The plane of polarization can be found by observing the polarized light through a second Nicol prism which can be rotated (the two prisms constitute the essential parts of a *polarimeter*). Maximum brightness occurs when the axis of the second crystal is aligned with the plane of polarization, and complete

darkness when they are at right angles. If a solution of a *dextro-rotatory* (*d*—) substance is placed between the two prisms, the plane of polarization is rotated in a clockwise (+) direction. If the dissolved substance is *lævo-rotatory* (*l*—), the direction of rotation is anti-clockwise (--).

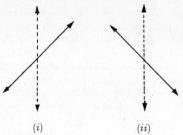

Effect on plane of polarization of (*i*) a dextro-rotatory solution, (*ii*) a lævo-rotatory solution

OPTICAL ISOMERISM

Theory of the Tetrahedral Carbon Atom

Both 2-hydroxypropanoic acid (lactic acid) and 2,3-dihydroxy-butanedioic acid (tartaric acid) exist in forms which are chemically similar, but which differ in their ability to rotate the plane of polarized light. These forms are called *optical isomerides*. The explanation of the phenomenon was first given by van't Hoff and Le Bel (1874).

If the structural formula of 2-hydroxypropanoic acid is examined it will be seen that the molecule contains a carbon atom joined to four different atoms or groups. A carbon atom joined to four different atoms or groups is called an *asymmetric* carbon atom. *All compounds which contain one or more asymmetric carbon atoms in the molecule possess optical isomerides.* Another example is 2-aminopropanoic acid (alanine).

$$CH_3$$
$$H—C—OH$$
$$COOH$$

2-hydroxypropanoic acid
(lactic acid)

$$CH_3$$
$$H—C—NH_2$$
$$COOH$$

2-aminopropanoic acid
(alanine)

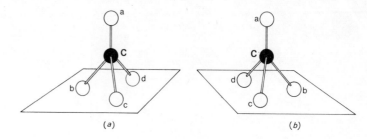

Fig. 17-1. Enantiomorphous forms of a molecule containing an asymmetric carbon atom

In general a molecule containing one asymmetric carbon atom can be represented by the formula $Cabcd$, in which a, b, c, and d stand for the four different atoms or groups attached to the carbon atom. According to van't Hoff and Le Bel a, b, c, and d are to be considered as situated at the corners of a tetrahedron with the carbon atom at the centre. The four valencies of the carbon atom are then geometrically disposed in space so that the directions in which they act form the same angle (109° 28′) with each other. With this arrangement two forms are possible for the molecule, the two being related to each other in the same way as is an object to its image in a mirror. If models of the two forms are constructed with coloured balls they will be found to have the appearance shown in Fig. 17-1.

The two forms of molecule illustrated in Fig. 17-1 are not identical. This can be seen if one imagines oneself to be sitting on each carbon atom in turn and looking down at the three atoms or groups below. The order of these, read in a clockwise direction, is bdc in model A, while in model B the order is bcd. The occurrence

of molecules in two forms with an 'object and image' relationship is described as *enantiomorphism*, and the forms are said to be *enantiomorphs* of each other. There are many examples in everyday life of objects which show a similar relationship—a right shoe is the mirror image of a left shoe, a right-handed glove of a left-handed glove, etc.

The two forms of a compound $Cabcd$, whether in solution or in the solid state, invariably affect the plane of polarized light in opposite directions. If one form is dextro-rotatory the other is lævo-rotatory to the same extent. When the compound is crystalline a crystal of one form is the mirror image of a crystal of the other. Otherwise the physical and chemical properties of the different forms are similar.

Molecules of the types Ca_4, Ca_3b, and Ca_2bc have only one arrangement of the atoms or groups in space no matter how these are distributed. This can be shown by constructing models of these molecules as described previously. Furthermore, with molecules of these types it is always possible to draw a *plane of symmetry* which divides the molecule into enantiomorphously

227

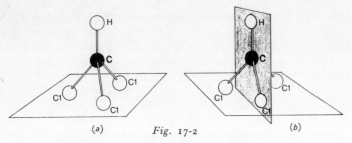

Fig. 17-2

(a) Model of a trichloromethane molecule;
(b) Division of the molecule by a plane of symmetry

related halves. Fig. 17-2a shows the tetrahedral arrangement of the atoms in a molecule of trichloromethane ($CHCl_3$) which belongs to the type Ca_3b. Fig. 17-2b shows how the molecule is divided into 'object and image' halves by a plane of symmetry (shaded) passing through the hydrogen atom, the carbon atom, and one of the chlorine atoms. Clearly there are two other planes of symmetry (passing through the other chlorine atoms respectively) which could be used for the same purpose. No molecule with a plane of symmetry is optically active.

A molecule containing an asymmetric carbon atom has no plane of symmetry. This fact is expressed in the term 'asymmetric,' the Greek prefix *a-* meaning 'not' or 'without.' Although the term is usually applied to the carbon atom, it more fittingly describes the whole molecule, since it is the latter which cannot be divided by a plane into enantiomorphously related halves.

STEREOISOMERISM OF 2-HYDROXYPROPANOIC ACID

2-Hydroxypropanoic acid (lactic acid) is typical of compounds containing one asymmetric carbon atom in the molecule. Three forms of the acid are known. One form is dextro-rotatory, one lævo-rotatory, while the third has no effect on the plane of polarized light. The first two are readily accounted for by the theory of space isomerism. Since the molecule contains an asymmetric carbon atom there should be two enantiomorphously related optical isomerides. If one of these is arbitrarily assumed to be dextro-rotatory (+), the other will be lævo-rotatory (−). The space formulæ of the enantiomorphs are as follows:

(+)2-hydroxypropanoic acid   (−)2-hydroxypropanoic acid

In these formulæ (and similar space formulæ given in this chapter) valency bonds shown by ordinary lines are to be taken as being in the plane of the paper; heavier lines represent bonds directed outward from the carbon atom above the plane of the paper, while dotted lines stand for bonds directed below the plane of the paper.

More convenient methods of representing the space formula of a molecule are often used. One method is to draw a tetrahedron, at the corners of which are placed the atoms or groups attached to the asymmetric carbon atom. The latter is assumed to be at the centre of the tetrahedron and is not included. Thus the two optical isomerides of 2-hydroxypropanoic acid are indicated by the *tetrahedral formulæ* now given.

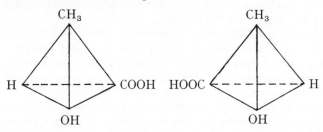

The simplest method, however, of showing the relative configurations of two enantiomorphs—and the one most frequently used—is to project the space formulæ on to one plane, when the corresponding *projection formulæ* are obtained:

$$
\begin{array}{cc}
\text{CH}_3 & \text{CH}_3 \\
| & | \\
\text{H}-\text{C}-\text{COOH} & \text{HOOC}-\text{C}-\text{H} \\
| & | \\
\text{OH} & \text{OH}
\end{array}
$$

Projection formulæ are particularly convenient for the representation of enantiomorphs in the case of more complex molecules. They must not be confused with Kekulé, or structural, formulæ, which they resemble in appearance. Projection formulæ differ from Kekulé formulæ in depicting the relative positions of the atoms or groups joined to the asymmetric carbon atom.

The third form of 2-hydroxypropanoic acid is optically neutral. It consists of equal amounts of the dextro-rotatory and lævo-rotatory forms combined together, so that the optical activity of one form is neutralized by that of the other. Such a modification is known as a *racemic compound*, a *racemate*, or a *dl-compound*

and is said to be *externally compensated*. Originally the name 'racemic acid' (Latin *racemus* = a grape) was given to the externally compensated form of 2,3-dihydroxybutanedioic acid (tartaric acid), the latter being the first example known of this type of compound. Nowadays externally compensated compounds in general are described by the term 'racemic'.

The composition of optically neutral 2-hydroxypropanoic acid can be demonstrated in two ways. Firstly, a solution of the racemic compound can be separated, or 'resolved,' into the two optically active forms as described presently. Secondly, if equal amounts of the *d*- and *l*-forms are dissolved separately in water and the solution, after mixing, is crystallized out only one set of crystals is obtained—those of the *dl*-compound. Crystals formed by a racemic compound are generally different from those of the *d*- and *l*-forms. When a racemic compound is dissolved in water it is largely dissociated again into the optically active isomerides.

(±)2-hydroxypropanoic acid ⇌ (+)2-hydroxypropanoic acid + (−)2-hydroxypropanoic acid

In some cases there is only a slight tendency for the optically active isomerides to combine, and if a mixed solution of the two is evaporated separate crystals are again deposited. The resulting mixture of enantiomorphs is then called a *racemic mixture*.

Preparation and Properties of the Stereoisomers of 2-Hydroxypropanoic Acid

(+)2-Hydroxypropanoic acid is found naturally in muscle tissue. Its common name is sarcolactic acid. It can be prepared from meat extract. 2-Hydroxypropanoic acid obtained from sour milk is the racemic compound. (−)2-Hydroxypropanoic acid does not occur naturally, and can only be prepared from the racemic compound (see later). 2-Hydroxypropanoic acid obtained by laboratory methods (e.g. by synthesis from ethanal) is also the racemic compound. It is a general rule that when a compound containing an asymmetric carbon atom is prepared from an optically inactive substance the racemic compound (or mixture) results. This is to be expected, since if a compound $Cabc_2$ is converted into a compound $Cabcd$ there is an even chance of either of the atoms or groups represented by c being replaced by d. Hence equal amounts of the dextro-rotatory and lævo-rotatory forms are produced.

Whereas the d- and l-forms of 2-hydroxypropanoic acid both melt at 26°C, the racemic compound melts at 18°C. The racemic compound also shows slight differences in chemical properties—chiefly in the number of molecules of water of crystallization occurring in some of the salts. Thus the zinc(II) salt of the racemic compound contains three molecules of water of crystallization, while the zinc(II) salts of the d- and l-forms of the acid contain only two such molecules.

STEREOISOMERISM OF 2,3-DIHYDROXYBUTANEDIOIC ACID

The constitution of 2,3-dihydroxybutanedioic acid (tartaric acid) is represented by the Kekulé formula now given, from which it will be seen that the molecule contains two asymmetric carbon atoms.

$$HO-\underset{\underset{COOH}{|}}{\overset{\overset{H}{|}}{C}}-\underset{\underset{COOH}{|}}{\overset{\overset{H}{|}}{C}}-OH$$

Four different forms of the acid are known. One form is dextro-rotatory, and one lævo-rotatory; the remaining two do not affect the plane of polarization of light.

The two optically active forms are accounted for in the same manner as those of 2-hydroxypropanoic acid; that is, there are two possible arrangements in space of the atoms or groups attached to the asymmetric carbon atoms, so that one arrangement is a mirror image of the other. These arrangements are most easily understood by considering first the atoms or groups in one half of the molecule. The three atoms or groups joined to one of the asymmetric carbon atoms can have two possible configurations, these being related as an object and its mirror image. The configurations can be illustrated by models, as shown in Fig. 17-3.

If the effect of arrangement A on the plane of polarized light is dextrorotatory, that of B will be lævo-rotatory. If to the dextro-rotatory half-molecule represented by A a similar half-molecule is attached, the complete molecule will be dextro-rotatory. The projection formula given below for (+)2,3-dihydroxy-butanedioic acid is thus obtained. In each half of the molecule

Fig. 17-3. Enantiomorphous forms of a half-molecule of 2,3-dihydroxybutanedioic acid

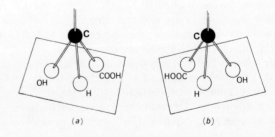

(a)　　　　　　(b)

the order of the atoms or groups, read from the asymmetric carbon atom in a clockwise direction, is H, —OH, —COOH.

$$\text{HOOC—C—OH}$$
$$\text{HO—C—COOH}$$

(+)2,3-dihydroxybutanedioic acid

$$\text{HO—C—COOH}$$
$$\text{HOOC—C—OH}$$

(−)2,3-dihydroxybutanedioic acid

The molecule of the lævo-rotatory form of the acid is derived by combining two half-molecules, each of which has a lævo-rotatory effect on the plane of polarized light. In this case the order of the atoms or groups in each half of the molecule is H, —COOH, —OH, when read from the asymmetric carbon atom in a clockwise direction. If the formulæ of the d- and l-forms of the acid are compared, it will be seen that the two molecules are enantiomorphs.

A third possibility is to combine a dextro-rotatory half-molecule with a lævo-rotatory half-molecule. This gives the form of the acid known as (meso)2,3-dihydroxybutanedioic acid.

$$\text{HO—C—COOH}$$
$$\text{HO—C—COOH}$$

(meso)2,3-dihydroxybutanedioic acid

The optical effects of the two half-molecules in the meso-form are equal and opposite, so that the molecule as a whole is optically neutral. Since the neutralization of the optical effects takes place within the molecule, the meso-form is said to be internally compensated. There is no method by which internally compensated compounds can be separated into optically active isomerides.

The fourth form of 2,3-dihydroxybutanedioic acid is also optically neutral. Its existence is explained by external compensation. It consists of equal amounts of the dextro-rotatory and lævo-rotatory isomerides, and, as only one kind of crystals is deposited when a solution is crystallized, it is a racemic compound. In fact it is commonly called racemic acid, although systematically it is designated as (±)2,3-dihydroxybutanedioic acid. It can be separated into the d- and l-forms by the methods shortly to be described.

The molecules of the d- and l-forms of the acid do not possess a plane of symmetry, that is, a plane which will divide the molecule into 'object and image' halves. They therefore agree with the rule that no optically active molecule has a plane of symmetry.

Preparation and Properties of the Stereoisomers of 2,3-Dihydroxy-butanedioic Acid

The common form of this acid, prepared from argol as described in Chapter 16, is the d-form. If it is heated under pressure with a little water in a sealed tube, it undergoes an intra-molecular change and yields a mixture of the meso-form and racemic form. These can be separated by means of the difference in their solubilities. (−)2,3-Dihydroxybutanedioic acid can only be prepared by separation of the d- and l-forms present in the racemic form (see next section).

231

The two optically active forms of the acid closely resemble each other in properties. Both have the same density, melting point, solubility, and dissociation constants. Apart from the 'object and image' relationship of their crystals and their opposite effects on the plane of polarized light the only differences between them are of a biological character. Thus the *d*-form is more readily attacked and decomposed by certain kinds of yeast and moulds. The *meso*-form differs from the optically active isomers in density, melting point, solubility, and crystalline form; moreover, the crystals deposited from solution contain water of crystallization. Similar points of difference are found in the case of the racemic compound. The four forms of 2,3-dihydroxybutanedioic acid are briefly compared in Table 17-1, the solubilities given being at 20°C.

Table 17-1. OPTICAL ISOMERIDES OF 2,3-DIHYDROXY-BUTANEDIOIC ACID

Form of acid	Formula	Melting point/°C	Optical effect	Solubility/ mol kg^{-1}
d-Form	$C_4H_6O_6$	170	(+)	9·27
l-Form	$C_4H_6O_6$	170	(−)	9·27
Meso-form	$C_4H_6O_6 . H_2O$	140 (anhydrous)	Neutral	7·44
Racemic form	$(C_4H_6O_6)_2 . 2H_2O$	206 (anhydrous)	Neutral	0·62

Resolution of Racemic Compounds and Mixtures

By suitable treatment it is possible to separate, or *resolve*, racemic modifications (that is externally compensated compounds or mixtures) into their dextro-rotatory and lævo-rotatory constituents. Several methods are available, but the three now described have been chiefly used. All of these were devised by Pasteur.

(*i*) *By Picking out Crystals*. In the case of a racemic *mixture* two kinds of crystals are formed, and these have the 'object and image' relationship. If the crystals are sufficiently large and well-defined they can be separated by picking out the individual crystals with a pair of tweezers. The method requires monumental patience, because the crystals deposited are usually small and difficult to distinguish even with the aid of a lens. Pasteur used the method, however, to resolve the sodium(I) ammonium salt of the racemic form of 2,3-dihydroxybutanedioic acid into its enantiomorphous forms. The method cannot be applied directly to the acid, because, as stated previously, the latter crystallizes from solution as a racemic compound. Pasteur, therefore, converted the acid into its sodium(I) ammonium salt, and crystallized out the latter. He then picked out the two kinds of crystals from the mixture, redissolved them separately, and demonstrated that the two solutions were dextro-rotatory and lævo-rotatory respectively. Furthermore, by treating the separate crystals with dilute sulphuric(VI) acid he was able to prepare the *d*- and *l*-forms of the organic acid. This method of resolution has only a limited application.

(*ii*) *By Selective Fermentation*. It has already been stated that certain kinds of yeast and moulds attack the *d*-form of 2,3-dihydroxybutanedioic acid in preference to the *l*-form. Thus, if the mould *Penicillium glaucum* is allowed to act on a solution of the ammonium salt of the racemic form of the acid, the dextro-rotatory constituent is fermented and decomposed, while the lævo-rotatory constituent is largely unaffected. The latter can be precipitated by concentrating the remaining solution and adding ethanol. The *l*-form of the acid can then be obtained from the ammonium salt by treatment with dilute sulphuric(VI) acid. Selective fermentation can be used in relatively few cases.

(*iii*) *Chemical Separation*. Racemic modifications are usually resolved nowadays by chemical methods such as the one now described. If a solution of the racemic form of 2,3-dihydroxy-butanedioic acid is treated with an optically active organic base, two salts are formed. One is derived from combination of the base with the *d*-form of the acid, the other from combination

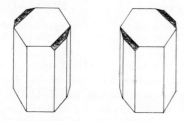

Fig. 17-4. Enantiomorphous forms of crystals of sodium(I) ammonium 2,3-dihydroxybutanedioate (distinctive facets are shaded)

of the base with the *l*-form. A suitable base is (+)cinchonine, an alkaloid found naturally along with quinine in Peruvian bark. With the racemic form of the acid this base yields the salts (+)cinchonine (+)2,3-dihydroxybutanedioate and (+)cinchonine (−)2,3-dihydroxybutanedioate. The two salts differ in solubility, and can be separated by fractional crystallization They are

then converted into the respective acids with a dilute mineral acid.

GEOMETRICAL ISOMERISM

It was seen in Chapter 5 that two configurations of a molecule are equivalent if they differ only as a result of rotation about a single bond. If this rule did not hold, there would be an infinite number of forms of, say, 1,2-dichloroethane (CH_2Cl—CH_2Cl), each differing in the relative positions in space of the chlorine and hydrogen atoms in the two halves of the molecule. In fact, only one form of this compound exists.

The situation is different in substituted ethene compounds of the type $CHX{=}CHX$. A simple compound of this type is 1,2-dichloroethene ($CHCl{=}CHCl$). Since both carbon atoms of the molecule are attached to only three other atoms, all six atoms are in the same plane. As noted in Chapter 6, the $C{=}C$ bond consists of a σ (sigma) bond and a π bond, the electrons of the latter occupying a banana-shaped orbital above, and below, the plane of the molecule. Rotation of the two halves of the molecule in opposite directions about the double bond involves rupture of the π bond, which requires a considerable amount of energy. Thus 'free rotation' cannot take place about the double bond.

In the above circumstances two isomers are possible because the chlorine atoms can be situated either on the same side of the double bond or on opposite sides. The two isomers, which are known as the *cis*-form and *trans*-form respectively of

1,2-dichloroethene, have the following structures:

$$
\begin{array}{cc}
H & Cl \\
\diagdown\!\diagup & \\
C & \\
\| & \\
C & \\
\diagup\!\diagdown & \\
H & Cl \\
\end{array}
\qquad
\begin{array}{cc}
H & Cl \\
\diagdown\!\diagup & \\
C & \\
\| & \\
C & \\
\diagup\!\diagdown & \\
Cl & H \\
\end{array}
$$

cis-form *trans*-form

The above type of isomerism is called *geometrical isomerism*. A simpler method of representing the geometrical isomers of 1,2-dichloroethene is by means of the projection formulæ now shown.

$$
\begin{array}{c}
H—C—Cl \\
\| \\
H—C—Cl \\
\end{array}
\qquad
\begin{array}{c}
H—C—Cl \\
\| \\
Cl—C—H \\
\end{array}
$$

cis-form *trans*-form

Both isomers of 1,2-dichloroethene are produced as colourless liquids when ethyne combines with a limited amount of chlorine under the conditions described in Chapter 7. Their chemical properties are similar, but they differ in physical properties. Thus the *cis*-form boils at 60°C, the *trans*-form at 48°C.

It should be noted that there is a third form of dichloroethene. This is 1,1-dichloroethene (CH_2=CCl_2). The latter, however, cannot form geometrical isomers because it does not fulfil the condition that each carbon atom joined by the double bond must be attached to two different atoms or groups.

234

Geometrical Isomers of Butenedioic Acid

This acid, which has the structural formula HOOC—CH=CH—COOH, also exists in *cis*- and *trans*-forms. In the first the two carboxyl groups are on the same side of the double bond, while in the second they are on opposite sides. In this case the geometrical isomers differ chemically as well as physically, and this led to the use of different trivial names for them. The *cis*-form of butenedioic acid is commonly called maleic acid, and the *trans*-form fumaric acid.

$$
\begin{array}{c}
H—C—COOH \\
\| \\
H—C—COOH \\
\end{array}
\qquad
\begin{array}{c}
H—C—COOH \\
\| \\
HOOC—C—H \\
\end{array}
$$

cis-butenedioic acid *trans*-butenedioic acid
(maleic acid) (fumaric acid)

Both forms of the acid are crystalline solids. As usually happens with geometrical isomers, the *cis*-form has the lower melting point, but a higher solubility in water. The *cis*-form melts at 130°C, and dissolves readily in water. The *trans*-form melts at 287°C, and is only slightly soluble in water. An important chemical difference is that *cis*-butenedioic acid is readily converted into its anhydride by heating at 100°C under reduced pressure.

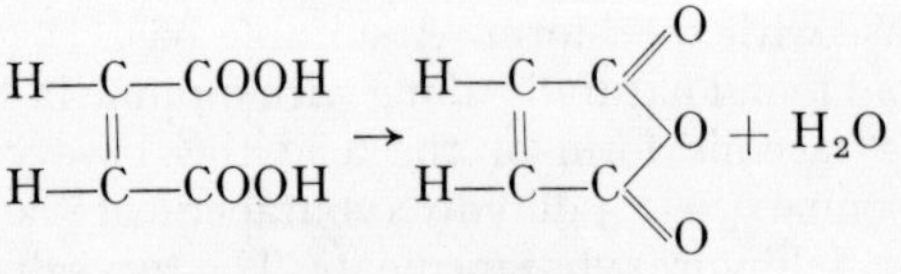

$$
\begin{array}{c}
H—C—COOH \\
\| \\
H—C—COOH \\
\end{array}
\rightarrow
\begin{array}{c}
H—C—C \\
\| \\
H—C—C \\
\end{array}
\!\!\!\diagdown\!\!O + H_2O
$$

cis-butenedioic anhydride
(maleic anhydride)

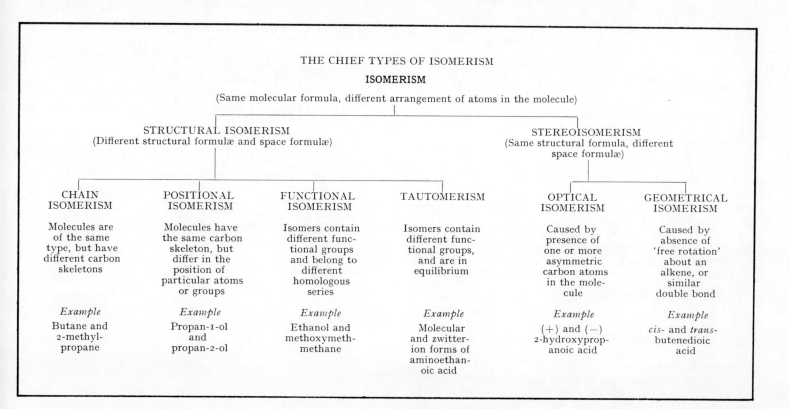

THE CHIEF TYPES OF ISOMERISM

ISOMERISM

(Same molecular formula, different arrangement of atoms in the molecule)

STRUCTURAL ISOMERISM
(Different structural formulæ and space formulæ)

STEREOISOMERISM
(Same structural formula, different space formulæ)

CHAIN ISOMERISM	POSITIONAL ISOMERISM	FUNCTIONAL ISOMERISM	TAUTOMERISM	OPTICAL ISOMERISM	GEOMETRICAL ISOMERISM
Molecules are of the same type, but have different carbon skeletons	Molecules have the same carbon skeleton, but differ in the position of particular atoms or groups	Isomers contain different functional groups and belong to different homologous series	Isomers contain different functional groups, and are in equilibrium	Caused by presence of one or more asymmetric carbon atoms in the molecule	Caused by absence of 'free rotation' about an alkene, or similar double bond
Example Butane and 2-methylpropane	*Example* Propan-1-ol and propan-2-ol	*Example* Ethanol and methoxymethane	*Example* Molecular and zwitter-ion forms of aminoethanoic acid	*Example* (+) and (−) 2-hydroxypropanoic acid	*Example* *cis*- and *trans*-butenedioic acid

The *trans*-form does not form a corresponding anhydride, although under drastic conditions (heating with phosphorus(V) oxide) it can be made to yield the same anhydride as the *cis*-form. The readiness with which the latter yields an anhydride is strong evidence that this form of the acid has the *cis*-configuration. The correctness of the conclusion has been confirmed by X-ray analysis of the two acids.

Optical Isomerism

1. State which of the following compounds have optical isomerides:

(a) $CH_3.CHBr.COOH$.

(b) $(CH_3)_2CH.CH_2OH$.

(c) $(CH_3)(C_2H_5)CH.CH_2Cl$.

(d) $HOOC.CH(OH).CH_2.COOH$.

2. (a) Name the stereoisomers of 2-aminopropanoic acid; (b) write projection formulæ to show the relationship between the forms which are optically active.

3. Give the terms used to describe the following: (a) a carbon atom attached to four different atoms or groups, (b) light in which the vibrations are in one plane only, (c) two molecules which have an 'object and image' relationship, (d) a compound consisting of equal amounts of d- and l-forms.

4. Which of the following statements are true?

(a) The chloroethanoic molecule ($CH_2Cl.COOH$) has a plane of symmetry.

(b) ($\pm$) 2-Hydroxypropanoic acid is externally compensated.

(c) 2-Aminopropanoic acid (alanine) synthesized from ethanal is optically neutral.

(d) The chemical properties of optical isomerides are identical.

5. (a) How many asymmetric carbon atoms are present in the molecule $HOOC.CHBr.CHBr.COOH$? (b) Write projection formulæ to show the relationship between the forms which are optically active.

6. Specify the optically neutral forms of the compound represented in question 5, and state whether they are internally or externally compensated.

Geometrical Isomerism

7. State which of the following compounds you would expect to have geometrical isomerides:

(a) $CH_3.CH{=}CH.CH_3$.

(b) $CH_3.CH{=}CH.COOH$.

(c) $CH_3.CH_2.CH{=}CH_2$.

(d) $CH_3.CBr{=}CBr.CH_3$.

(e) $(CH_3)_2C{=}CH.CH_3$.

8. Which of the following statements are true of the cis- and trans-forms of butenedioic acid?

(a) They have the same structural formula.

(b) All the carbon atoms are in the same plane.

(c) They melt at the same temperature.

(d) The cis-form readily gives an anhydride.

(e) They are both dextro-rotatory.

18   Carbohydrates

Carbohydrates *are a class of compounds containing carbon, hydrogen, and oxygen, the hydrogen and oxygen being present in the same proportion as in water.* The general formula of carbohydrates can be written in the form $C_x(H_2O)_y$. The composition of these compounds is indicated by the name 'carbohydrate,' which is an abbreviation of 'carbon hydrate.' It must be emphasized, however, that the name merely summarizes the chemical composition, and, in fact, carbohydrates are not hydrates of the element carbon.

Carbohydrates include many substances of great biological and industrial importance. They are divided into the following groups according to the number of carbon atoms in the molecule:

(*i*) *Monosaccharides.* These are carbohydrates with six, or less than six, carbon atoms in the molecule. Examples are glucose and fructose, both of which have the molecular formula $C_6H_{12}O_6$.

(*ii*) *Disaccharides.* Disaccharides always contain twelve carbon atoms in the molecule and have the molecular formula $C_{12}H_{22}O_{11}$. They include sucrose (cane-sugar) and maltose (malt-sugar).

(*iii*) *Polysaccharides.* Polysaccharides have the formula $(C_6H_{10}O_5)_n$, in which n is a number varying with the particular compound. In all cases the value of n can be found only approximately. It may be as small as 25 or as large as 10 000. Starch and cellulose are the best known polysaccharides.

Monosaccharides and disaccharides are collectively called *sugars.* They have a sweet taste, dissolve readily in water and can be obtained from solution in crystalline form. The chemical names of sugars end in *-ose* (e.g. sucrose—cane-sugar, and maltose—malt-sugar). Polysaccharides are tasteless, and they are usually insoluble in water or form colloidal solutions with water. Monosaccharide molecules are the chemical units from which the molecules of disaccharides and polysaccharides are built up. This is expressed in the names of the different classes of carbohydrates (monosaccharides—one unit; disaccharides—two units; polysaccharides—many units).

MONOSACCHARIDES

The most important monosaccharides possess six carbon atoms in the molecule, but in some cases the number of carbon atom is less than six. An example is ribose $C_5H_{10}O_5$), which occurs in a combined form in ribonucleic acid (RNA), an important constituent of living cells. Monosaccharides are further classified according to the number of carbon atoms in the molecule. Ribose is described as a *pentose* sugar, while glucose and fructose are called *hexose* sugars. Another method of classification depends on whether an aldehyde group (—CHO) or a ketone group (—CO—) is present in the molecule. Thus glucose, which contains the aldehyde group, is called an *aldose,* and fructose, which contains the ketone group, is described as a *ketose.* The two systems are sometimes combined, glucose being classified as an *aldohexose* and fructose as a *ketohexose.*

GLUCOSE, $C_6H_{12}O_6$

Melting point: 146°C   *Boiling point:* (decomposes)

Glucose is also known as grape-sugar and dextrose. The name

'glucose', like the names of certain other sweet-tasting compounds, is derived from the Greek word *glukus*, meaning 'sweet'. The presence of glucose in ripe grapes accounts for the name 'grape-sugar,' while 'dextrose' refers to the fact that the naturally occurring compound is dextro-rotatory. Glucose is found in other fruits besides grapes, notably in apples, pears, and figs. It is also present in table syrup and honey, the latter containing 80 per cent of glucose and fructose in about equal proportions. The blood sugar of animals consists of glucose.

Glucose is manufactured by heating starch with very dilute sulphuric(VI) acid under pressure. As explained later, starch molecules are made up of glucose 'units' formed by glucose molecules condensing together. Hot mineral acids bring about hydrolysis of the starch as shown.

$$(C_6H_{10}O_5)_n + nH_2O \rightarrow nC_6H_{12}O_6$$

The chief application of glucose is as a sweetening agent in confectionery and soft drinks. It is not as sweet as ordinary sugar, but is more easily digested. In hospitals a dilute solution is used to 'drip-feed' patients who cannot be fed by mouth, the solution being run directly into the veins. Glucose is also used as a weak reducing agent.

Properties of Glucose

At ordinary temperatures glucose crystallizes from aqueous solution as colourless needle-shaped crystals, which contain one molecule of water of crystallization. These crystals melt at 86°C. Anhydrous glucose, obtained by crystallization above 36°C, melts at 146°C. Glucose is very soluble in water, but only sparingly soluble in ethanol.

In many of its chemical reactions glucose behaves either as a pentahydric alcohol or as an aldehyde. These reactions are in accordance with the Kekulé formula

The chief way in which glucose acts as an alcohol is in the formation of esters. Thus, when glucose is refluxed with ethanoic, or acetic, anhydride in the presence of anhydrous zinc(II) chloride as a catalyst, all five hydroxyl groups in the molecule undergo ethanoylation, or acetylation, and glucose penta-ethanoate, or penta-acetate, is formed. Glucose also forms esters with nitric(V) acid and phosphoric(V) acid under suitable conditions.

Some of the reactions of the simpler aldehydes are not given by glucose; for example, glucose does not produce a violet colour with Schiff's reagent, nor are addition compounds formed with ammonia and sodium(I) hydrogensulphate(IV) (sodium bisulphite). Glucose, however, yields an addition compound with hydrogen cyanide.

As in the case of simple aldehydes, the —CHO group can be oxidized to a —COOH group and reduced to a —CH$_2$OH group. Mild oxidizing agents such as Fehling's solution and Tollens's reagent convert glucose to *gluconic acid*.

$$CH_2OH.(CHOH)_4.CHO + O \rightarrow CH_2OH.(CHOH)_4.COOH$$

gluconic acid

Fehling's solution is used for the detection of glucose. Another reagent used for the same purpose is *Benedict's solution*. This is a greenish-blue solution containing copper(II) sulphate(VI), sodium(I) carbonate, and sodium(I) citrate. Like Fehling's solution, it gives a yellow precipitate (which turns red) of copper(I) oxide when boiled with glucose or other reducing sugars.

Glucose is reduced by hydrogen and a nickel catalyst to *sorbitol*, a hexahydric alcohol.

$$CH_2OH.(CHOH)_4.CHO + H_2$$
$$\rightarrow CH_2OH.(CHOH)_4.CH_2OH$$
sorbitol

Reduction with hydriodic acid, a strong reducing agent, converts glucose to hexane.

Glucose has two reactions with phenylhydrazine. In the cold and with an equimolecular proportion of phenylhydrazine it yields a phenylhydrazone in the same way as ethanal.

$$CH_2OH.(CHOH)_4.CHO + H_2N.NH.C_6H_5 \rightarrow$$
$$CH_2OH.(CHOH)_4.CH:N.NH.C_6H_5 + H_2O$$

The phenylhydrazone of glucose, like those formed by other reducing sugars, is of little use for identification purposes, because it is very soluble in water. If glucose is heated with excess of phenylhydrazine, however, a further reaction occurs. The secondary alcohol group adjacent to the aldehyde group is first oxidized to a carbonyl group with the formation of ammonia and phenylamine. The carbonyl group then undergoes condensation with another molecule of phenylhydrazine in the manner typical of ketones. Omitting that part of the glucose molecule which takes no part in the changes, the further reactions are as follows:

$$\begin{array}{l} | \\ CHOH \\ | \\ CH:N.NH.C_6H_5 \end{array} + H_2N.NH.C_6H_5 \rightarrow \begin{array}{l} | \\ CO \\ | \\ CH:N.NH.C_6H_5 \\ \quad + NH_3 + C_6H_5NH_2 \end{array}$$

$$\begin{array}{l} | \\ CO \\ | \\ CH:N.NH.C_6H_5 \end{array} + H_2N.NH.C_6H_5 \rightarrow \begin{array}{l} | \\ C:N.NH.C_6H_5 \\ | \\ CH:N.NH.C_6H_5 + H_2O \end{array}$$

The product, *glucosazone*, is obtained as a yellow precipitate. Many other sugars yield osazones with phenylhydrazine, and these provide a valuable means of identifying the parent sugars. The sugars themselves are often difficult to crystallize, whereas the osazones are sparingly soluble and crystalline. Furthermore, the different osazones crystallize in distinctive forms, which can be readily recognized when the crystals are examined under a microscope. The preparation of glucosazone is described below.

An aqueous solution of glucose undergoes fermentation by yeast with the production of ethanol and carbon dioxide (Chapter 8).

Chemical Tests for Glucose

The following tests are given not only by glucose but by carbohydrates in general:

(i) *Decomposition by Heat*. Heat a little glucose in a dry tube. The glucose melts and turns brown, giving off an odour of burnt sugar. Moisture is deposited in the cooler part of the tube. Finally a black residue of carbon is left in the tube.

(ii) Concentrated Sulphuric(VI) Acid. Warm about 0·5 g of glucose gently with a few drops of concentrated sulphuric(VI) acid. The glucose is dehydrated and a puffy black mass of carbon is formed.

(iii) Molisch's Test. To 5 cm³ of a dilute solution of glucose add two drops of a 20 per cent solution of naphthalen-1-ol (α-naphthol) in ethanol. Then slope the tube and carefully pour concentrated sulphuric(VI) acid down the side so as to obtain a lower layer of the acid. A violet colour is formed at the junction of the two liquids.

The following tests are given by glucose and fructose, but not by sucrose or starch.

(iv) Sodium(I) Hydroxide Solution. Add about 0·5 g of glucose to 5 cm³ of dilute sodium(I) hydroxide solution, and boil the liquid. The latter turns first yellow and then brown, and an odour of caramel is evolved.

(v) Fehling's Solution and Tollens's Reagent. The reactions with these reagents, which give a yellow precipitate of copper(I) oxide and a silver mirror respectively, are carried out as described for ethanal (Chapter 10).

(vi) Osazone Formation. Dissolve 2 g of glucose in 10 cm³ of water, and add a solution of 4 cm³ of phenylhydrazine in 4 cm³ of glacial ethanoic acid. Heat the mixture in a beaker of boiling water. A yellow precipitate of glucosazone is formed in about 15 minutes.

Note. Phenylhydrazine hydrochloride can be used instead of phenylhydrazine itself, but in this case sodium(I) ethanoate (about 5 g) must be added to diminish the acidity. If the solution is too acidic the osazone undergoes hydrolysis.

Structure of Glucose

X-ray analysis does not support the open-chain structure which was used in describing the reactions of glucose. It shows instead that in the anhydrous crystals the glucose molecule has a ring structure. Moreover, if glucose is precipitated by adding glacial ethanoic acid to a boiling aqueous solution, a second form of crystals can be obtained. This form (called β-glucose) also has a

ring structure which differs slightly, however, from that of the first form (α-glucose). There are also small differences in properties (e.g. melting point).

The molecules of both α-glucose and β-glucose contain a ring of five carbon atoms and one oxygen atom. The structures are represented in I and II, below by means of 'Haworth perspective formulæ' (named after W. N. Haworth, an expert in the field of sugar chemistry). This type of formula was devised to show the configuration of sugar molecules in the clearest possible manner. The carbon atoms in the ring are omitted. The three heavier lines in the ring stand for bonds above the plane of the paper, and the other three lines for bonds below this plane. By convention the oxygen atom is placed in the position shown, and the carbon atoms in the ring are specified by numbering them in a clockwise direction, starting from the one after the oxygen atom. The formula is not a picture of the molecule in space. The ring is not flat, but puckered, the valency bonds forming the usual tetrahedral angle with each other.

The only difference between the structures of α-glucose and β-glucose is in the relative positions of the H atom and —OH group attached to carbon-1. In α-glucose the hydrogen atom is above the ring and the hydroxyl group below, while in β-glucose their positions are reversed. The difference appears to be trivial, but it is very important from the biochemical point of view.

When either α-glucose or β-glucose is dissolved in water the ring structure opens out, and a small amount of the open-chain structure III results. After a time an equilibrium is established, in which all three forms are present. An aqueous solution of glucose thus provides an example of 'tautomerism.' At ordinary temperatures the equilibrium mixture contains about 38 per cent of α-glucose, 62 per cent of β-glucose, and less than 1 per cent of the open-chain compound. Nevertheless, in aqueous solution glucose reacts chiefly in accordance with structure III, the only structure which contains the aldehyde group. α-Glucose and β-glucose can only change into each other through intermediate formation in solution of the open-chain structure.

FRUCTOSE, $C_6H_{12}O_6$

Melting point: 95°C   *Boiling point:* (decomposes)

Fructose is also known as fruit-sugar and lævulose. It is found in many fruits, usually in association with glucose. The name 'lævulose' signifies that the naturally occurring compound is lævo-rotatory.

Fructose is obtained by hydrolysing inulin (a polysaccharide present in the tubers of the dahlia plant) with boiling dilute sulphuric(VI) acid.

$$(C_6H_{10}O_5)_n + nH_2O \rightarrow nC_6H_{12}O_6$$

The reactions of fructose closely resemble those of its isomer glucose, and agree generally with the following open-chain structure:

$$CH_2OH-CHOH-CHOH-CHOH-\underset{\underset{O}{\|}}{C}-CH_2OH$$

Fructose shows the reactions of a pentahydric alcohol and a ketone. Like glucose it forms a penta-ethanoate with ethanoic anhydride, and yields an addition compound with hydrogen cyanide. It also reduces Fehling's solution and Tollens's reagent. This may appear surprising in view of the fact that simple ketones (e.g. propanone) are not reducing agents. The ability to act as reducing agents, however, is a particular feature of 2-hydroxy-ketones such as fructose.

Fructose, like glucose, reacts with phenylhydrazine to form both a phenylhydrazone and an osazone. The former is obtained by treating fructose in the cold with an equimolecular proportion of phenylhydrazine. With excess of the latter at 100°C fructose yields the same osazone as glucose.

Fructose also resembles glucose in having two cyclic structures (α-fructose and β-fructose), in which there are hexagonal rings composed of five carbon atoms and one oxygen atom. Again the aqueous solution contains the two cyclic forms in equilibrium with a small proportion of the open-chain form. In the solid state, however, fructose exists in only one form, β-fructose.

A third cyclic structure (γ-fructose) is found in the fructose units which enter into the composition of sucrose and inulin. This form contains a five-membered ring of four carbon atoms and one oxygen atom, as shown by the Haworth perspective formula below.

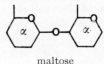

γ-fructose

γ-Fructose is too unstable to exist in the free state, although it may be present in trace amounts in an aqueous solution of fructose. When sucrose (cane-sugar) is hydrolysed to glucose and fructose the latter appears, not as the γ-form, but as a mixture of the α and β forms.

DISACCHARIDES

A disaccharide molecule is derived by condensation of two monosaccharide molecules, which may be similar or different. Condensation takes place by an —OH group of one molecule reacting with an —OH group of the second molecule, a molecule of water being eliminated. The monosaccharide residues or units are thus joined together by an oxygen atom. To show the nature of the monosaccharide units concisely use is sometimes made of special symbols. These indicate the type of ring structure present and the situation of any carbon atoms which may be attached to the ring. Examples are given below.

α-glucose   β-glucose   γ-fructose

Only three disaccharides occur as such in Nature. These are maltose (malt sugar), sucrose (cane-sugar), and lactose (milk sugar).

MALTOSE, $C_{12}H_{22}O_{11}$

Maltose is isomeric with cane-sugar. It is produced when the hydrolysing enzyme diastase, which occurs in malt, acts upon starch.

$$2(C_6H_{10}O_5)_n + nH_2O \rightarrow nC_{12}H_{22}O_{11}$$

starch   maltose

A molecule of maltose consists of two α-glucose units joined together in 'head to tail' fashion, as now shown:

maltose

When maltose is boiled with dilute sulphuric(VI) acid hydrolysis takes place through addition of a water molecule at the oxygen linkage. The latter is broken, and two molecules of glucose are formed.

$$C_{12}H_{22}O_{11} + H_2O \rightarrow 2C_6H_{12}O_6$$

Hydrolysis occurs similarly with the yeast enzyme maltase, and by further action of the enzyme zymase the glucose is converted into ethanol. This is utilized in making beer from malt.

Many of the properties of maltose are similar to those of glucose. An aqueous solution of maltose is dextro-rotatory, as might be expected from its composition. Maltose is a reducing sugar, and reduces both Fehling's solution and Tollens's reagent. It can be distinguished from glucose and other sugars by the characteristic form of its osazone crystals.

SUCROSE, $C_{12}H_{22}O_{11}$

Melting point: 160°C   *Boiling point:* (decomposes)

Sucrose, commonly known as 'sugar,' occurs in most forms of plant life. It is responsible for the sweet taste of many fruits, including oranges, bananas, dates, peaches, and strawberries. The sugar-cane and sugar-beet are outstanding for their ability to store sucrose, and these form the chief sources of the commercial product. The sugar is first partially purified in factories near the sugar plantations, and refining of the 'raw' sugar is then completed in the consuming countries. An important by-product is *molasses*, the brownish-black mother-liquor left after the crystallization of raw sugar. The liquor is used in various fermentation processes, including the manufacture of penicillin.

Structure and Properties of Sucrose

Hydrolysis of sucrose with a boiling mineral acid shows that its molecule is composed of one glucose unit and one fructose unit. Further evidence shows that these have the α-form and γ-form respectively.

sucrose

Sucrose forms monoclinic crystals, which are very soluble in water, but only slightly soluble in ethanol. The high solubility in water is due to the presence of eight hydroxyl groups in the molecule (four in each monosaccharide unit).

From the commercial standpoint the most important property of sucrose is its sweet taste. Many attempts have been made to relate this property to some feature of molecular architecture, but without success. Chemists have prepared a number of compounds which are many times sweeter than cane-sugar. The best known of these is saccharin (Chapter 22), which is 675 times sweeter than sucrose.

When sucrose is heated it melts at about 160°C with slight decomposition. At 190°–200°C the melted sugar loses water and changes to a brown amorphous solid called *caramel*. This is used for colouring confectionery, gravy, and rum. At still higher temperatures decomposition occurs, and a black mass of carbon (sugar charcoal) remains.

The chemical reactions of sucrose differ in many ways from those of glucose and fructose. This is explained by the inability of the two monosaccharide rings to open out in aqueous solution to give open-chain structures with an aldehyde group and a ketone group. Thus sucrose is not a reducing sugar and does not react with Fehling's solution or Tollens's reagent. It does not form an addition compound with hydrogen cyanide, nor does it yield a phenylhydrazone or an osazone with phenylhydrazine. When sucrose is refluxed with ethanoic anhydride and anhydrous zinc(II) chloride (catalyst) it forms an octa-ethanoate.

If sucrose is boiled with dilute sulphuric(VI) acid or dilute hydrochloric acid it is hydrolysed to glucose and fructose in equimolecular proportions.

$$C_{12}H_{22}O_{11} + H_2O \rightarrow C_6H_{12}O_6 + C_6H_{12}O_6$$
$$\text{glucose} \qquad \text{fructose}$$

Fructose is more strongly lævo-rotatory than glucose is dextro-rotatory. The solution obtained by hydrolysis is therefore lævo-rotatory, whereas the original solution of cane-sugar is dextro-rotatory. For this reason the hydrolysis is often described as the *inversion* of cane-sugar, and the resulting mixture of glucose and fructose as *invert sugar*. The inversion can also be brought about by the enzyme invertase, as noted previously in connection with the fermentation of sucrose by yeast (Chapter 8).

Like other sugars, sucrose undergoes dehydration when warmed with concentrated sulphuric(VI) acid. It leaves a black porous mass of carbon.

$$C_{12}H_{22}O_{11} - 11H_2O \rightarrow 12C$$

Chemical Tests for Sucrose

Sucrose gives the three general tests for carbohydrates described previously for glucose. It differs from glucose and fructose in not forming an osazone with phenylhydrazine; also, its aqueous solution does not turn brown when it is heated with sodium(I) hydroxide solution.

The usual method of recognizing sucrose is to show that its aqueous solution will reduce Fehling's solution after, but not before, it has been boiled with dilute sulphuric(VI) acid. After boiling, the solution contains the reducing sugars glucose and fructose. Before the liquid is tested with the reagent it should be made slightly alkaline with sodium(I) hydroxide solution.

POLYSACCHARIDES

The most important polysaccharides are starch, dextrin, glycogen, and cellulose. All of these have the molecular formula $(C_6H_{10}O_5)_n$, and are composed of chains of glucose units condensed together. In the first three compounds the units consist of α-glucose, while in cellulose they are of β-glucose. The basic structures can be represented as follows:

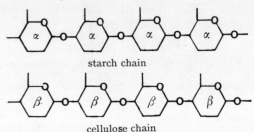

starch chain

cellulose chain

The lengths of the molecular chains vary not only for different polysaccharides, but also for the same one. If the relative molecular mass is determined by measuring osmotic pressures of solutions, different values are obtained for n in the formula $(C_6H_{10}O_5)_n$ according to the source of the compound. The value may be several hundred or many thousands. In some polysaccharides branching of the molecular chains occurs.

STARCH, $(C_6H_{10}O_5)_n$

The starch of cereals and potatoes forms one of the chief foodstuffs of the human race. In plants starch acts as a reserve food material, being readily reconverted by enzyme action into glucose. Wheat, corn, rice, and potatoes are all sources of commercial starch. The latter is used in laundries and in the textile industry as a 'sizing' material. Starch is also employed in the manufacture of glucose, dextrin, and ethanol (in brewing).

Properties of Starch

Starch is a white, tasteless, amorphous powder. Natural starch consists of two distinct varieties. One form dissolves in water to give a colloidal solution, while the other is insoluble. The soluble compound is called *amylose* (Latin *amylum* = starch), and constitutes about 20 per cent of natural starch. It consists of unbranched chains of α-glucose units, and has a relative molecular mass of the order of half a million. Amylose is the form of starch which gives the well-known blue colour with iodine solution.

The insoluble variety of starch is called *amylopectin*, and is composed of branched chains of some thousands of α-glucose units. An aqueous suspension gives a violet colour with iodine solution. The same colour is obtained with 'soluble starch solution which has been kept for a long time. This appears to be caused by slow conversion of amylose to amylopectin, which is gradually precipitated.

Starch has no reducing action on Fehling's solution or Tollens's reagent, and it does not form an osazone. Its most important reaction is that of hydrolysis. This can be brought about either by enzyme action (with diastase) or by heating starch with dilute mineral acids. Hydrolysis results in progressive simplification of the molecular chains as follows:

$$\text{Starch} \rightarrow \text{dextrin} \rightarrow \text{maltose} \rightarrow \text{glucose}$$

Hydrolysis of starch by diastase normally yields a mixture of dextrin and maltose, but if the reaction is prolonged the dextrin also is converted into maltose. The products obtained by hydrolysis with mineral acids vary with the concentration of the acid, the temperature, and the duration of the reaction. Thus, by adjusting the conditions, dextrin, maltose, or glucose can be obtained as the chief product. If starch is boiled with dilute hydrochloric acid for about twenty minutes glucose is formed in quantitative yield.

$$(C_6H_{10}O_5)_n + nH_2O \rightarrow nC_6H_{12}O_6$$

Chemical Tests for Starch

The tests now described are carried out with starch 'solution,' which is best prepared by shaking about 1 g of starch with cold water in a test-tube and pouring the suspension into about 100 cm³ of boiling water in a beaker.

(*i*) *Reaction with Iodine*. To a portion of the *cold* solution add one or two drops of iodine solution. A deep blue colour is formed.

(*ii*) *Hydrolysis*. To 30 cm³ of the solution in a boiling-tube add 2 cm³ of concentrated hydrochloric acid. Place the tube in a beaker of water, and boil the water. From time to time pour off a small portion of the solution into a test-tube, cool it, and test with iodine solution. After a few minutes a red colour is obtained instead of a blue one. The red colour is given by dextrin. After several minutes of further boiling the liquid gives no colour reaction at all with iodine. When this stage is reached cool a portion of the liquid and neutralize it with dilute sodium(I) hydroxide solution. Confirm the presence of glucose in the liquid by showing that the latter now reduces Fehling's solution.

Dextrin. Dextrin is manufactured by heating starch at about 200°C in the absence of air. The product is an amorphous powder known as 'British gum.' It dissolves in water to form a viscous solution, which is used as an adhesive.

Glycogen

This compound occurs chiefly in the liver and muscle tissue of animals, where it acts as a reserve material for supplying energy. (Glycogen is sometimes called 'animal starch.') Glycogen can be prepared from liver as a white amorphous powder. With water it forms a colloidal solution which is dextro-rotatory and which gives a brownish-red colour with iodine solution. Mineral acids hydrolyse glycogen to glucose on boiling.

Like amylopectin glycogen consists of branched chains of α-glucose units, but the method of branching is different. The relative molecular mass of glycogen has been estimated at between one million and two million.

CELLULOSE, $(C_6H_{10}O_5)_n$

Cellulose forms a large proportion of the framework of plants. The walls of the individual cells consist chiefly of cellulose, and in the plant stem some of the cells take the form of fibres which help to keep the stem erect. Several of the tougher kinds of fibre (e.g. linen, jute, and hemp) are used in the textile industry. Another valuable source of cellulose fibres is provided by the seed-hairs of certain plants (e.g. cotton).

Several branches of industry are based on the cellulose of wood. In wood the cellulose fibres are cemented together by a brown resinous compound called lignin, and this must be removed before the cellulose can be used. The wood is cut into chips, and heated under pressure either with a solution of calcium(II) hydrogensulphate(IV) (calcium bisulphite) or with sodium(I) hydroxide solution. The lignin dissolves, leaving the cellulose as a fibrous mass called 'wood pulp,' which is then compressed into sheets. Vast quantities of wood pulp are used to make newsprint and viscose rayon (Chapter 26).

Properties of Cellulose

Cellulose is insoluble in all the common solvents. It can be dissolved in a solution of anhydrous zinc(II) chloride in concentrated hydrochloric acid or in a solution of copper(II)hydroxide in aqueous ammonia. It can also be brought into solution by converting it into a derivative, such as cellulose nitrate(V), which can then be dissolved in an appropriate solvent.

Cellulose molecules consist of unbranched chains of β-glucose units varying in number from about 600 to about 3 000. Each unit contains three hydroxyl groups, and these are responsible for most of the chemical activity of the compound. The insolubility of cellulose in water may appear surprising in view of the large number of hydroxyl groups. This is explained by the fact that a cellulose fibre is composed of a number of parallel molecular chains, and hydrogen bonds are formed between the —OH groups in adjacent chains, which therefore cohere strongly. Some hydrogen bonds are formed with water molecules, but they are relatively few and unable to bring about solution.

Most carnivorous animals, including man, are unable to utilize cellulose as a food material. The enzymes present in their digestive systems can break down the linkages between the α-glucose units in starch and glycogen, but not those between the β-glucose units in cellulose. This is an illustration of how a minor difference in the space structure of molecules can have a profound biological effect.

1. **Reaction with Sodium(I) Hydroxide.** When cotton or paper is treated with a fairly concentrated solution of sodium(I) hydroxide it becomes gelatinous. The fibres, which normally have a flat ribbon-like appearance under the microscope, swell and assume a smooth cylindrical shape, while at the same time they contract in length. In *mercerizing* cotton (a process invented by John Mercer in 1844) the cloth is kept taut to prevent contraction of the fibres and is passed through a bath of sodium(I) hydroxide solution. This treatment gives cotton the lustrous appearance of silk. The physical change is accompanied by a chemical one,

which can be represented in a simplified form as follows:

$$R-OH + NaOH \rightarrow R-ONa + H_2O$$

In this equation R represents the portion of the cellulose molecule which does not react. The precise composition of the product, sodium(I) cellulose, depends on the extent to which the hydroxyl groups of the cellulose react. This in turn depends on the concentration of the alkali and the length of immersion.

2. Hydrolysis by Sulphuric(VI) Acid. If cellulose is heated with dilute sulphuric(VI) acid under pressure it is hydrolysed to glucose.

$$(C_6H_{10}O_5)_n + nH_2O \rightarrow nC_6H_{12}O_6$$

When a piece of filter-paper (which consists of nearly pure cellulose) is left for a few seconds in concentrated sulphuric(VI) acid and then washed, it is found that the surface of the paper has become tough and horny. This change, which is due to partial hydrolysis of the cellulose, is used in the manufacture of *parchment paper*. The latter is employed for legal documents, diplomas, etc., because of its durability.

3. Formation of Esters. Cellulose behaves as an alcohol in forming esters with certain acids. When cotton is immersed in a mixture of fuming nitric(V) acid and concentrated sulphuric(VI) acid for about an hour most of the —OH groups of the glucose units are replaced by nitrate(V) groups and a yellowish solid called *gun-cotton* is formed. The composition of this approximates to cellulose trinitrate(V), $C_6H_7O_2(NO_3)_3$.

In making *cordite*, the propellant for rifle cartridges, a mixture of gun-cotton and nitroglycerine (Chapter 13) is dissolved in propanone, and the solution is mixed with petroleum jelly. The pasty mass is forced through small holes, giving filaments, or cords, from which the propanone evaporates. The cords are then cut into the required lengths.

Under suitable conditions lower nitrates of cellulose can be obtained (cellulose nitrates in general are known commercially as *nitrocellulose*). The lower nitrates are soluble in various solvents, including ethyl ethanoate, 3-methylbutyl ethanoate (isoamyl acetate), and propanone. The solutions are used as lacquers, a common example being nail varnish.

Another important ester of cellulose is cellulose ethanoate, which is used both as a plastic and as a textile fibre (Chapter 26).

EXERCISE 18

1. Classify the following as monosaccharides, disaccharides, or polysaccharides: (*a*) sucrose, (*b*) fructose, (*c*) starch, (*d*) maltose, (*e*) glucose.

2. State the nature of X and Y in the following open-chain formula for glucose: $X-(CHOH)_4-Y$.

3. Which of the following are correct for glucose only, for sucrose only, for both, or for neither?

(*a*) Its aqueous solution is dextro-rotatory.

(*b*) It reduces Fehling's solution.

(*c*) It forms an ethanoyl derivative with ethanoic anhydride.

(*d*) It forms an osazone with phenylhydrazine.

(*e*) Its aqueous solution turns brown when boiled with aqueous sodium(I) hydroxide.

4. Give the names used to describe:

(a) A carbohydrate which has six carbon atoms in its molecule and behaves as an aldehyde in aqueous solution.

(b) The hydrolysis of sucrose, which is dextro-rotatory, to a mixture of glucose and fructose, which is lævo-rotatory.

(c) The type of isomerism shown by glucose in aqueous solution.

(d) The reagent made from copper(II) sulphate(VI), sodium(I) carbonate, and sodium(I) citrate.

(e) The final condensation product of a sugar with excess of phenylhydrazine.

5. Which of the following statements are correct?

(a) A monosaccharide is a sugar with six carbon atoms in the molecule.

(b) Fructose is sweeter than cane-sugar.

(c) Dextrin is another name for glucose.

(d) Diastase converts starch to maltose.

(e) Nitrocellulose is not a nitro-compound.

6. Disaccharide and polysaccharide molecules are composed of monosaccharide units such as α-glucose, β-glucose, etc. State the units present in (a) maltose, (b) sucrose, (c) starch, (d) cellulose, (e) glycogen.

7. Which of the following carbohydrates are hydrolysed by boiling dilute sulphuric(VI) acid: (a) maltose, (b) glucose, (c) sucrose, (d) starch, (e) cellulose?

8. Write equations for the hydrolysis reactions in question 7.

part three

AROMATIC COMPOUNDS

19   Benzene—General Properties of Aromatic Compounds

Amongst the naturally occurring compounds investigated in the early days of organic chemistry were a number to which no satisfactory structure could be assigned. These included turpentine, oil of winter-green, oil of cloves, etc., and, as the compounds had a pleasant aroma, they were classified together as 'aromatic' compounds. Aroma can no longer be regarded, however, as the distinctive feature of aromatic compounds. Nowadays *aromatic organic chemistry is the chemistry of benzene and related compounds.* Just as methane (CH_4) can be looked upon as the parent substance of aliphatic compounds, so benzene (C_6H_6) can be regarded as the starting-point for making aromatic compounds, although in practice many of the latter are obtained from other sources. Benz*ene* must not be confused with benz*ine*, which is a petroleum fraction boiling at 70°–90°C. Both benzene and benzine are used as solvents.

In contrast with aliphatic hydrocarbons, which have an open-chain structure, benzene has a closed-chain, or ring, structure. The ring consists of six carbon atoms, to each of which a hydrogen atom is attached, as shown in (*i*). For simplicity the molecule may be represented, as shown in (*ii*), by a regular hexagon with alternate single and double lines to indicate single and double bonds between the carbonatoms. As will be seen shortly, however, the carbon-carbon bonding is not as simple as this, and other methods of formulation are also used.

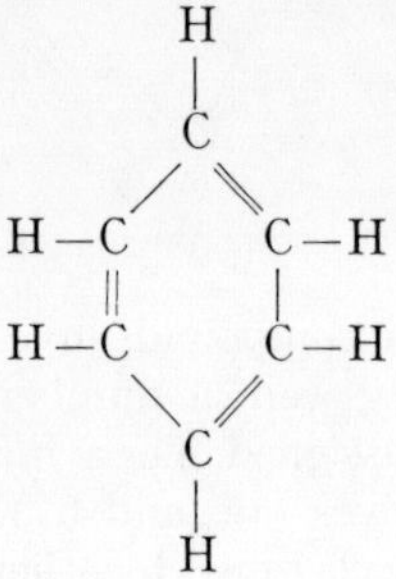

$$H$$
$$|$$
$$C$$
$$H-C \quad C-H$$
$$H-C \quad C-H$$
$$C$$
$$|$$
$$H$$

(i)  (ii)

Derivatives of benzene are formed by substituting other atoms or groups for the hydrogen atoms in the molecule. Thus chlorobenzene has the formula C_6H_5Cl, dichlorobenzene $C_6H_4Cl_2$, etc. Homologues of benzene, e.g. methylbenzene (toluene), $C_6H_5.CH_3$, and dimethylbenzene (xylene), $C_6H_4(CH_3)_2$, are obtained by the introduction of alkyl radicals. Although many benzene derivatives contain more than six carbon atoms in the molecule, the ring of six carbon atoms, which is commonly known as the *benzene nucleus*, is present in all of them. No derivative exists which has less than six carbon atoms in the molecule. Even if hydrogen atoms have been replaced by other atoms or groups, the hexagon formula is still used to represent the benzene nucleus. Thus, the structural formula for methylbenzene can be written simply in either of the ways shown (it is immaterial which of the six hydrogen atoms is substituted by the methyl radical).

(i) (ii)

methylbenzene (toluene) $C_6H_5.CH_3$

BENZENE, C_6H_6

Boiling point: 80·2°C   *Density:* 0·88 g cm^{-3}

Benzene and its homologues occur to varying extents in petroleum. In 1845 Hofmann identified benzene as a constituent of coaltar, and it is now manufactured from this source as shortly described. Benzene is seldom prepared in the laboratory, although there are several methods by which it can be obtained from other compounds. The best known of these methods are the following:

(i) *From Ethyne (Acetylene).* When ethyne is heated to a temperature of about 400°C it polymerizes to benzene (Chapter 7).

$$3C_2H_2 \rightarrow C_6H_6$$

(ii) *From Sodium(I) Benzoate and Soda-lime.* By heating sodium(I) benzoate (or benzoic acid) with soda-lime, the carboxyl group of the former is removed and benzene is formed.

$$C_6H_5.COONa + NaOH \rightarrow C_6H_6 + Na_2CO_3$$

This reaction is analogous to the decarboxylation of sodium(I) ethanoate by soda-lime to give methane (Chapter 5).

Experiment

Mix together a small amount of benzoic acid and twice the amount of soda-lime. Heat the mixture in a dry test-tube. The benzene vapour which is evolved can be ignited at the mouth of the tube. It burns with a very smoky flame owing to the high percentage of carbon which it contains.

(*iii*) *From Phenol and Zinc.* If the vapour of phenol (carbolic acid) is passed over strongly heated zinc powder the latter reduces the phenol to benzene.

$$C_6H_5OH + Zn \rightarrow C_6H_6 + ZnO$$

Manufacture and Uses of Benzene

(*i*) *From Coal.* When coal is subjected to destructive distillation at about 1200°C (*high temperature carbonization*) benzene and other aromatic compounds are given off as vapours. These are partially condensed by cooling and form part of the gas-tar which collects in the tar-well. Most of the benzene and other hydrocarbons of low boiling point, however, escape condensation. They are removed from the coal-gas by 'scrubbing' with a heavy oil such as gas oil (Chapter 5). The resulting solution is distilled in steam to drive out the dissolved hydrocarbons. The latter are condensed and separated from the lower aqueous layer, giving 'crude benzol,' which consists mostly of benzene. For further purification the crude benzol is combined with 'light oil' obtained from coal-tar as now described.

The liquid in the tar-well separates into two layers, an upper layer of watery ammoniacal liquor and lower layer of black oily coal-tar. The two layers are separated, and the coal-tar is fractionally distilled to give five fractions, as shown in the following table:

Table 19-1. FRACTIONS OBTAINED BY DISTILLATION OF COAL-TAR

Name of fraction	Boiling point range /C°	Chief constituents
1. Light oil	Up to 170	Benzene, methyl-benzenes
2. Middle oil, or carbolic oil	170–230	Phenol (C_6H_5OH), naphthalene ($C_{10}H_8$)
3. Heavy oil, or creosote oil	230–270	Cresols ('creosote')
4. Anthracene oil	Above 270	Anthracene ($C_{14}H_{10}$)
5. Pitch	Residue in still	

The light oil (so called because it floats on water) is a mixture of compounds of the types now shown.

(*i*) Hydrocarbons-benzene, methylbenzene, and dimethyl-benzene.

(*ii*) Acidic compounds—chiefly phenol (carbolic acid).

(*iii*) Basic compounds—chiefly phenylamine (aniline), $C_6H_5NH_2$, and pyridine, C_5H_5N.

As crude benzol and light oil have a similar composition they are purified together. The mixture is first given a chemical treatment, which consists of agitating it in turn with aqueous

251

sodium(I) hydroxide and medium sulphuric(VI) acid. The alkali removes acidic impurities, while the acid combines with basic compounds. The purified oil is washed with water and then fractionally distilled to give benzene (b.p. 80°C), methylbenzene (b.p. 110°C), and a higher-boiling fraction consisting of isomeric dimethylbenzenes and ethylbenzene.

Both the benzene and methylbenzene fractions still contain small amounts of undesirable impurities such as sulphur compounds, which have an objectionable odour. These are removed by agitation with concentrated sulphuric(VI) acid, after which the liquids are again washed with sodium(I) hydroxide solution. Finally redistillation yields commercial benzene ('benzol') and methylbenzene ('toluol'). About 90 per cent of the benzene obtained is derived from the crude benzol.

The other coal-tar fractions are also sources of important compounds. The middle oil, which has a density about equal to that of water, is treated to recover phenol and naphthalene. The heavy oil, which sinks in water, contains compounds called cresols, which are closely related to phenol. The oil is used as a preservative ('creosote') for timber and for making commercial disinfectants. Anthracene, obtained from the anthracene oil, is used to prepare alizarin dyes. The pitch left behind in the still is employed in surfacing roads and in making roofing felt, paint, and carbon electrodes.

(ii) *From Petroleum Hydrocarbons.* Aromatic hydrocarbons are produced from naphtha, a light petroleum fraction, by the *catalytic reforming process.* This consists of passing the vaporized naphtha under pressure over a heated platinum catalyst (see Chapter 5). Some of the cycloalkanes and straight-chain alkanes undergo simultaneous dehydrogenation and aromatization, as illustrated by the following equations:

$$C_6H_{12} \rightarrow C_6H_6 + 3H_2$$
cyclohexane

$$C_7H_{16} \rightarrow C_6H_5 . CH_3 + 4H_2$$
heptane

Benzene is employed as a fuel (in petrol), as a solvent (for oils, fats, and resins). and as a starting-point in chemical manufacture. Some 80 per cent of the world's output of benzene is used in making other compounds such as phenylethene (styrene), phenol, and nitrobenzene. These in turn are used to make plastics, insecticides, dyes, drugs, and many other useful substances.

Properties of Benzene

Benzene is a colourless volatile liquid with a characteristic smell. The vapour is toxic, and, when inhaled in quantity, induces unconsciousness. The liquid freezes at 5·5°C. It is immiscible with water, but mixes with ethanol, ethoxyethane, and methylbenzene in all proportions. Besides being a good solvent for oils and fats, benzene also dissolves iodine and sulphur. Benzene is highly flammable, and burns in air with a smoky flame owing to incomplete combustion. Complete combustion is represented by the equation given.

$$2C_6H_6 + 15O_2 \rightarrow 12CO_2 + 6H_2O$$

The provisional structural formula with three double bonds given earlier for benzene does not agree very well with the chemical behaviour of the hydrocarbon. A molecule with three double bonds might be expected to be highly unsaturated and to form addition compounds very readily with electrophilic reagents (as in the case of ethene). Although benzene has a few addition reactions (and is therefore unsaturated), it does not give the usual two tests for unsaturation in hydrocarbons; that is, it does not immediately decolorize bromine, nor does it reduce Baeyer's reagent. Also, it does not react additively with hydrogen halides, nor with concentrated sulphuric(VI) acid. Most of the characteristic reactions of benzene and its homologues are *substitutions*, in which electrophilic reagents replace one or more hydrogen atoms by other atoms or groups. These apparent anomalies will be discussed after the reactions of benzene have been described.

(A) Addition Reactions of Benzene

1. Addition of Hydrogen. If a mixture of benzene vapour and hydrogen is passed over a nickel catalyst at 180°C, direct combination occurs, and cyclohexane is formed. The same reaction takes place at ordinary temperatures if hydrogen is bubbled into benzene containing colloidal platinum. In the addition of hydrogen to benzene no intermediate stages (C_6H_8 and C_6H_{10}) occur.

$$C_6H_6 + 3H_2 \rightarrow C_6H_{12}$$

Or,

cyclohexane

Cyclohexane is a colourless flammable liquid (b.p. 81°C), which is used mainly as a solvent. It is not an aromatic compound, but a cycloalkane. Its molecule has a puckered-ring structure, the valency bonds between the carbon atoms forming the usual tetrahedral angle of 109·5° with each other. This contrasts with the benzene molecule, which has a flat-ring structure (see later).

2. Addition of Chlorine and Bromine. Benzene combines slowly with chlorine and bromine at ordinary temperatures in the presence of direct sunlight or ultraviolet light. The product is 1,2,3,4,5,6-hexachlorocyclohexane (benzene hexachloride). Again no intermediate compounds, such as $C_6H_6Cl_2$ or $C_6H_6Cl_4$, are formed.

$$C_6H_6 + 3Cl_2 \rightarrow C_6H_6Cl_6$$

The addition compound of benzene and chlorine is a white solid, which is used as an insecticide ('Gammexane') against locusts, cockroaches, cattle-ticks, etc.

3. Addition of Ozone. When ozonized oxygen is bubbled into benzene at ordinary temperatures an unstable triozonide of formula $C_6H_6(O_3)_3$ is produced.

4. Addition of Alkenes. By far the most important addition reactions of benzene from the industrial point of view are those which take place with ethene and propene. These reactions, which yield homologues of benzene, are described in the next chapter.

(B) Substitution Reactions of Benzene

1. Substitution by Halogens (Halogenation). If chlorine is passed into benzene at ordinary temperatures in the presence of a 'chlorine carrier' (iodine, red phosphorus, finely divided iron, etc.) substitution of hydrogen by chlorine occurs, and chlorobenzene is formed.

$$C_6H_6 + Cl_2 \rightarrow C_6H_5Cl + HCl$$

Or,

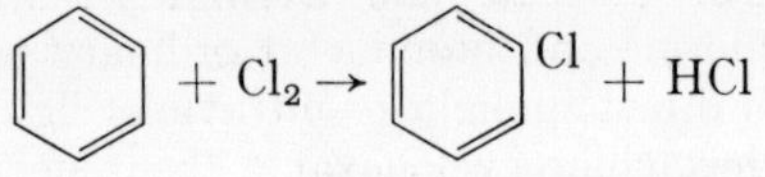

chlorobenzene

The reaction should be carried out preferably in the absence of sunlight, because in the presence of sunlight some of the benzene is converted into the addition compound. If the chlorination is continued, further substitution occurs and dichlorobenzene ($C_6H_4Cl_2$), trichlorobenzene ($C_6H_3Cl_3$), etc., are produced. By prolonged treatment it is possible to substitute all six hydrogen atoms of the benzene molecule and obtain eventually hexachlorobenzene (C_6Cl_6).

Bromine brings about substitution in benzene under the same conditions as chlorine, but reacts more slowly.

Experiment

To 2 cm³ of benzene add two or three drops of bromine. No reaction occurs. Now add some clean iron filings or a little powdered iodine. Steamy fumes of hydrogen bromide are soon evolved. Test the fumes with damp blue litmus paper and by blowing them across a bottle of concentrated ammonia.

Iodine scarcely reacts with benzene even in the presence of a halogen carrier. This is because hydrogen iodide is a strong reducing agent and tends to reconvert any iodobenzene formed to benzene.

$$C_6H_6 + I_2 \rightleftharpoons C_6H_5I + HI$$

Iodine brings about substitution in benzene, however, if an oxidizing agent (e.g. nitric(V) acid) is added to remove the hydrogen iodide formed.

1. Substitution by Nitric(V) Acid (Nitration). Very little reaction occurs between benzene and concentrated nitric(V) acid alone even on heating, although a small amount of nitrobenzene is produced. If, however, a mixture of equal volumes of

$$C_6H_6 + HNO_3 \rightarrow C_6H_5NO_2 + H_2O$$

Or,

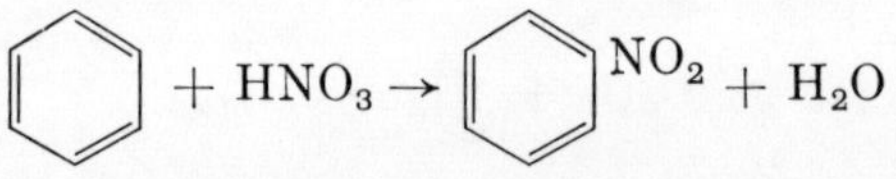

nitrobenzene

concentrated nitric(V) acid and concentrated sulphuric(VI) acid is added to benzene at ordinary temperatures considerable heat is evolved and nitrobenzene is rapidly formed. The part played by the sulphuric(VI) acid in promoting the reaction is discussed in Chapter 22. The process of substituting a hydrogen atom by the nitro group (—NO_2) is described as *nitration*. The mixture of equal volumes of concentrated nitric(V) acid and concentrated sulphuric(VI) acid is often called 'nitrating mixture'.

It is possible to substitute up to three hydrogen atoms of the benzene molecule by nitro groups, and in this way dinitrobenzene, $C_6H_4(NO_2)_2$, and trinitrobenzene, $C_6H_3(NO_2)_3$, are obtained.

3. Substitution by Sulphuric(VI) Acid (Sulphonation).

Benzene is attacked only slowly by concentrated sulphuric(VI) acid even on heating. Substitution occurs when the two compounds are refluxed together for about twelve hours, benzenesulphonic acid being formed. In this case the reaction is reversible.

$$C_6H_6 + H_2SO_4 \rightleftharpoons C_6H_5SO_3H + H_2O$$
benzenesulphonic acid

Substitution takes place rapidly, and without heating, if fuming sulphuric(VI) acid (oleum) is used. The introduction of a sulphonic group (—SO_3H) into a molecule in place of a hydrogen atom is known as *sulphonation*.

Two other types of substitution undergone by benzene are *alkylation* and *acylation*, which are described in the next chapter. In these reactions a hydrogen atom is replaced by an alkyl radical and an acyl group respectively.

Chemical Tests for Aromatic Hydrocarbons

Benzene and its homologues can be distinguished from liquid *alkanes*

(e.g. hexane) by the tests now given.

(*i*) *Nitration.* Mix together in a boiling-tube 2 cm³ of concentrated nitric(V) acid and 2 cm³ of concentrated sulphuric(VI) acid. Add drop by drop 1 cm³ of benzene, *shaking the tube well and cooling it under the tap.* Nitrobenzene, which has a smell of almonds, separates as a liquid layer above the acid mixture. Pour the contents of the tube into a beaker of water. The nitrobenzene forms yellow oily drops at the bottom of the beaker.

No reaction occurs with a liquid alkane under similar conditions.

(*ii*) *Sulphonation.* To 2 cm³ of benzene in a boiling-tube add carefully two or three drops of fuming sulphuric(VI) acid and shake the tube. Heat is evolved, and one homogeneous liquid is obtained. If the experiment is repeated with a liquid alkane no reaction takes place. Then hydrocarbon forms a separate layer above the acid.

It is necessary to use other tests to distinguish aromatic hydrocarbons from liquid alkenes such as hex-1-ene

$$(CH_3.CH_2.CH_2.CH_2.CH{=\!=}CH_2).$$

Liquid alkenes react with both nitrating mixture and fuming sulphuric(VI) acid. These hydrocarbons, however, give the usual reactions of alkenes with a solution of bromine in trichloromethane and with Baeyer's reagent, whereas benzene and its homologues are not immediately affected.

Isomerism of Benzene Derivatives

Derivatives of benzene are obtained by substituting the hydrogen atoms in the molecule by univalent atoms or groups such as —Cl, —Br, —CH_3, —NO_2, etc. The compounds resulting from the substitution of up to three hydrogen atoms by chlorine atoms are shown below.

CHLORO-SUBSTITUTION DERIVATIVES OF BENZENE

Monochlorobenzene, C_6H_5Cl

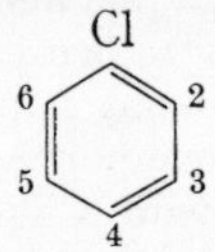

Dichlorobenzenes, $C_6H_4Cl_2$

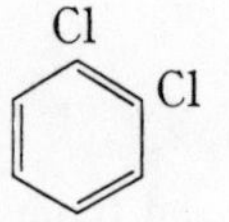

| 1,2-, or ortho-(o-) dichlorobenzene | 1,3-, or meta-(m-) dichlorobenzene | 1,4-, or para-(p-) dichlorobenzene |

Trichlorobenzenes, $C_6H_3Cl_3$

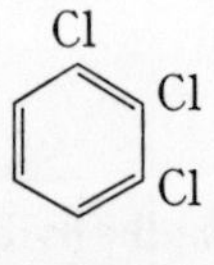

| 1,2,3- Trichlorobenzene (adjacent) | 1,2,4- Trichlorobenzene (asymmetrical) | 1,3,5- Trichlorobenzene (symmetrical) |

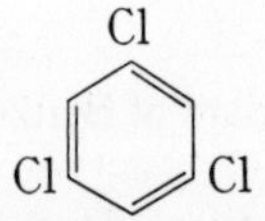

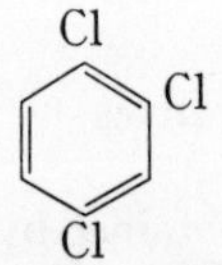

The substitution of a single hydrogen atom results in only one derivative, but further substitution gives rise to a type of isomerism known as *positional isomerism*. In this the isomers differ according to the different relative positions taken up by the chlorine atoms. These positions can be denoted by numbering the carbon atoms of the nucleus from 1 to 6 in a clockwise direction. The number 1 position is regarded as occupied by the first atom introduced, and the relative positions of further substituting atoms are then indicated by the appropriate numbers.

There are three dichloro-substitution derivatives of benzene. These can be specified as 1,2-dichlorobenzene, 1,3-dichlorobenzene, and 1,4-dichlorobenzene. Often, however, a different method of naming is used in the case of disubstitution derivatives, the 1,2-compound being described as *ortho-(o-)*, the 1,3-compound as *meta-(m-)*, and the 1,4-compound as *para-(p-)*. If the two atoms or groups introduced are different the number of isomers is still three.

The substitution of three hydrogen atoms by chlorine atoms again produces three isomers: 1,2,3-trichlorobenzene, 1,2,4-trichlorobenzene, and 1,3,5-trichlorobenzene. Occasionally the terms 'adjacent', 'asymmetrical', and 'symmetrical' are used to describe these three isomers. If the substituting atoms or groups are not identical the number of possible isomers is larger.

With four substituting chlorine atoms three isomers are again obtained. The possible arrangements are easily deduced, since only two unsubstituted hydrogen atoms remain, and these can occupy only the 1,2-, 1,3-, and 1,4-positions. The substitution of five or six hydrogen atoms by chlorine atoms yields only one compound. There are few cases in which more than three hydrogen atoms can be substituted by similar atoms or groups.

The chief differences between positional isomers are found in physical properties such as melting point and boiling point, although chemical differences sometimes occur as well.

Structure of Benzene

The theory that the benzene molecule has a ring structure of six carbon atoms, each with an attached hydrogen atom, was first put forward by the German chemist Kekulé in 1865. It was accepted long before it was proved because it explained a number of facts in the chemistry of benzene. Thus it accounted for the stability of the group of six carbon atoms. In most of the reactions of benzene and its derivatives the nucleus of six carbon atoms persists unchanged. Some derivatives contain more than six carbon atoms in the molecule, but by suitable treatment they can be broken down into compounds which again possess the group of six carbon atoms. Usually further breaking down can be accomplished only by drastic methods, such as combustion. Again, if one of the hydrogen atoms in the benzene molecule is substituted by, say, a chlorine atom, only one monochlorobenzene (C_6H_5Cl) is formed. This shows that all six hydrogen atoms are in a similar state of combination, which would not be the case if the molecule had an open-chain structure such as

$$CH\equiv C-CH_2-CH_2-C\equiv CH.$$

Modern physical methods of investigating molecular structure confirm the correctness of the ring structure. An electron density 'map' of a benzene crystal (see Fig. 3-5) shows that all twelve atoms of the benzene molecule lie in the same plane, that the carbon atoms are equidistant from each other, and that the bond

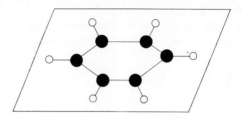

Fig. 19-1. Model of the flat benzene molecule

angles are 120° (Fig. 19-1). Thus the molecule is not only flat, but also symmetrical. This illustrates again the general rule that when a carbon atom is joined to only three other atoms all four atoms are in the same plane, and the angles formed by the bonds between them are approximately 120°.

Electronic Structure. Three of the four valencies of each carbon atom in the ring can be assigned to the formation of single covalent bonds with a hydrogen atom and the two adjacent carbon atoms. Kekulé resolved the question of the disposal of the fourth valency by placing a double bond between alternate pairs of carbon atoms, thus giving the molecule three double bonds. The formula obtained in this way is known as the *Kekulé formula*. It accounts for the addition to the molecule of six hydrogen atoms, six chlorine atoms, or three molecules of ozone. The degree of unsaturation implied by the Kekulé formula is too great, however, to agree with the other chemical properties. Thus the formula is inconsistent with the fact that benzene reacts with nitric(V) acid and sulphuric(VI) acid by substitution and not by addition, and that the hydrocarbon does not react at all with Baeyer's reagent or hydrogen chloride.

According to modern theory the benzene molecule is a resonance hybrid with the *two* Kekulé structures shown as its chief contributing structures. This means that the valency bonds between

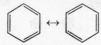

the carbon atoms are neither single nor double, but an intermediate type of bond. The electronic nature of this is most readily understood from the following. In Chapter 6 it was seen that in the flat ethene molecule the C=C bond consists of a σ bond and a π bond, the latter being formed by overlapping of atomic orbitals above, and below, the plane of the molecule (see Fig. 6-3). In the resulting molecular orbital, which consists of two banana-shaped figures, the two π electrons are *localized*; that is, they are restricted to the two carbon atoms joined by the double bond. The same applies to the π electrons in the molecules of propene (CH_3—CH=CH_2) and but-2-ene (CH_3—CH=CH—CH_3).

The situation is different when a number of carbon atoms are joined (in the conventional formula) by alternate single and double bonds. In this type of system, and providing there are at least two double bonds, the bonds undergo 'conjugation.' For example, in the flat butadiene molecule (CH_2=CH—CH=CH_2) the atomic orbitals of the π electrons of the two middle carbon atoms not only overlap with those of the two end carbon atoms, but also overlap with each other (Fig. 19-2a). This results in a composite molecular orbital embracing the whole length of the carbon chain (Fig. 19-2b). Since the four π electrons can move anywhere within this orbital, and are not restricted to two particular carbon atoms, they are described as *delocalized* electrons.

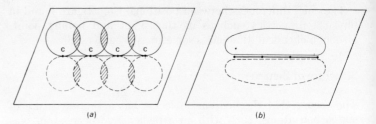

Fig. 19-2. (*a*) Overlapping of atomic orbitals of π electrons in butadiene; (*b*) resulting delocalized molecular orbital

A similar situation exists in the benzene molecule, except that the carbon atoms form a ring. In this case the delocalized molecular orbital of the six π electrons extends completely round the ring. It takes the form of two tubes, one immediately above, and one immediately below, the carbon atoms of the benzene ring (Fig. 19-3). In this orbital the π electrons have freedom of movement, so that an electrical effect in one part of the ring can be readily transmitted to another part. This is important in connection with the mechanism of substitution, as explained shortly.

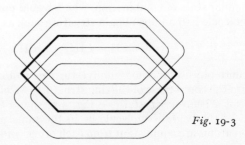

Fig. 19-3

Thus the intermediate type of bond between the carbon atoms in benzene consists of a σ bond and a special kind of extended π bond. Since the true nature of the bonding cannot be shown by conventional means, other ways must be used to portray it. In the method used so far the benzene nucleus has been represented simply by the Kekulé formula or the corresponding hexagonal symbol with alternate single and double bonds. An alternative symbol now in common use is a plain hexagon with a circle inscribed as shown below, the circle representing the delocalized molecular orbital of the six π electrons (the 'aromatic sextet'). This symbol, like the first, may be used both for benzene and benzene derivatives.

benzene   chlorobenzene

Neither the Kekulé symbol nor its alternative is a completely satisfactory method of representing the benzene nucleus. In the rest of this book the newer symbol will be used except where the Kekulé method of formulation is more advantageous.

Experimental Evidence of Resonance in Benzene

(i) *Carbon-carbon Bond Lengths*. As seen earlier, the distances between the carbon atoms in the benzene ring are the same. This shows that the bonds between the atoms are similar, and not alternate single and double bonds. The measured bond length (0·139 nm) is intermediate between the length of a C—C bond (0·154 nm) and that of a C=C bond (0·133 nm). This agrees with the bonds having partial single bond, and partial double bond, character.

(ii) *Heat of Hydrogenation of Benzene*. It was seen in Chapter 4 that if a compound is a resonance hybrid more energy per mole is evolved in forming it from its elements than if it had the conventional structure shown by its formula. The difference in energy (the 'resonance energy') represents the extra extent to which the compound is stabilized by resonance. Benzene has a large resonance energy. This is shown by the fact that the heat of hydrogenation of benzene to cyclohexane is considerably *less* than that calculated for a Kekulé structure. The two values for the heat evolved are 210 kJ mol^{-1} and 360 kJ mol^{-1}, giving a resonance energy for benzene of 150 kJ mol^{-1}. This can be understood more clearly from the following. 72 g of carbon and 12 g of hydrogen can be theoretically converted to 84 g of cyclohexane by first forming either the Kekulé structure or the resonance hybrid of benzene, and then hydrogenating these. Since the total energy change is independent of the path followed (Hess's law) the separate amounts of heat evolved (in kJ mol^{-1}) would be as shown in the following diagram:

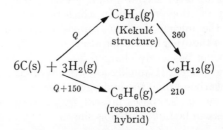

General Mechanism of Substitution in the Benzene Ring

The delocalized π electrons of the benzene ring play a fundamental part in the substitution of hydrogen atoms by electrophilic reagents. Investigation has shown that substitution is initiated by an ion X^+ or a molecule $\overset{\delta+}{X}—\overset{\delta-}{Y}$ which can split off an ion X^+. In nitration X^+ may be $NO_2{}^+$, in chlorination Cl^+, in sulphonation SO_3H^+, and in alkylation a positively charged alkyl radical R^+. It will be seen later how these intermediate ions come to be formed under the experimental conditions used.

An electrophilic reagent seeks a point of high electron density for its attack, but, since the benzene molecule is symmetrical and electrically neutral, there is no reason why the attack should be made at one carbon atom in preference to another. However, when an X^+ ion approaches one of the carbon atoms it attracts the mobile π electrons in the ring towards that atom (this is said to be due to the *electromeric effect* of the ion). At the moment of combination the carbon atom, now negatively charged, donates two electrons to form a co-ordinate covalent bond with X^+. This gives an addition complex, in which the remaining four π electrons of the nucleus have a composite molecular orbital extending over the other five carbon atoms in the ring, this portion of the ring being positively charged as a whole. The addition complex is short-lived. It quickly loses the hydrogen atom from the attacked carbon atom, the hydrogen atom being removed as a proton by one of the bases present (e.g. a water molecule). These changes can be represented as shown.

GENERAL PROPERTIES OF AROMATIC COMPOUNDS

Physical Properties

The general rule found for aliphatic compounds that melting points and boiling points rise with increase in size of the molecules holds also for aromatic compounds. Since even the simplest aromatic compounds are fairly large, all these compounds are either liquids or solids at ordinary temperatures.

The phenyl radical ($C_6H_5—$) is more strongly hydrophobic than the lower alkyl radicals, and therefore the solubilities of benzene derivatives in water are less than those of methyl and ethyl compounds of similar type. If, however, the phenyl radical is compared with the hexyl radical ($C_6H_{13}—$), which contains the same number of carbon atoms, it is found that they do not differ greatly in hydrophobic character. Thus both phenol (C_6H_5OH) and hexan-1-ol ($C_6H_{13}OH$) dissolve only slightly in water.

Chemical Properties

(*i*) *Stability of the Benzene Ring*. This feature has been mentioned previously in connection with the constitution of benzene. Aromatic compounds undergo many different types of reaction, but nearly all of these involve changes in the atoms or groups attached to the benzene ring, and not in the ring itself.

(*ii*) *A Weak Degree of Unsaturation*. Benzene and some of its derivatives show a certain amount of unsaturated character. Benzene, methylbenzene, phenol, phenylmethanol (benzyl alcohol), and benzoic acid all form addition compounds with hydrogen under the catalytic influence of finely divided nickel, and in each case three molecules of hydrogen combine with one molecule of the aromatic compound. Only benzene, however, forms addition compounds with chlorine and bromine. A few benzene derivatives, like benzene itself, form unstable ozonides.

(*iii*) *Tendency to Undergo Substitution with Electrophilic Reagents*. Benzene and most of its derivatives undergo substitution in the nucleus with chlorine, bromine, nitrating mixture, and concentrated sulphuric(VI) acid.

(*iv*) *Special Behaviour of Atoms or Groups Attached to the Nucleus*. If the properties of aliphatic and aromatic compounds of similar type (e.g. C_2H_5X and C_6H_5X) are compared the chemical behaviour of the atom or group X is often different in the two cases. Reactions which are characteristic of X in aliphatic compounds may be absent in the corresponding aromatic compounds, or, on the other hand, new reactions may take place in the case of the aromatic compounds. In the latter the behaviour of X is modified by the presence of the benzene nucleus.

Because of the differences in properties between aromatic compounds and parallel aliphatic compounds, many derivatives of benzene were named differently from derivatives of alkanes. For example, the compound C_6H_5OH, which has few of the properties of an aliphatic alcohol, was not called phenyl alcohol, but phenol. This is a trivial name, which has now been retained for common use.

EXERCISE 19

1. Name the following compounds: (*a*) $C_6H_5 . CH_3$, (*b*) C_6H_5Cl, (*c*) C_6H_5OH, (*d*) $C_6H_5NO_2$, (*e*) $C_6H_5SO_3H$.

2. Give the names used for (*a*) the benzene-containing fraction obtained in distilling coal-tar, (*b*) commercial benzene, (*c*) the method of obtaining aromatic hydrocarbons from cyclic and straight-chain alkanes with the aid of a platinum catalyst, (*d*) substitution of a hydrogen atom by (*i*) the —NO_2 group, (*ii*) the —SO_3H group.

3. Which is the odd one out of the following: benzene, methylbenzene, cyclohexane, chlorobenzene, dichlorobenzene?

4. State whether addition compounds are formed with the following by benzene only, by hex-1-ene only, by both, or by neither: (*a*) hydrogen, (*b*) bromine, (*c*) hydrogen bromide, (*d*) concentrated sulphuric(VI) acid, (*e*) ozone.

5. Give the formulæ and names of the addition compounds formed by benzene with any of the reagents in question 4.

6. Write the structural formulæ of the three dibromobenzenes, labelling the positions of the substituent atoms.

7. Which of the following statements are true?

(*a*) Benzene is unsaturated.

(*b*) All the electrons forming the carbon–carbon bonds in benzene are delocalized.

(*c*) The carbon–carbon bonds in benzene are shorter than those in ethene.

(*d*) If benzene had a Kekulé structure less heat per mole would be evolved in forming it from its elements than in the case of the actual structure.

(*e*) Benzene undergoes substitution with electrophilic reagents.

20 Homologues of Benzene—Orientation and Reactivity of Benzene Derivatives

HOMOLOGUES OF BENZENE

Two series of homologous compounds are derived from benzene as now shown.

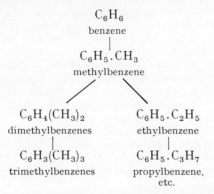

C_6H_6
benzene

$C_6H_5 . CH_3$
methylbenzene

$C_6H_4(CH_3)_2$
dimethylbenzenes

$C_6H_3(CH_3)_3$
trimethylbenzenes

$C_6H_5 . C_2H_5$
ethylbenzene

$C_6H_5 . C_3H_7$
propylbenzene,
etc.

Both series of homologues possess the general formula C_nH_{2n-6}. After benzene, methylbenzene (toluene) is the most important member. The following are used industrially: methylbenzene, the three dimethylbenzenes (xylenes), ethylbenzene, and (1-methylethyl)benzene (isopropylbenzene).

An alkyl radical attached to the benzene nucleus is called a *side-chain*. The term is not restricted to alkyl radicals, but is applied also to other aliphatic groups

$$(—CH_2OH, —CHO, —COOH, etc.)$$

which are directly linked to the nucleus *by means of a carbon atom*. The term is not used for groups like —OH, —NH$_2$, —SO$_3$H, etc., which are not attached to the nucleus by a carbon atom.

As seen earlier, monovalent radicals such as CH_3— and C_2H_5—, which are derived from alkane molecules by removal of a hydrogen atom, are called alkyl radicals. Corresponding radicals derived from aromatic hydrocarbons by removal of a hydrogen atom *from the nucleus* are known as *aryl* radicals, and are represented by Ar. Thus from benzene the phenyl radical is obtained,

C_6H_5—
phenyl

C_6H_4
CH$_3$
methylphenyl

and from methylbenzene the methylphenyl radical. Methylbenzene also gives rise to the phenylmethyl radical. This, however, is not an aryl radical, but a substituted alkyl radical.

$C_6H_5 . CH_2$—
phenylmethyl

Synthesis of Homologues of Benzene

Friedel-Crafts' Reactions. In 1877 a French chemist Friedel and an American chemist Crafts, working in collaboration, discovered that under the catalytic influence of *anhydrous* aluminium(III) chloride alkyl halides bring about substitution of a hydrogen atom in benzene by an alkyl radical. Thus chloromethane and benzene yield methylbenzene.

$$C_6H_6 + CH_3Cl \xrightarrow{AlCl_3} C_6H_5.CH_3 + HCl$$

Other aromatic hydrocarbons can be used instead of benzene and other alkyl halides in place of chloromethane. Hence the reaction can be expressed in the general form shown.

$$ArH + RX \xrightarrow{AlCl_3} ArR + HX$$

Contrary to the usual rule, in this type of reaction alkyl chlorides are more reactive than bromides, and bromides more reactive than iodides. Introduction of an alkyl radical into a molecule in place of a hydrogen atom is described as *alkylation*.

Experimental evidence indicates that the alkylation reaction depends on the halogen atom of the alkyl halide undergoing co-ordination with the aluminium atom of the catalyst to give an intermediate complex (e.g. R—Cl → AlCl$_3$). The latter attacks the benzene ring through the alkyl radical R, which has a partial positive charge (hence the attack is electrophilic). The alkyl radical is added to the nucleus as a carbonium ion R$^+$ by means of two of π electrons of the benzene ring, as explained in the la This is followed by removal of the hydrogen atom as a pr base AlCl$_4{}^-$.

$$ArH + \overset{\delta+}{R}\overset{\delta-}{(AlCl_4)} \rightleftharpoons \overset{+}{Ar}\overset{\displaystyle H}{\underset{\displaystyle R}{\big<}} + AlCl_4{}^-$$

$$\rightleftharpoons ArR + AlCl_3 + HCl$$

The original type of Friedel-Crafts reaction has now little practical importance, although it is sometimes used for making higher alkylbenzenes. The scope of the term 'Friedel-Crafts reaction' has, however, been extended. For example, it now includes substitution in the benzene ring brought about by acyl chlorides and also addition of alkenes to the benzene ring. Both of these reactions are catalysed by anhydrous aluminium(III) chloride. With acyl chlorides benzene yields aromatic ketones. Thus ethanoyl, or acetyl, chloride and benzene give rise to phenylethanone (acetophenone).

$$C_6H_6 + CH_3.COCl \rightarrow C_6H_5.CO.CH_3 + HCl$$
$$\text{phenylethanone}$$

Addition of ethene and propene to benzene under the influence of heat, pressure, and anhydrous aluminium(III) chloride gives ethylbenzene and (1-methylethyl)benzene respectively. These

...e important homologues of benzene. Ethylbenzene is an intermediate in making the plastic poly(phenylethene) (Chapter 26), and (1-methylethyl)benzene is used in the manufacture of phenol (Chapter 24).

$$C_6H_6 + CH_2{=}CH_2 \rightarrow C_6H_5{-}C_2H_5$$
$$\text{ethylbenzene}$$

$$C_6H_6 + CH_3{-}CH{=}CH_2 \rightarrow C_6H_5{-}CH \underset{CH_3}{\overset{CH_3}{\big\langle}}$$
$$\text{(1-methylethyl)benzene}$$

METHYLBENZENE, $C_6H_5.CH_3$

Boiling point: 110°C   *Density:* 0·87 g cm^{-3}

Methylbenzene (toluene) is seldom prepared in the laboratory as it is readily available from industrial sources. The manufacture of the commercial compound ('toluol') from coal-tar and petroleum has been described in Chapter 19. Toluol is chiefly used for blending with motor-spirit and as a solvent. It is also used as a starting-point in the manufacture of other compounds.

Properties of Methylbenzene

Methylbenzene is a colourless liquid with a smell resembling that of benzene. The vapour is much less toxic than that of benzene.

The freezing point of methylbenzene ($-93°C$) is considerably lower than the freezing point of benzene. Methylbenzene is immiscible with water, but mixes with ethanol and ethoxyethane in all proportions. It dissolves oils, fats, and rubber, and also many naturally occurring and synthetic gums and resins. Like benzene, methylbenzene burns with a smoky flame.

The chemical reactions of methylbenzene fall into two groups: (*i*) those which affect the side-chain; and (*ii*) those which involve the benzene nucleus. The chief reactions of the first group are those in which the $CH_3{-}$ radical undergoes oxidation or in which the hydrogen atoms of this radical are substituted by halogen atoms. Reactions of the second group consist mainly of further substitution of the hydrogen atoms of the nucleus. Theoretically

when methylbenzene is converted into a derivative by introduction of another atom or group X 1,2-(*o*-), 1,3-(*m*-), and 1,4-(*p*-) isomers should be obtained. In practice, the product consists almost entirely of the first and third isomers with only traces of the second. As will be seen later, the presence of the methyl group in the molecule plays a very important part in determining the positions taken up by the incoming atom or group. Besides yielding substitution derivatives, methylbenzene shows a weak degree of unsaturation in forming an addition compound with hydrogen. It does not form addition compounds with chlorine and bromine, however, as benzene does.

(A) Reactions Involving the Methyl Radical

1. Oxidation. When methylbenzene is refluxed with dilute nitric(V) acid or an alkaline solution of potassium(I) manganate(VII) (potassium permanganate) the methyl radical is slowly oxidized to a carboxyl group.

$$C_6H_5.CH_3 + 3O \rightarrow C_6H_5.COOH + H_2O$$
$$\text{benzoic acid}$$

It should be noted that any alkyl radical attached to the nucleus is converted to a carboxyl group by the above oxidizing agents, the remainder of the hydrocarbon chain being oxidized to carbon dioxide and water; e.g.

$$C_6H_5.C_2H_5 + 6O \rightarrow C_6H_5.COOH + CO_2 + 2H_2O$$

2. Halogenation. Chlorine substitutes hydrogen either in the side-chain or in the nucleus according to the experimental conditions. *In the absence of a chlorine carrier, in sunlight, and with boiling* methylbenzene substitution occurs in the side-chain to give (chloromethyl)benzene (benzyl chloride).

$$C_6H_5.CH_3 + Cl_2 \rightarrow C_6H_5.CH_2Cl + HCl$$
$$\text{(chloromethyl)benzene}$$

If the chlorination is continued, further substitution takes place until all three hydrogen atoms of the methyl radical have been replaced. (Dichloromethyl)benzene ($C_6H_5.CHCl_2$) and (trichloromethyl) benzene ($C_6H_5.CCl_3$) are produced in turn.

Bromine substitutes hydrogen in the same way as chlorine, but the reactions occur less readily.

(B) Reactions Involving the Nucleus

1. Hydrogenation. Methylbenzene forms an addition compound with hydrogen under the same conditions as benzene. Thus, if a mixture of methylbenzene vapour and hydrogen is passed over finely divided nickel at 180°C or platinum black at ordinary temperatures, methylcyclohexane is produced. This compound is a colourless liquid.

$$C_6H_5.CH_3 + 3H_2 \rightarrow C_6H_{11}.CH_3$$
$$\text{methylcyclohexane}$$

2. Alkylation. One of the features of alkylbenzenes is that they undergo substitution in the nucleus more readily than benzene itself. If methylbenzene is treated with chloromethane in the presence of anhydrous aluminium(III) chloride, the nucleus is further alkylated, and a mixture of 1,2- and 1,4-dimethyl-benzenes (xylenes) is formed.

$$C_6H_5.CH_3 + CH_3Cl \rightarrow C_6H_4(CH_3)_2 + HCl$$
$$\text{dimethylbenzene}$$

3. Halogenation. *In the presence of a chlorine carrier, in the absence of sunlight, and at ordinary temperatures* chlorine brings about substitution in the nucleus. A mixture of chloro-2-methylbenzene (*o*-chlorotoluene) and chloro-4-methylbenzene (*p*-chlorotoluene) is obtained. Suitable chlorine carriers to use are iron filings, anhydrous iron(III) chloride, and an aluminium-mercury couple.

$$C_6H_5.CH_3 + Cl_2 \rightarrow C_6H_4 \begin{smallmatrix} CH_3 \\ \\ Cl \end{smallmatrix} + HCl$$

chloro-2-methylbenzene

chloro-4-methylbenzene

If the chlorination is continued, further substitution occurs in the nucleus, and firstly dichloromethylbenzenes and then trichloromethylbenzenes are produced.

Again bromine brings about similar substitution, but reacts more slowly than chlorine.

The naming of disubstitution derivatives of benzene requires a special word of explanation. As in the case of substitution in alkanes (Chapter 5), substituents are specified in alphabetical order. The first substituent mentioned is then regarded as occupying the number 1 position in the benzene ring. The number 1, however, is not included in the name unless both substituents are of the same kind. The position of the second substituent is shown by one of the numbers 2, 3, or 4 as appropriate. These rules are illustrated by the names

chloro-2-methylbenzene (really 1-chloro-2-methylbenzene), methyl-4-nitrobenzene, and 1,2-dichlorobenzene.

2-Nitrophenol and 4-methylphenylamine are also disubstitution derivatives of benzene, but are not named as such. Instead, they are named as monosubstitution derivatives of phenol (C_6H_5OH) and phenylamine ($C_6H_5NH_2$) respectively. Here the number 1 position in the ring is that occupied by the group already present (that is, —OH or —NH$_2$). The relative position taken up by the second substituting group is indicated by the number in front of its name.

4. Nitration. When a mixture of equal volumes of concentrated nitric(V) acid and concentrated sulphuric(VI) acid (nitrating mixture) is carefully added to cold methylbenzene a mixture of methyl-2-nitrobenzene and methyl-4-nitrobenzene is formed.

$$C_6H_5.CH_3 + HNO_3 \rightarrow C_6H_4 \begin{smallmatrix} CH_3 \\ \\ NO_2 \end{smallmatrix} + H_2O$$

methylnitrobenzene

As in the nitration of benzene considerable heat is evolved. By further nitration first methyldinitrobenzene and then methyltrinitrobenzene can be obtained.

5. Sulphonation. If methylbenzene is refluxed with concentrated sulphuric(VI) acid, or treated with fuming sulphuric(VI) acid at ordinary temperatures, substitution again takes place in the nucleus. A mixture of 2- and 4-methylbenzenesulphonic acids is produced. In this case the reaction is reversible, but under the conditions used the equilibrium point in the equation shown is far to the right.

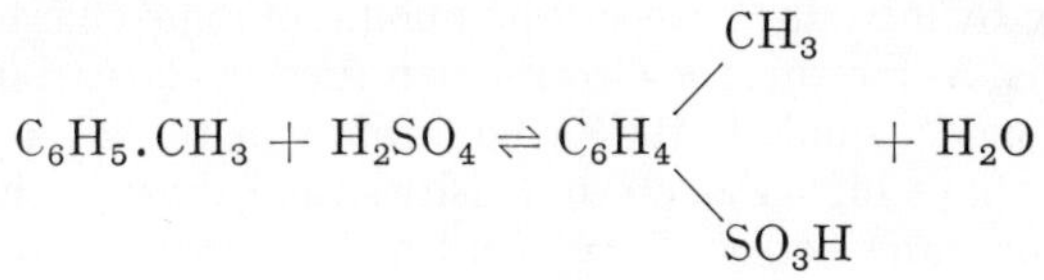

$$C_6H_5.CH_3 + H_2SO_4 \rightleftharpoons C_6H_4 \Big\langle{}^{\displaystyle CH_3}_{\displaystyle SO_3H} + H_2O$$

methylbenzenesulphonic acid

DIMETHYLBENZENES, $C_6H_4(CH_3)_2$

Dimethylbenzene(xylene) exists in three forms, which are positional isomers. These have the names and structures shown.

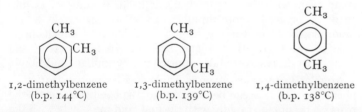

1,2-dimethylbenzene   1,3-dimethylbenzene   1,4-dimethylbenzene
(b.p. 144°C)   (b.p. 139°C)   (b.p. 138°C)

The dimethylbenzenes are colourless flammable liquids resembling methylbenzene. Like the latter they are obtained from coal-tar and petroleum. A mixture of the three isomers is used as an industrial solvent ('xylol'). The isomers cannot be readily separated by fractional distillation owing to the closeness of their boiling points. However, the freezing point (13°C) of 1,4-dimethylbenzene is appreciably higher than those of the other isomers, and therefore 1,4-dimethylbenzene can be obtained as a crystalline solid by cooling. The other two isomers can then be separated by careful fractional distillation. Large amounts of 1,4-dimethylbenzene are used to make benzene-1,4-dicarboxylic, $C_6H_4(COOH)_2$, which is employed in the manufacture of Terylene (Chapter 26).

ORIENTATION AND REACTIVITY OF BENZENE DERIVATIVES

Orientating Effect of a Substituent in the Nucleus

Orientation (the relative positions of the substituting atoms or groups in benzene derivatives) is an important feature of aromatic substitution. If a monosubstitution derivative of benzene is converted into a disubstitution derivative, the three possible isomers are produced unequally. The product consists either of the 1,2- and 1,4-isomers (usually with a small amount of the 1,3-compound) or of the 1,3-isomer (usually with small amounts of the 1,2- and 1,4-compounds). The nature of the product is determined by the atom or group *already present* in the nucleus. There are several empirical rules which indicate the types of isomers formed. One of the simplest is Vorländer's rule, which may be stated as follows:

If there is an atom or group X already united with the nucleus the relative positions taken up by a further substituting atom or group depend on the nature of X. If X contains a double or triple bond the product is chiefly the 1,3-compound.

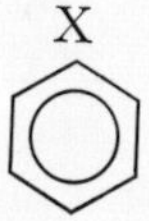

267

If X does not contain a double or triple bond the product consists mostly of the 1,2- and 1,4-compounds. The rule also holds for the introduction of two similar atoms or groups in one operation.

Some common cases in which Vorländer's rule applies are the following:

Nature of X	*Chief Product(s)*
—NO$_2$, —SO$_3$H, —CHO, —COOH, —CN, —COCH$_3$	1,3-
—Cl, —Br, —I, —OH, —NH$_2$, —CH$_3$ (or other alkyl group)	1,2- and 1,4-

Although Vorländer's rule does not hold rigidly, it is a useful guide to the types of isomers obtained when the simpler monosubstitution derivatives of benzene are halogenated, nitrated, or sulphonated. Table 20-1 shows the relative amounts of 1,2-, 1,3-, and 1,4-isomers formed when methylbenzene, chlorobenzene, phenol, and nitrobenzene undergo nitration. (The —NO$_2$ group of nitrobenzene contains a double bond.)

Table 20-1. PERCENTAGES OF ISOMERS FORMED BY NITRATION OF SOME MONOSUBSTITUTED BENZENES C$_6$H$_5$X

	Methylbenzene (C$_6$H$_5$.CH$_3$)	Chlorobenzene (C$_6$H$_5$Cl)	Phenol (C$_6$H$_5$OH)	Nitrobenzene (C$_6$H$_5$NO$_2$)
Per cent 1,2-	41	30	40	6·4
Per cent 1,3-	3	1	0	93·2
Per cent 1,4-	56	69	60	0·3

According to the electronic theory the reason for the orientating effect of a substituent already present in the nucleus is that it disturbs the symmetrical distribution of electrons in the benzene ring by introducing a certain amount of ionic character into the bonds. As a result the electron densities on the carbon atoms are no longer equal. If the carbon atoms in the 2-, 4-, and 6- positions have higher electron densities than those in the 3- and 5- positions, an electrophilic reagent may be expected to attack the nucleus preferentially at the former positions. Conversely, if the carbon atoms in the 3- and 5- positions have the highest electron densities, these will be the main points of attack.

Activation and Deactivation of the Nucleus

Closely bound up with the orientating effect of a substituent X is the *rate* at which further substitution occurs in C$_6$H$_5$X in comparison with the rate for benzene itself. In some cases substitution proceeds more rapidly than with benzene, in others more slowly. This can be seen from Table 20-2, which shows relative rates of nitration and bromination for benzene and some of its derivatives under approximately similar conditions.

No relative rates are given in Table 20-2 for mononitration of phenol or nitrobenzene because, under the conditions used, the reaction occurs too rapidly with the first, and too slowly with the second, for accurate comparisons to be made.

If an electrophilic reagent reacts more rapidly with a monosubstituted benzene than with benzene itself, it is inferred that the substituent X increases the electron density on the nucleus as a whole; that is, it *activates* the nucleus. Conversely, if the reaction takes place more slowly than with benzene, it is inferred that the substituent has decreased the electron density on the

Table 20-2. RELATIVE RATES OF SUBSTITUTION OF BENZENE
AND SOME DERIVATIVES C_6H_5X

Compound	Mononitration	Monobromination
C_6H_5OH	(very rapid)	$1\cdot1 \times 10^{11}$
$C_6H_5.CH_3$	$24\cdot5$	$3\cdot4 \times 10^2$
C_6H_6	$1\cdot00$	$1\cdot00$
C_6H_5Cl	$0\cdot033$	$0\cdot11*$
$C_6H_5NO_2$	(very slow)	$1\cdot8 \times 10^{-6}$

* This relative rate is for chlorination, not bromination.

nucleus as a whole. In this case it has *deactivated* the nucleus.
Practical evidence that these inferences are correct is obtained
from measurement of electric dipole moments of monosubstituted
benzenes. It is found that the directions of the dipole moments
agree with a displacement of electrons towards the nucleus when
the nucleus is activated, and a displacement in the opposite
direction when the nucleus is deactivated. The directions of the
displacements are represented by means of an arrow with a plus
sign at the end to show the positive end of the dipole.

$$\longleftarrow\!\!\oplus \qquad\qquad \oplus\!\!\longrightarrow$$

$$C_6H_5\!-\!CH_3 \qquad\qquad C_6H_5\!-\!NO_2$$

In methylbenzene the   In nitrobenzene the
nucleus is activated   nucleus is deactivated

Activation of the nucleus is accompanied by 1,2- and 1,4-
orientation, and, except in one group of compounds, deactivation
by 1,3-orientation. The exceptions include chlorobenzene, which,
as already seen, yields 1,2- and 1,4-isomers. The behaviour of
chlorobenzene will be explained presently.

Permanent ionic character in monosubstituted benzenes can
be caused both by mesomerism and by the inductive effect of the
substituent. Usually both operate, and the permanent ionic
character of the molecule depends on the directions and magni-
tudes of the electron displacements due to each. This is illustrated
in the next section.

Combination of Mesomeric and Inductive Effects

(*i*) *Phenol, Phenylamine, and Chlorobenzene.* The two effects are
most clearly explained by using the Kekulé formula for the benzene ring. All three compounds mentioned have mesomeric forms,
two of which are Kekulé structures analogous to those of benzene.
In the case of phenol the two Kekulé structures can be represented as shown.

Mesomerism between these structures merely results in a
symmetrical distribution of electrons in the nucleus as in the case
of benzene. The structures can be ignored as regards further
substitution.

269

When, however, the atom joined to the nucleus has a pair of unshared electrons, as in the case of the oxygen atom of phenol, mesomerism can give rise to three further contributing structures. The oxygen atom can donate a lone pair of electrons to form a double bond with the carbon atom C. This gives C_1 ten electrons, which is not permissible. C_1, however, can dispose of two electrons by transferring one of the two pairs shared with C_2 (see structure I) completely to the latter. In this way the double bond between C_1 and C_2 is converted into a single bond (structure III). The oxygen atom, having lost a share in two electrons, now has a positive charge, while C_2, having gained complete control of two electrons previously shared, now has a negative charge.

Electron transfer can be applied in exactly the same way to structure II. In this case the negative charge is acquired by C_6 instead of by C_2. This accounts for the contributing structure IV.

To derive structure V the process of electron transfer is continued round the ring. C_2 (in structure III) can get rid of its negative charge by sharing its lone pair of electrons with C_3, thereby establishing a double bond and transferring the negative charge to C_3. But C_3 would then have ten electrons, which again is not permissible. C_3, however, can solve the problem in the same way as C_1 previously. It can give up its share in two of the four electrons of the double bond with C_4, thus changing the double bond to a single bond. C_4, having now sole possession of two electrons previously shared, has acquired a negative charge. What has happened in effect is that the negative charge has been relayed from the 2- position to the 4- position, as shown in structure V.

If the process of electron transfer is continued round the ring from structure IV instead of from structure III, structure V is still arrived at. C_5 cannot retain the negative charge for the same reason as C_3, and the charge is passed on to C_4. Thus *the negative charge can be located on the carbon atoms in the 2-, 4-, and 6-positions, but not on those in the 3- and 5- positions.*

The phenol molecule is a resonance hybrid of all five contributing structures. The ionic structures III, IV, and V produce partial negative charges on the carbon atoms in the 2-, 4-, and 6- positions, as represented below.

The ionic structures also increase the electron density on the nucleus as a whole (at the expense of the hydroxyl group). Thus in phenol the mesomeric effect is such as to *activate* the nucleus and facilitate substitution by an electrophilic reagent. The effect is therefore described as a $+M$ effect.

Superimposed on the mesomeric effect in phenol is the inductive effect of the —OH group. As the latter has a $-I$ (electron-attracting) effect it tends to deactivate the nucleus. The mesomeric effect and the inductive effect thus oppose each other. Since,

(III)   (IV)   (V)

however, phenol undergoes nitration and halogenation more rapidly than benzene, the resultant effect must be activation of the nucleus. Hence in phenol the $+M$ effect is greater than the $-I$ effect, the latter merely decreasing the partial negative charge on the carbon atoms in the 2-, 4-, and 6- positions. This agrees with the fact that in monosubstitution of phenol by an electrophilic reagent the product consists almost entirely of the 1,2- (or 1,6-) and 1,4-isomers. (The 1,6-isomer is not specified separately, but is included with the 1,2-compound.)

In phenylamine and chlorobenzene (or bromobenzene) the atom attached to the nucleus again has a pair of unshared electrons which can be used to form a double bond with the nucleus, giving once more a $+M$ effect. This again is combined with a $-I$ effect due to the substituent. In phenylamine, as in phenol, the $+M$ effect predominates, the nucleus is activated, and the monosubstitution derivatives are mainly 1,2- and 1,4-isomers.

Chlorobenzene and bromobenzene are special cases. In these the $+M$ and $-I$ effects roughly balance, the $-I$ effect being somewhat larger as shown by the fact that the compounds undergo nitration more slowly than benzene. Here the orientation of the monosubstitution derivatives is determined by the *electromeric effect* of the reagent. This effect has been explained in Chapter 19 in connection with substitution in benzene. An electrophilic reagent attracts the electrons of the ring towards the carbon atom being attacked, the mobile π electrons being mainly affected. Displacement of the π electrons takes place through the mesomerism mechanism. Therefore the electromeric effect of the reagent can increase the electron density when the reagent approaches closely to a carbon atom in either the 2- or 4- position, but not when the approach is made to one in the 3- position. Thus, in spite of deactivation of the nucleus as a whole, the chief products are the 1,2- and 1,4-isomers.

The electromeric effect operates in *all* cases of electrophilic substitution in benzene derivatives, but is usually negligible in comparison with the mesomeric and inductive effects. Only when the latter are approximately equal and opposite does it become the deciding factor in orientation.

(*ii*) *Benzenecarbaldehyde, Benzenecarooxylic Acid, and Nitrobenzene.* (The first two compounds may also be called by their trivial names benzaldehyde and benzoic acid.) In these compounds the mesomeric effect operates in the opposite manner to that in phenol. Here the effect is associated with the presence of a double bond adjacent to the bond which links the substituting group to the nucleus. In benzenecarbaldehyde mesomerism gives rise to the five contributing structures now shown.

(2 forms)
(VI)   (VII)   (VIII)   (IX)

Structures VII, VIII, and IX are again derived from the Kekulé structures VI by electron transfer, but this time the extra bond between the nucleus and the attached atom is formed by *withdrawal* of two electrons from the nucleus ($-M$ effect).

This leaves the nucleus with a positive charge. By reasoning similar to that used in the case of phenol (but in reverse) it can be shown that the positive charge can be located on the 2-, 4-, and 6- positions, but not on those in the 3- and 5- positions.

$$\overset{\delta-}{\underset{\delta+}{\text{CHO}}}$$

Since the substituting groups in benzenecarbaldehyde, benzenecarboxylic acid, and nitrobenzene are electron-attracting groups, the $-M$ effect is combined with a $-I$ effect. The result is strong deactivation of the nucleus and a much slower rate of reaction than benzene with electrophilic reagents. Also, since the carbon atoms in the 3- and 5- positions have the highest electron density, the 1,3- (or 1,5-) isomer is the main product.

(*iii*) *Methylbenzene*. In methylbenzene mesomerism cannot arise in either of the ways illustrated by phenol and benzene-carbaldehyde, and orientation of monosubstitution derivatives is determined by the $+I$ effect of the CH_3- radical. The inductive effect activates the nucleus, and accordingly methylbenzene undergoes nitration some 25 times faster than benzene. The relaying of the inductive effect again occurs mainly through displacement of the mobile π electrons by the mesomerism mechanism. Thus partial negative charges are created on the carbon atoms in the 2-,4-, and 6- positions. Hence the substitution products formed with electrophilic reagents are chiefly the 1,2- and 1,4-isomers.

EXERCISE 20

1. Which (if any) of the following are not homologues of benzene: (*a*) $C_6H_5.C_2H_5$, (*b*) $C_6H_4(CH_3)_2$, (*c*) C_6H_{12}, (*d*) $C_6H_5.C_6H_5$, (*e*) $C_6H_5.CH(CH_3)_2$?

2. Give the names and formulæ of the organic compounds formed when the following react together in the presence of anhydrous aluminium(III) chloride as a catalyst: (*a*) benzene and chloroethane, (*b*) benzene and ethene, (*c*) benzene and propene, (*d*) benzene and 1-chlorohexane, (*e*) methylbenzene and chloromethane.

3. Specifying, where appropiate, the orientation of the main products, name the monosubstitution compounds formed when methylbenzene is (*a*) treated with nitrating mixture, (*b*) boiled with bromine, (*c*) treated in the cold with bromine and iron filings, (*d*) refluxed with concentrated sulphuric(VI) acid.

4. Write structural formulæ for (*a*) (chloromethyl)benzene (*b*) bromo-4-methylbenzene, (*c*) 2-methylbenzenesulphonic acid, (*d*) 1,3-dimethylbenzene.

5. Using chemical formulæ and showing the reagents employed, indicate the stages by which methylbenzene could be converted to benzene.

6. State whether you would expect the following conversions to result chiefly in 1,2- and 1,4- isomers or chiefly in the 1,3- derivative: (*a*) nitrobenzene to chloronitrobenzene, (*b*) chlorobenzene to chloronitrobenzene, (*c*) phenol to nitrophenol, (*d*) benzene to dinitrobenzene, (*e*) benzene to dibromobenzene.

7. Which of the following statements are true?

(*a*) In sunlight methylbenzene forms an addition compound with chlorine.

(*b*) $C_6H_5.CH_2$— is not an aryl radical.

(*c*) When a monosubstitution derivative of benzene is converted into a disubstitution derivative the orientation of the latter is governed by the nature of the group being introduced.

(*d*) An increase in the electron density of the benzene nucleus makes the nucleus more reactive towards electrophilic reagents.

(*e*) Methylbenzene reacts with electrophilic reagents more rapidly than benzene does under similar conditions.

8. State whether you would expect the reactions of the following with nitrating mixture to be faster or slower than that of benzene under similar conditions: $C_6H_5.C_2H_5$, (*b*) C_6H_5F, (*c*) $C_6H_5.COOH$, (*d*) $C_6H_5.CHCl_2$, (*e*) 1,4-$C_6H_4(CH_3)_2$.

21   Halogen Substitution Derivatives

It was seen in Chapter 19 that chlorine and bromine react with benzene to form both addition compounds (e.g. $C_6H_6Cl_6$) and nuclear substitution compounds (e.g. C_6H_5Cl, $C_6H_4Cl_2$, etc.). Methylbenzene (toluene) yields only substitution compounds, but substitution may occur in either the nucleus or the side-chain according to the experimental conditions. Marked differences are found in the properties of the derivatives according to whether substitution has taken place in the nucleus or the side-chain. For this reason the two kinds of derivatives are considered separately. Compounds formed by substitution of a hydrogen atom in the nucleus of an aromatic hydrocarbon by a halogen atom are called *aryl halides*.

NUCLEAR SUBSTITUTION DERIVATIVES

The more important aryl halides derived from benzene and methylbenzene are listed in Table 21-1 together with their boiling points and densities. With the exception of bromo-4-methylbenzene (*p*-bromotoluene) which is a solid, the compounds are colourless liquids with pleasant smells (their vapours are toxic). As usual boiling points show a progressive rise with increase in relative molecular mass. As in the case of alkyl halides (Chapter 9), densities increase in the order: chloro-derivatives < bromo-derivatives < iodo-derivatives. The liquids are immiscible with water, but mix with ethoxyethane and benzene in all proportions.

Laboratory Preparation of Aryl Halides

Unlike alkyl halides aryl halides are not prepared by treating the corresponding hydroxy-compounds with the hydrogen halide or a phosphorus halide. This is because hydroxy-compounds having the —OH group attached to the nucleus do not react with hydrogen halides, and the yields obtained with phosphorus halides are poor because of side-reactions. Aryl halides are usually made by the method given on the next page.

Table 21-1. BOILING POINTS AND DENSITIES OF ARYL HALIDES

Compound	Formula	Boiling point/°C	Density/ g cm^{-3}
Chlorobenzene	C_6H_5Cl	132	1·11
Bromobenzene	C_6H_5Br	156	1·50
Iodobenzene	C_6H_5I	189	1·83
Chloro-2-methylbenzene		159	1·08
Chloro-3-methylbenzene	$C_6H_4(CH_3)Cl$	162	1·07
Chloro-4-methylbenzene		162	1·07
Bromo-2-methylbenzene		182	1·43
Bromo-3-methylbenzene	$C_6H_4(CH_3)Br$	184	1·41
Bromo-4-methylbenzene		184	(solid)

Direct Halogenation of the Hydrocarbon. Benzene and methylbenzene undergo chlorination and bromination in the nucleus at ordinary temperatures in the presence of a halogen carrier.

$$C_6H_6 + Cl_2 \rightarrow C_6H_5Cl + HCl$$

$$C_6H_5 . CH_3 + Cl_2 \rightarrow C_6H_4 \begin{matrix} CH_3 \\ \diagup \\ \diagdown \\ Cl \end{matrix} + HCl$$

The reactions should be carried out in the absence of direct sunlight because the latter leads to the formation of 1,2,3,4,5,6-hexachlorocyclohexane (benzene hexachloride) in the first case and side-chain derivatives of methylbenzene in the second.

For the laboratory preparation of aryl halides the halogen carrier may be an aluminium-mercury couple, anhydrous iron(III) chloride, or iron filings (which forms iron(III) chloride or bromide with the halogen). There is evidence that the action of the catalyst depends on its ability to react with the halogen and form a small proportion of positively charged halogen ions (*chloronium* ions or *bromonium* ions). These are stronger electrophiles than the molecular halogens. Thus, if iron or iron(III) chloride is used as the carrier, chloronium ions are thought to be produced in accordance with the following equation.

$$Cl-Cl + FeCl_3 \rightleftharpoons Cl^+ + FeCl_4^-$$

The mechanism of the chlorination can then be represented as follows:

$$ArH + Cl^+ \rightarrow Ar \begin{matrix} H \\ \diagup \\ + \\ \diagdown \\ Cl \end{matrix} \xrightarrow{FeCl_4^-} ArCl + FeCl_3 + HCl$$

When an aluminium-mercury couple is used as the carrier anhydrous aluminium(III) chloride is formed, and this probably acts in the same way as iron(III) chloride.

To prepare chlorobenzene a known mass of benzene is placed in a flask, a small amount of chlorine carrier is added, and the flask is attached to a reflux condenser. The reaction is exothermic and, since a rise of temperature leads to formation of dichlorobenzene, the flask is cooled in a water-bath. Chlorine is obtained by dropping concentrated hydrochloric acid on to crystals of potassium(I) manganate(VII) (potassium permanganate). The gas is passed into the flask until the theoretical increase in mass (calculated from the equation) has taken place. This stage is ascertained by weighing the flask on a rough balance before chlorination and weighing it from time to time during the chlorination. The chlorobenzene is purified as described below for bromobenzene.

In the laboratory bromobenzene is more convenient to prepare than chlorobenzene. The preparation of bromobenzene is carried out as now described.

Experiment

Set up the apparatus shown in Fig. 21-1, including the device for absorbing hydrogen bromide in water (an alternative method is shown in Fig. 9-1). Put into the 50 cm³ pear-shaped flask 12 cm³ of dry benzene and about 0.5 g of aluminium-mercury couple (prepared as described in Chapter 5). Place 8 cm³ of bromine in the dropping-funnel (this should be done very carefully in the fume-cupboard

275

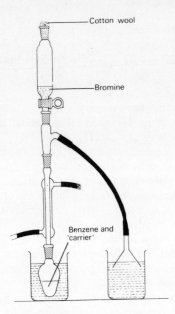

Fig. 21-1. Preparation of bromobenzene

before assembling the apparatus), and insert a plug of cotton wool in the top to reduce evaporation. Add the bromine in small portions to the flask, cooling the latter in the water-bath and waiting for the reaction to subside before adding a fresh portion. When all the bromine has been added put a tripod and gauze under the water-bath, and heat the flask in the latter, keeping the temperature at 65°–70°C for twenty minutes. Bromobenzene is left in the flask.

Purification. The crude product contains small amounts of dibromobenzene, hydrogen bromide, and unchanged benzene and bromine.

Allow the liquid to cool, and pour it into 25 cm³ of 'bench' sodium(I) hydroxide solution in a separating-funnel. Shake the funnel well, and let the contents separate into two layers. Run off the lower layer of bromobenzene. Repeat the washing with another portion of sodium(I) hydroxide solution and then with distilled water. Dry the product with anhydrous calcium(II) chloride in a stoppered flask. Finally separate the bromobenzene from benzene and dibromobenzene by fractional distillation, collecting the fraction which passes over at 150°–160°C. The yield is about 8 cm³ (12 g).

The monochloro- and monobromo-derivatives of methylbenzene are not prepared in the laboratory by direct halogenation of the hydrocarbon. This is because the boiling points of the isomers produced are too close to permit separation by fractional distillation (see Table 21-1). They are made from the corresponding methylnitrobenzenes by the reactions described in Chapter 23 (see Sandmeyer's reactions).

Manufacture and Uses of Aryl Halides

The monochloro- and monobromo-derivatives of benzene and methylbenzene are obtained on a large scale by chlorination or bromination of the hydrocarbons, iron borings being used as the halogen carrier. In the case of methylbenzene derivatives special apparatus is available for separating the isomers of nearly similar boiling point. The halogen derivatives are employed as solvents and as intermediates in making other compounds.

(A) Reactions of Aryl Halides Involving the Halogen Atom

The most important reactions of aryl halides can be divided into those involving only the halogen atom and those in which the

276

nucleus takes part. The derivatives of methylbenzene and other alkylbenzenes have additional reactions owing to the presence of the alkyl radical.

The outstanding feature of aryl halides is *the firmness with which the halogen atom is attached to the nucleus*. In this respect they offer a striking contrast to alkyl halides (Chapter 9). The halogen atom of the latter is readily substituted by nucleophilic reagents such as aqueous alkalies, ammonia, CN^- ions, and anions of silver(I) salts. Similar replacements in aryl halides are difficult to bring about, and require high temperatures and pressures. The greater stability of aryl halides is due to resonance, which arises in the same way as in phenol (Chapter 20). Thus chlorobenzene, like phenol, has five contributing structures. Two of these are Kekulé structures, and three are ionic structures.

$$C_6H_5-\overset{..}{\underset{..}{Cl}}: \leftrightarrow C_6\overset{-}{H_5}=\overset{+}{Cl}:$$

2 structures   3 structures

Evidence of resonance in chlorobenzene is provided by the C—Cl bond length, which is only 0·169 nm as compared with 0·176 nm for the same bond in chloromethane. The shortening of the bond in chlorobenzene agrees with the presence of a certain amount of double bond character in the linkage.

In spite of its lower reactivity the halogen atom enters into a number of reactions with nucleophilic reagents. These reactions can be illustrated by means of the benzene derivatives.

1. **Reduction.** Chlorobenzene and bromobenzene can be reduced to benzene by passing a mixture of the vapour and hydrogen over finely divided nickel at 270°C.

$$C_6H_5Cl + H_2 \rightarrow C_6H_6 + HCl$$

2. **Fittig's Reaction.** Chlorobenzene or, preferably, bromobenzene (which is more reactive) can be used to synthesize alkylbenzenes. A mixture of bromobenzene and an alkyl bromide or iodide is warmed with metallic sodium.

$$C_6H_5Br + RBr + 2Na \rightarrow C_6H_5R + 2NaBr$$

Some diphenyl ($C_6H_5-C_6H_5$) and an alkane (R—R) are formed as impurities by the sodium reacting separately with the aryl halide and alkyl halide.

3. **Reaction with Sodium(I) Hydroxide Solution.** Although chlorobenzene is unaffected by aqueous sodium(I) hydroxide under ordinary conditions, reaction occurs under more drastic conditions. Thus, at a temperature of 300°C and a pressure of 150 atmospheres the 'inactive' chlorine atom becomes 'active,' and firstly phenol and then sodium(I) phenoxide are produced.

$$C_6H_5Cl + NaOH \rightarrow C_6H_5OH + NaCl$$

$$C_6H_5OH + NaOH \rightarrow C_6H_5ONa + H_2O$$
sodium(I)
phenoxide

This is one of the methods used to manufacture phenol (Chapter 24).

4. **Formation of Grignard Reagents.** Bromobenzene and iodobenzene (but not the less reactive chlorobenzene) form Grignard reagents when they are dissolved in ethoxyethane and the solutions are treated with magnesium. Thus bromobenzene gives phenylmagnesium(II) bromide in solution. As in the preparation of aliphatic Grignard reagents (Chapter 9), all the

materials used must be rigorously dried. The Grignard reagents are used in the synthesis of aromatic compounds.

(B) Reactions Involving the Nucleus

Chlorobenzene and bromobenzene can be further halogenated, nitrated, and sulphonated in the nucleus. As explained in Chapter 20, the rates at which these reactions occur are slower than for benzene owing to deactivation of the nucleus by the halogen atom.

1. Halogenation. Chlorination or bromination of benzene in the presence of a halogen carrier can be continued beyond the stage of monosubstitution to obtain disubstitution derivatives. Thus on further chlorination monochlorobenzene gives mainly a mixture of 1,2- and 1,4-dichlorobenzenes.

$$C_6H_5Cl + Cl_2 \rightarrow C_6H_4\begin{array}{c} Cl \\ \\ Cl \end{array} + HCl$$

The isomers can be readily separated because the 1,2-compound is a liquid (b.p. 179°C), while the 1,4-compound is a solid (m.p. 53°C). 1,4-Dichlorobenzene is widely used as an insect repellant (e.g. against midges, mosquitoes, and moths).

Dichlorobenzenes and dibromobenzenes can be still further halogenated to produce more highly substituted derivatives.

2. Nitration. Chlorobenzene and bromobenzene undergo nitration in the cold with a mixture of concentrated nitric(V) acid and concentrated sulphuric(VI) acid. Heat is evolved in the

reaction. In the case of chlorobenzene the product consists entirely of chloro-2-nitrobenzene and chloro-4-nitrobenzene.

$$C_6H_5Cl + HNO_3 \rightarrow C_6H_4\begin{array}{c} Cl \\ \\ NO_2 \end{array} + H_2O$$

chloro-2-nitro-
benzene
(31 per cent)

chloro-4-nitro-
benzene
(69 per cent)

The introduction of the nitro group into the 2- and 4- positions with respect to the chlorine atom has a remarkable effect on the reactivity of the halogen atom. The latter can now be readily removed by means of boiling aqueous alkalies, ethanolic ammonia, etc. With alkali chloro-2-nitrobenzene yields 2-nitrophenol.

$$C_6H_4\begin{array}{c} Cl \\ \\ NO_2 \end{array} + OH^- \rightarrow C_6H_4\begin{array}{c} OH \\ \\ NO_2 \end{array} + Cl^-$$

Similar 'activation' of the chlorine atom does not occur in the case of chloro-3-nitrobenzene. This isomer cannot be obtained by nitration of chlorobenzene. It is prepared by chlorination of nitrobenzene, as described in the next chapter.

The activating effect on the halogen atom of a nitro group in the 2- or 4- position is caused by the $-M$ and $-I$ effects of the $-NO_2$ group. These strongly deactivate the 2- and 4- positions relative to the group, but leave the 3- position largely unaffected. Thus the carbon atoms in the 2- and 4- positions have low electron densities, which make them vulnerable to attack by *nucleophilic* reagents such as NH_3, OH^-, and other negative ions.

3. Sulphonation. Aryl halides undergo further substitution in the nucleus when they are refluxed with concentrated sulphuric(VI) acid or treated in the cold with fuming sulphuric(VI) acid. In the case of chlorobenzene the product consists almost wholly of 4-chlorobenzenesulphonic acid.

$$C_6H_5Cl + H_2SO_4 \rightarrow C_6H_4\begin{smallmatrix} Cl \\ \\ SO_3H \end{smallmatrix} + H_2O$$

(C) Reactions Involving Alkyl Radicals

1. Oxidation. If one of the isomeric chloromethylbenzenes is refluxed with dilute nitric(V) acid or an alkaline solution of potassium(I) manganate(VII), the methyl radical is oxidized to a carboxyl group, and the corresponding chlorobenzenecarboxylic acid (chlorobenzoic acid) is slowly formed.

$$C_6H_4\begin{smallmatrix} CH_3 \\ \\ Cl \end{smallmatrix} + 3O \rightarrow C_6H_4\begin{smallmatrix} COOH \\ \\ Cl \end{smallmatrix} + H_2O$$

2. Substitution in the Side-chain. When chlorine is passed into boiling chloromethylbenzene, in the absence of a chlorine carrier, substitution occurs in the methyl radical. Thus chloro-2-methylbenzene yields chloro-2-(chloromethyl)benzene.

$$C_6H_4\begin{smallmatrix} CH_3 \\ \\ Cl \end{smallmatrix} + Cl_2 \rightarrow C_6H_4\begin{smallmatrix} CH_2Cl \\ \\ Cl \end{smallmatrix} + HCl$$

Tests for Aryl Halides

Aryl halides are most easily recognized by their infrared absorption spectra, which are characteristic for the individual compounds. There are no specific chemical tests for aryl halides, but the tests now given help to identify a compound as a member of this particular group of compounds.

(*i*) *The Lassaigne Test*. This test, which is carried out as described in Chapter 3, indicates the presence of a halogen and also shows its nature.

(*ii*) *Nitration*. The aromatic character of the compound is shown by the fact that it can be nitrated. Put into a boiling-tube 1 cm³ of chlorobenzene, 2 cm³ of concentrated nitric(V) acid and 2 cm³ of concentrated sulphuric(VI) acid. As a precaution hold the tube in a clip. Shake the mixture well for three or four minutes, and then pour it into a large amount of water in a beaker. A pale-yellow crystalline precipitate, consisting of the 1,2- and 1,4-isomers of chloronitrobenzene, is formed. This smells of almonds.

(*iii*) *Failure to Undergo Hydrolysis*. The nitration test does not distinguish between an aryl halide like chlorobenzene and a compound like (chloromethyl) benzene, in which the halogen is present in the side-chain. If the compound has the halogen in the side-chain, it is hydrolysed by boiling aqueous alkalies which do not hydrolyse aryl halides.

Boil a few drops of an ethanolic solution of chlorobenzene in a boiling-tube with 'bench' sodium(I) hydroxide solution, shaking the tube well and using a very small flame. After a few minutes cool the tube under the tap, acidify the contents with dilute nitric(V) acid, and add silver(I) nitrate(V) solution. No precipitate of silver(I) chloride is formed. If the experiment is repeated with (chloromethyl)benzene in place of chlorobenzene, a precipitate is obtained owing to liberation of Cl⁻ ions by hydrolysis.

SIDE-CHAIN SUBSTITUTION DERIVATIVES

As was seen in Chapter 20, chlorination of methylbenzene under suitable conditions brings about substitution of the three hydrogen atoms in the side-chain, giving in turn (chloromethyl)-benzene (benzyl chloride), $C_6H_5.CH_2Cl$, (dichloromethyl)-benzene (benzylidene dichloride), $C_6H_5.CHCl_2$, and (trichloromethyl)benzene (benzotrichloride), $C_6H_5.CCl_3$. The dimethyl-benzenes and ethylbenzene also undergo chlorination in their side-chains under the same conditions as methylbenzene, but yield mixtures of isomers. (Chloromethyl)benzene may be taken as representative of the monosubstitution derivatives.

(CHLOROMETHYL)BENZENE, $C_6H_5.CH_2Cl$

Boiling point: 179°C   *Density:* 1·10 g cm⁻³

Preparation

(Chloromethyl)benzene (benzyl chloride) is prepared both in the laboratory and in industry by passing chlorine into boiling methylbenzene in the absence of a chlorine carrier and, preferably, in direct sunlight. Ultraviolet light may be used in place of sunlight. In the laboratory chlorination is continued until the theoretical increase in mass (in accordance with the equation) is obtained. The product is purified as described earlier for bromobenzene.

$$C_6H_5.CH_3 + Cl_2 \rightarrow C_6H_5.CH_2Cl + HCl$$

As seen earlier, chlorination in the nucleus probably depends on formation of chloronium (Cl^+) ions under the influence of a catalyst like iron(III) chloride. In contrast (as shown by the effect of light), chlorination in the side-chain takes place by a free-radical mechanism involving homolytic fission of the Cl—Cl bond. The three steps of the mechanism are similar to those which occur in the chlorination of methane (Chapter 4).

Properties of (Chloromethyl)benzene

(Chloromethyl)benzene is a heavy colourless liquid. It is immiscible with water, but miscible with ethanol and ethoxyethane in all proportions. Its vapour has an extremely irritating effect on the nose and throat, and rapidly produces a flow of tears even in minute concentration. This is characteristic of aromatic compounds containing a halogen atom in the side-chain.

The reactions of (chloromethyl)benzene are partly those of the chloromethyl group (—CH₂Cl) and partly those of the nucleus. A halogen atom present in a side-chain is always 'active,' and can be replaced by groups such as —OH, —NH₂, etc., by treatment with appropriate nucleophilic reagents. In this respect (chloromethyl)benzene resembles an alkyl halide. It differs from an alkyl halide, however, in undergoing nuclear substitution reactions.

(A) Reactions Involving the Chloromethyl (—CH₂Cl) Group

(*i*) (Chloromethyl)benzene is slowly hydrolysed by boiling water to phenylmethanol (benzyl alcohol)

$$C_6H_5.CH_2Cl + H_2O \rightleftharpoons C_6H_5.CH_2OH + HCl$$
phenylmethanol

Hydrolysis occurs more rapidly with boiling aqueous sodium(I) hydroxide, but this gives rise to side-reactions.

(*ii*) If (chloromethyl)benzene is refluxed with an *ethanolic* solution of sodium(I) hydroxide, it forms an ether, (ethoxymethyl)benzene (benzyl ethyl ether). This is a colourless liquid (b.p. 185°C).

$$C_6H_5.CH_2Cl + C_2H_5O^- \rightarrow C_6H_5.CH_2.O.C_2H_5 + Cl^-$$

(*iii*) The halogen atom of (chloromethyl)benzene can be replaced by an amino group by heating with an ethanolic solution of ammonia in a sealed vessel. (Phenylmethyl)amine ($C_6H_5.CH_2NH_2$) is produced.

(*iv*) When (chloromethyl)benzene is refluxed with potassium(I) cyanide in aqueous-ethanolic solution phenylethanonitrile, or phenylacetonitrile, is obtained.

$$C_6H_5.CH_2Cl + CN^- \rightarrow C_6H_5.CH_2CN + Cl^-$$
phenylethanonitrile

Phenylethanonitrile can be converted by hydrolysis with a mineral acid into phenylethanoic acid ($C_6H_5.CH_2.COOH$).

(*v*) (Chloromethyl)benzene reacts with an ethanolic solution of silver(I) ethanoate when the two are heated together. Silver(I) chloride is precipitated and an ester, phenylmethyl ethanoate (benzyl acetate), $CH_3COOCH_2.C_6H_5$, is formed.

(*vi*) (Chloromethyl)benzene can be reduced to methylbenzene by adding sodium to an ethanolic solution of the compound.

$$C_6H_5.CH_2Cl + 2H \rightarrow C_6H_5.CH_3 + HCl$$

(*vii*) A Grignard reagent is formed when (chloromethyl)benzene (or the corresponding bromo-compound) is dissolved in dry ethoxyethane and the solution is treated with magnesium.

(B) Reactions Involving the Nucleus

(Chloromethyl)benzene shows typical aromatic character in undergoing nuclear substitution. It can be chlorinated and nitrated under the same conditions as chlorobenzene. The products are the 1,2- and 1,4-isomers of the compounds shown.

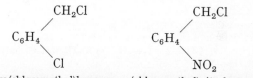

chloro(chloromethyl)benzene   (chloromethyl)nitrobenzene

(Chloromethyl)benzene does not undergo sulphonation with concentrated sulphuric(VI) acid. Instead, it is partly hydrolysed to phenylmethanol and partly converted to complex resinous products.

Chemical Tests for (Chloromethyl)benzene

The tests now described, when taken together, distinguish (chloro-methyl)benzene (or the corresponding bromo- compound) from chlorobenzene (or bromobenzene) and alkyl halides.

(i) *Hydrolysis*. As seen earlier, (chloromethyl)benzene can be distinguished from chlorobenzene by warming with 'bench' sodium(I) hydroxide solution for a few minutes. The tube is then cooled under the tap. When the contents are acidified with dilute nitric(V) acid and silver(I) nitrate(V) solution is added a white precipitate of silver(I) chloride is obtained in the case of (chloromethyl)benzene, but not in that of chlorobenzene.

(ii) *Nitration*. Nitrate (chloromethyl)benzene in the manner described earlier for chlorobenzene. When the product is poured into water the 1,2- and 1,4-isomers of (chloromethyl)nitrobenzene separate as a lower oily layer, which changes in a few minutes into a pale-yellow solid. (Crystallization is helped by scratching the inside of the tube with a glass rod.) This test distinguishes (chloro-methyl)benzene from an alkyl halide, but not from chlorobenzene.

EXERCISE 21

1. State which of the following are aryl halides:
(a) C_6H_5Br, (b) CH_3—C_6H_4—Br, (c) $C_6H_5.CH_2Cl$, (d) $C_6H_4Cl_2$, (e) Cl—C_6H_4—NO_2.

2. (a) In a preparation of bromobenzene 12 cm³ of benzene were treated with 8 cm³ of liquid bromine. Did these quantities correspond to an excess of benzene or an excess of bromine? (The densities of benzene and bromine are 0·88 g cm⁻³ and 3·12 g cm⁻³ respectively; H = 1, C = 12, Br = 80.)

(b) Suggest two reasons why the particular reactant should be used in excess.

3. State which of the following react with bromobenzene only, with bromoethane only, with both, or with neither: (a) hydrogen (nickel catalyst), (b) boiling aqueous potassium(I) hydroxide, (c) ammonia, (d) silver(I) ethanoate, (e) magnesium (in the presence of ethoxyethane).

4. Write equations for the reactions (if any) which occur between (a) bromobenzene and nitrating mixture, (b) chloro-4-methylbenzene and aqueous sodium(I) hydroxide, (c) chloro-4-nitrobenzene and aqueous sodium(I) hydroxide, (d) bromo-2-methylbenzene and boiling dilute nitric acid.

5. Name the organic products of the reactions which occur in question 4, and specify the orientations of the compounds where appropriate.

6. Which of the following statements are true?

(a) The C—Cl bond length is shorter in chlorobenzene than in chloromethane.

(b) A halogen atom attached to the benzene nucleus cannot be removed.

(c) Chlorobenzene undergoes nitration more slowly than benzene.

(d) Chlorination of the benzene nucleus involves heterolytic fission of the Cl—Cl bond.

(e) Chlorination of methylbenzene in the nucleus is assisted by direct sunlight.

7. What are the names of the following: (a) Cl^+, (b) $C_6H_5.CCl_3$, (c) $CH_3.COOCH_2.C_6H_5$, (d) $C_6H_5.CH_2NH_2$, (e) $C_6H_5.CH_2OH$?

22 Nitro- and Sulphonic-Derivatives

Nitration

Nitration is the process of substituting one or more hydrogen atoms in a molecule by the corresponding number of nitro (—NO$_2$) groups.

Up to three hydrogen atoms of the benzene molecule can be substituted by nitro groups to give nitrobenzene, $C_6H_5NO_2$, dinitrobenzene, $C_6H_4(NO_2)_2$, and trinitrobenzene, $C_6H_3(NO_2)_3$. Nitration is usually carried out with a mixture of nitric(V) acid and sulphuric(VI) acid. The extent to which nitration occurs depends on:

(*i*) the concentration of the acids;
(*ii*) the proportions of the reactants;
(*iii*) the temperature;
(*iv*) the duration of the reaction.

For the introduction of a single nitro group into the benzene molecule a mixture of concentrated nitric(V) acid and concentrated sulphuric(VI) acid ('nitrating mixture') is used. A little of the disubstitution derivative is formed at the same time, but dinitrobenzene cannot be satisfactorily prepared unless a mixture of fuming nitric(V) acid and concentrated sulphuric(VI) acid is employed. The replacement of the third hydrogen atom requires still more drastic conditions. To obtain trinitrobenzene it is necessary to heat dinitrobenzene with a mixture of fuming nitric(V) acid and fuming sulphuric(VI) acid for several days.

It should be noted that nitrobenzene is not an ester and the process of nitration is not reversible.

$$C_6H_6 + HNO_3 \rightarrow C_6H_5NO_2 + H_2O$$

Experiments have shown that the effective agent in nitration is the *nitryl* (nitronium) cation NO_2^+. This is formed by inter-action of concentrated nitric(V) acid and concentrated sul-phuric(VI) acid as shown.

$$HNO_3 + 2H_2SO_4 \rightleftharpoons NO_2^+ + H_3O^+ + 2HSO_4^-$$

The mechanism of nitration consists of addition of the nitryl cation to the nucleus, followed by removal of a hydrogen atom as a proton by the basic HSO_4^- ion (or a water molecule).

$$ArH + NO_2^+ \rightarrow \overset{+}{Ar}\!\!\begin{array}{c} \diagup H \\ \diagdown NO_2 \end{array} \xrightarrow{HSO_4^-} ArNO_2 + H_2SO_4$$

Since the sulphuric(VI) acid is re-formed during the reaction it functions as a catalyst. It also acts as a dehydrating agent. Concentrated nitric(V) acid contains about 30 per cent of water, which causes dissociation of the acid into hydrogen ions and nitrate(V) ions. By absorbing the water the concentrated sulphuric(VI) acid greatly reduces the ionic dissociation, and the bulk of the nitric(V) acid is kept in the form of HNO_3 molecules.

The presence of nitryl cations in nitrating mixture can be demon-strated spectroscopically. Their importance in nitration is shown by the fact that, if nitrating mixture is progressively diluted until the ions can no longer be detected spectroscopically, nitration ceases. Further evidence of the conversion of nitric(V) acid to nitryl cations is obtained from cryoscopic measurements on solutions of nitric(V) acid in sulphuric(VI) acid. If small amounts of pure nitric(V) acid are dissolved in pure sulphuric(VI) acid, the depression of freezing point of the latter is four times the depression calculated for molecular nitric acid; that is, four times as many particles are present in solution. This agrees with complete conversion of the nitric(V) acid into NO_2^+ ions in accordance with the equation given earlier.

The existence of nitryl cations has also been shown by the preparation of crystalline nitryl salts such as nitryl nitrate(V), $NO_2^+NO_3^-$, and nitryl chlorate(VII) (nitronium perchlorate), $NO_2^+ClO_4^-$. The ionic structures of these compounds, which are good nitrating agents, have been proved by X-ray analysis.

The nitryl cation mechanism is not the only one by which nitration can take place. Some very reactive compounds (e.g. phenol) are nitrated by dilute nitric(V) acid, in which NO_2^+ ions are absent. The reaction mechanism in these cases has not yet been fully established.

NITROBENZENE, $C_6H_5NO_2$

Boiling point: 210°C   *Density:* 1·21 g cm^{-3}

Laboratory Preparation of Nitrobenzene

The amounts of benzene, nitric(V) acid, and sulphuric(VI) acid theoretically required for this preparation are in the ratio indicated by the formulæ

$$C_6H_6 : HNO_3 : 2H_2SO_4$$

Usually, however, an excess of nitric(V) acid is used, as the acid is not completely converted into nitryl cations. The preparation is carried out as now described.

Experiment

Put 10 cm^3 of concentrated nitric(V) acid into a dry 50 cm^3 pear-shaped flask. Add 10 cm^3 of concentrated sulphuric(VI) acid in small portions, swirling the flask round and cooling it in a water-bath. Add 8 cm^3 of benzene in portions of about 1 cm^3 to the cold acid mixture. During the addition of the benzene (which should take about ten minutes) again swirl the flask round to ensure mixing of the liquids, and cool it in the water-bath. *At no stage must the flask be allowed to become too hot to hold comfortably on the hand.* If the temperature rises above 55°C, the reaction tends to become violent and also considerable amounts of dinitrobenzene are formed. When all the benzene has been added attach a reflux condenser to the flask, and complete the nitration by heating the flask in the water-bath for twenty minutes, keeping the temperature of the water at about 60°C.

Purification. Allow the flask to cool, and pour the contents into about 100 cm^3 of water in a beaker. The nitrobenzene forms an oily lower layer, while most of the remaining acid dissolves in the upper layer of water. Stir the liquids with a glass rod to remove as much acid as possible from the nitrobenzene. Decant off most of the aqueous layer, and transfer the remaining liquid to a separating-funnel. Shake the liquid in turn with 20 cm^3 of water, 20 cm^3 of 'bench' sodium(I) hydroxide solution, and again 20 cm^3 of water, in each case running off the lower layer of nitrobenzene. Dry the latter for a few hours in a stoppered flask with anhydrous calcium(II) chloride.

Finally distil the nitrobenzene, using an air-condenser. (A water-condenser is liable to crack owing to the high temperature of the vapour.) Collect the liquid which passes over at 206°–211°C. There is danger of an explosion if the distillation is continued much above 215°C. The yield is about 9·5 g. A small residue of 1,3-dinitrobenzene remains in the flask.

Properties of Nitrobenzene

Nitrobenzene is usually seen as a yellowish liquid, although when it is carefully purified it is colourless. It has an almond-like odour, and freezes at 5·5°C. The liquid is almost insoluble in water, but mixes with ethanol, ethoxyethane, and benzene in all proportions. Nitrobenzene itself is a good solvent for many organic compounds. Both the liquid and the vapour are poisonous. Nitrobenzene burns with a smoky flame.

The nitro group of nitrobenzene, like the chlorine atom of chlorobenzene, is firmly attached to the nucleus and cannot be removed by any simple treatment, such as boiling nitrobenzene with sodium(I) hydroxide solution. The only important reaction of the nitro group is that of reduction. The benzene nucleus, however, is susceptible to attack, and under suitable conditions can be further nitrated, chlorinated, brominated, and sulphonated.

1. Reduction. Reduction of a nitro-compound in acid solution converts the nitro group into an amino group. This change is of great importance in the manufacture of dyestuffs. Nitrobenzene yields phenylamine.

$$C_6H_5NO_2 + 6H \rightarrow C_6H_5NH_2 + 2H_2O$$

Reduction can be carried out with hydrochloric acid and a suitable metal (tin, iron, or zinc).

2. Further Substitution in the Nucleus. Nitrobenzene undergoes substitution much more slowly than benzene. This is owing to strong deactivation of the nucleus by the nitro group, which has both a $-M$ effect and a $-I$ effect. Nitrobenzene is thus analogous to benzenecarbaldehyde, or benzaldehyde (Chapter 20).

The $-M$ effect arises as follows. The structure of the nitro group is represented by either formula A or B.

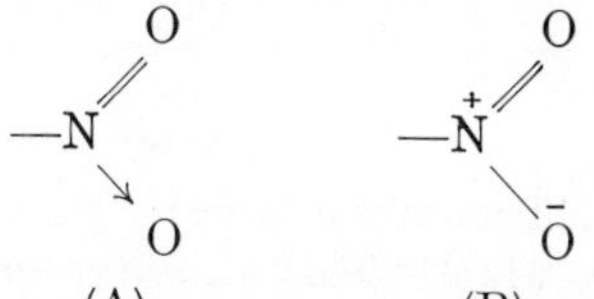

(A)   (B)

It will be seen that the nitro group (like the —CHO group in benzene-carbaldehyde) contains a suitably situated double bond which can conjugate with the bonds in the nucleus. Thus nitrobenzene is a resonance hybrid with five contributing structures. In addition to the usual two Kekulé structures there are three mesomeric structures of the form shown, this form being derived by withdrawal of two electrons from the nucleus.

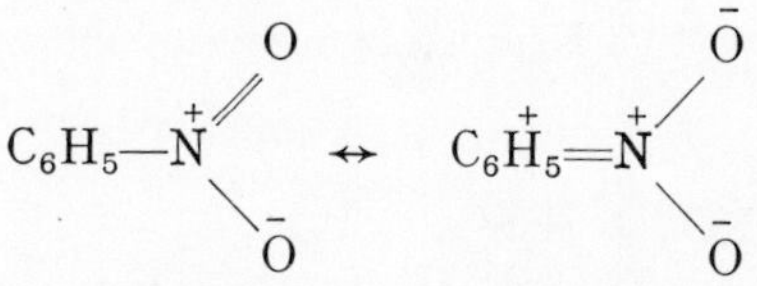

2 structures   3 structures

The resulting positive charge on the nucleus can be located on the carbon atoms in the 1,2- or 1,4- positions, but not on the one in the 1,3- position. Hence the latter has the highest electron density, and is preferentially attacked by an electrophilic reagent like Cl^+ or NO_2^+.

If nitrobenzene is treated with chlorine or bromine in the presence of a halogen carrier, a hydrogen atom in the 3-position is substituted. In the case of chlorine the product is chloro-3-nitrobenzene.

chloro-3-nitrobenzene

This reaction should be contrasted with the nitration of chlorobenzene (Chapter 21), which yields a mixture of the 1,2- and 1,4-isomers of chloronitrobenzene.

The introduction of a second nitro group into the nucleus requires more drastic conditions than that of the first. Although a little 1,3-dinitrobenzene, $C_6H_4(NO_2)_2$, is formed when nitrobenzene is heated with ordinary nitrating mixture, a good yield can be obtained only with a mixture of fuming nitric(V) acid and concentrated sulphuric(VI) acid.

Sulphonation of nitrobenzene occurs when the latter is heated with fuming sulphuric(VI) acid at 110°–115°C for about 45 minutes. The product is 3-nitrobenzenesulphonic acid.

Chemical Tests for Nitrobenzene

(*i*) *Reduction to Phenylamine.* Put a little granulated zinc and 3–4 cm³ of dilute hydrochloric acid into a test-tube. When hydrogen is freely evolved add one or two drops of nitrobenzene and shake the tube well for a few minutes. Then pour off the liquid into another test-tube, cool it under the tap, and make the liquid alkaline by adding sodium(I) hydroxide solution. Finally, add one or two drops of sodium(I) chlorate(I) (sodium hypochlorite) solution. With the latter the phenylamine which is present gives a violet colour.

(*ii*) *Nitration.* To 2 cm³ of concentrated nitric(V) acid in a boiling tube add 2 cm³ of concentrated sulphuric(VI) acid and 1 cm³ of nitrobenzene. Heat the mixture carefully (without boiling) for five minutes in the fume-cupboard, *keeping the mixture well shaken.* Pour the remaining liquid into a large amount of water in a beaker. A yellow precipitate of 1,3-dinitrobenzene is formed.

1,3-DINITROBENZENE, $C_6H_4(NO_2)_2$

Melting point: 90°C   *Boiling point:* 303°C

Laboratory Preparation of 1,3-Dinitrobenzene

Although 1,3-dinitrobenzene can be obtained directly from benzene, it is more conveniently prepared by further nitration of nitrobenzene by means of fuming nitric(V) acid and concentrated sulphuric(VI) acid.

Experiment

(The preparation should be carried out in the fume-cupboard, as poisonous fumes of nitrogen dioxide are given off). Gradually mix 5 cm³ of concentrated sulphuric(VI) acid with 4 cm³ of fuming nitric(V) acid in a 100 cm³ round-bottomed flask, and add a few fragments of porous pot. Support the flask in a water-bath, and add 3 cm³ of nitrobenzene in small portions, shaking the flask well after each addition. *It is most important that the liquids should be thoroughly mixed together.* Heat is evolved, and the contents of the flask darken in colour. When all the nitrobenzene has been added heat the flask in the water-bath, keeping the water boiling gently. During the heating shake the flask frequently to mix the contents. After half an hour withdraw a few drops of the liquid with a long glass tube and run them into some water in a test-tube. Repeat the sampling from time to time, and stop the heating when the drops solidify in the water. Allow the contents of the flask to cool partially, and then pour the liquid into 100 cm³ of cold water. The dinitrobenzene solidifies as a pale-yellow mass. Filter the solid at the pump, and wash it with cold water until the filtrate ceases to turn blue litmus red. The yield is about 3·5 g.

Purification. The crude product contains about 7 per cent of 1,2- and 1,4-dinitrobenzene. It can be purified from these by recrystallization from ethanol.

Properties of 1,3-Dinitrobenzene

The dinitro-derivative is usually pale yellow in colour, although when perfectly pure it is colourless. It is insoluble in water. It has only a small solubility in cold ethanol, but is fairly soluble in hot ethanol. Like nitrobenzene it is poisonous.

Both of the nitro groups in 1,3-dinitrobenzene are firmly attached to the nucleus, and are unaffected by boiling solutions of alkalies. The most important reaction of the compound, as in the case of nitrobenzene, is that of reduction. The two nitro groups can be reduced in stages by passing hydrogen sulphide into a mixture of ethanolic dinitrobenzene and concentrated ammonia.

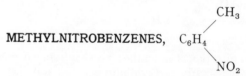

With tin and concentrated hydrochloric acid 1,3-dinitrobenzene is reduced directly to benzene-1,3-diamine. Both 3-nitrophenylamine and benzene-1,3-diamine are used to make dyestuffs.

METHYLNITROBENZENES, $C_6H_4 \begin{cases} CH_3 \\ NO_2 \end{cases}$

As noted in Chapter 20, methylbenzene undergoes nitration more rapidly than benzene owing to activation of the nucleus (particularly in the 2- and 4- positions) by the alkyl radical. If methylbenzene is treated with a mixture of concentrated nitric(V) acid and concentrated sulphuric(VI) acid, it yields a mixture of methyl-2-nitrobenzene (*o*-nitrotoluene) and methyl-4-nitrobenzene (*p*-nitrotoluene).

$$C_6H_5.CH_3 + HNO_3 \rightarrow C_6H_4 \begin{cases} CH_3 \\ NO_2 \end{cases} + H_2O$$

Methyl-2-nitrobenzene is a liquid of boiling point 222°C; methyl-4-nitrobenzene is a solid which melts at 52°C and boils at 238°C. The isomers resulting from nitration can therefore be separated either by cooling the mixture or by subjecting it to fractional distillation. Methyl-3-nitrobenzene can only be prepared indirectly.

The methylnitrobenzenes are 'intermediates' in the manufacture of dyestuffs. They undergo reduction in the same way as nitrobenzene, and yield the corresponding methylphenylamines (toluidines).

$$C_6H_4{\overset{CH_3}{\underset{NO_2}{}}} + 6H \rightarrow C_6H_4{\overset{CH_3}{\underset{NH_2}{}}} + 2H_2O$$

Further nitration of either the 1,2- or 1,4-isomer of methylnitrobenzene gives firstly methyldinitrobenzene and then the symmetrical methyl-2,4,6-trinitrobenzene. The latter is commonly called trinitrotoluene (TNT), and has the structure shown.

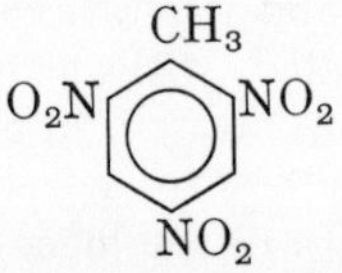

methyl-2,4,6-trinitrobenzene

TNT is a pale-yellow solid which melts at 81°C. It is quite stable under normal conditions, and when heated in air it burns with a smoky flame without explosion. If it is detonated, however, it explodes violently. The trinitroderivative of methylbenzene is actually a less powerful explosive than trinitrobenzene, but the latter has no commercial value because it is difficult to manufacture. When TNT explodes the carbon and hydrogen in the molecule undergo 'internal oxidation' by the oxygen present. The molecule contains insufficient oxygen, however, for complete oxidation. In 'high explosives' (e.g. Amatol) the deficiency is made good by incorporating ammonium nitrate(V), which is rich in oxygen.

SULPHONIC-DERIVATIVES

Sulphonation

Sulphonation is the process of substituting one or more hydrogen atoms in a molecule by the corresponding number of sulphonic (—SO₃H) groups.

As seen in Chapter 19, aromatic hydrocarbons differ from aliphatic hydrocarbons in undergoing sulphonation when they are treated with concentrated or fuming sulphuric(VI) acid. Thus, benzene readily yields benzenesulphonic acid with fuming sulphuric(VI) acid.

$$C_6H_6 + H_2SO_4 \rightleftharpoons C_6H_5SO_3H + H_2O$$

Unlike nitration, sulphonation is reversible, although the equilibrium point is well to the right in the presence of excess of sulphuric(VI) acid.

The sulphonic group is derived from sulphuric(VI) acid, and must not be confused with the hydrogensulphate(IV) ion (bisulphite ion), which is derived from sulphuric(IV) acid (sulphurous acid), H_2SO_3. Sulphuric(VI) acid and benzene-sulphonic acid have the structures now shown (the arrows represent co-ordinate covalent-bonds).

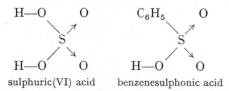

sulphuric(VI) acid   benzenesulphonic acid

It will be noticed that one of the two acidic hydrogen atoms of the sulphuric(VI) acid molecule is still present in the molecule of benzenesulphonic acid. This explains why the latter behaves as an acid comparable in strength with sulphuric(VI) acid itself. When benzenesulphonic acid is dissolved in water it is highly dissociated as shown.

$$C_6H_5SO_3H \rightleftharpoons C_6H_5SO_3^- + H^+$$

The replacement of the replaceable hydrogen of benzenesulphonic acid by metals results in the formation of salts such as sodium(I) benzenesulphonate ($C_6H_5SO_3^-Na^+$).

The usual agents employed in the sulphonation of aromatic compounds are the following:

(*i*) *Concentrated Sulphuric(VI) Acid.* It is usually necessary to heat the compound (often for long periods) with excess of concentrated sulphuric(VI) acid. Unless excess of the acid is used the reaction tends to slow down owing to dilution of the acid by the water formed. Iodine or kieselguhr is sometimes employed as a catalyst. In a few cases (e.g. phenol) concentrated sulphuric(VI) acid brings about sulphonation at ordinary temperatures.

(*ii*) *Fuming Sulphuric(VI) Acid (Oleum).* This is a much quicker sulphonating agent than concentrated sulphuric(VI) acid. It is particularly valuable when the compound is difficult to sulphonate, as in the case of nitrobenzene.

(*iii*) *Chlorosulphonic Acid.* Chlorosulphonic acid is greatly favoured as a sulphonating agent nowadays. It reacts rapidly, and, if used in the correct amount, yields a product which is easily purified.

$$C_6H_6 + ClSO_3H \rightarrow C_6H_5SO_3H + HCl$$

The mechanism of sulphonation is still uncertain. According to one theory the attack on the nucleus is made by the SO_3H^+ ion, which is believed to be present in small amounts in concentrated sulphuric(VI) acid and oleum. In this case the mechanism could be represented as follows:

$$ArH + SO_3H^+ \rightleftharpoons Ar\overset{+}{\underset{SO_3H}{\overset{H}{<}}} \xrightleftharpoons[]{HSO_4^-} ArSO_3H + H_2SO_4$$

The —SO_3H group is one of the most strongly hydrophilic of all groups. Most compounds containing this group are very soluble in water. In the manufacture of drugs and dyes the primary products obtained are often insoluble in water. They are rendered soluble ('solubilized') by preparing the sulphonic-derivatives. If the final products are required as neutral substances, the sulphonic-derivatives are converted into sodium(I) salts by neutralization.

The sulphonic group, like the nitro group, deactivates the nucleus, and therefore further substitution by sulphuric(VI) acid requiries more drastic conditions. When benzene is heated with excess of fuming sulphuric(VI) acid at 200°C, firstly benzene-1,3-disulphonic acid, $C_6H_4(SO_3H)_2$, and then benzenetrisulphonic acid, $C_6H_3(SO_3H)_3$, are formed. This is as far as sulphonation of benzene can be carried.

BENZENESULPHONIC ACID-$1\frac{1}{2}$-WATER

$$C_6H_5SO_3H . 1\tfrac{1}{2}H_2O$$

Melting point: 46°C (*approx.*)   *Boiling point: decomposes*

Preparation of Benzenesulphonic Acid

Benzenesulphonic acid can be prepared by boiling benzene with excess of concentrated sulphuric(VI) acid until the upper layer of benzene has been absorbed. This may take from six to twenty hours, according to how frequently the liquids are shaken together.

$$C_6H_6 + H_2SO_4 \rightleftharpoons C_6H_5SO_3H + H_2O$$

In practice the time required for sulphonation is reduced by employing a mixture of concentrated, and fuming, sulphuric(VI) acid. Benzenesulphonic acid itself is little used, and it is usually isolated in the form of its sodium(I) salt (which is used in the manufacture of phenol).

The sodium(I) salt is obtained by a process known as *liming*, which utilizes the solubility of the calcium(II) salt in water. The liquid remaining after sulphonation is diluted with water and neutralized with milk of lime. This gives a precipitate of calcium(II) sulphate(VI) (with the excess of sulphuric(VI) acid) and a solution of calcium(II) benzenesulphonate (with the benzenesulphonic acid).

$$2C_6H_5SO_3H + Ca(OH)_2 \rightarrow (C_6H_5SO_3)_2Ca + 2H_2O$$

After filtering, the solution of calcium(II) benzenesulphonate is treated with sodium(I) carbonate solution. A precipitate of calcium(II) carbonate is formed, and sodium(I) benzenesulphonate is left in solution.

$$(C_6H_5SO_3)_2Ca + Na_2CO_3 \rightarrow 2C_6H_5SO_3Na + CaCO_3 \downarrow$$

The sodium(I) salt is obtained by filtering and crystallizing out the filtrate.

Benzenesulphonic acid itself can be prepared by adding dilute sulphuric(VI) acid instead of sodium(I) carbonate to the calcium(II) benzenesulphonate solution. The precipitate of calcium(II) sulphate(VI) is filtered, and the filtrate is evaporated.

$$(C_6H_5SO_3)_2Ca + H_2SO_4 \rightarrow 2C_6H_5SO_3H + CaSO_4 \downarrow$$

Sulphonation of methylbenzene takes place more rapidly than that of benzene owing to activation of the nucleus by the CH_3— radical. A mixture of the 1,2- and 1,4-isomers of methylbenzenesulphonic acid (toluenesulphonic acid) is formed.

$$C_6H_5.CH_3 + H_2SO_4 \rightleftharpoons C_6H_4{\overset{\displaystyle CH_3}{\underset{\displaystyle SO_3H}{<}}} + H_2O$$

2-methylbenzenesulphonic acid is used in the manufacture of the sweetening agent saccharin.

Reactions of Benzenesulphonic Acid

The sulphonic group in arylsulphonic acids is not attached to the nucleus as firmly as the nitro group in nitro-derivatives. It can be replaced by other atoms or groups, although fairly drastic conditions are required.

I. Reversal of Sulphonation. When benzenesulphonic acid is heated with dilute sulphuric(VI) acid in a sealed vessel the sulphonic group is replaced by a hydrogen atom, and benzene is reformed.

$$C_6H_5SO_3H + H_2O \rightleftharpoons C_6H_6 + H_2SO_4$$

2. Reaction with Fused Sodium(I) Hydroxide. The sulphonic group of benzenesulphonic acid cannot be removed by boiling the acid with sodium(I) hydroxide solution. If the acid, or its sodium(I) salt, is fused with solid sodium(I) hydroxide, however, the sulphonic group is replaced by a hydroxyl group. Phenol is first produced, but this reacts with excess of the alkali to give sodium(I) phenoxide.

$$C_6H_5SO_3Na + NaOH \rightarrow C_6H_5OH + Na_2SO_3$$
$$C_6H_5OH + NaOH \rightarrow C_6H_5ONa + H_2O$$

3. Reaction with Fused Sodium(I) Cyanide. When sodium(I) benzenesulphonate is fused with sodium(I) cyanide the sulphonic group is replaced by a cyanide group to give benzenecarbonitrile, or benzonitrile.

$$C_6H_5SO_3Na + NaCN \rightarrow C_6H_5CN + Na_2SO_3$$

This reaction results in a carbon atom being attached to the benzene nucleus and therefore provides a means of synthesizing aromatic compounds. Thus, when hydrolysed, benzenecarbonitrile yields benzenecarboxylic, or benzoic acid ($C_6H_5.COOH$).

4. Reaction with Phosphorus Pentachloride. The molecule of benzenesulphonic acid contains a hydroxyl group and, as might be expected, this reacts with phosphorus pentachloride. If the anhydrous acid is heated with phosphorus pentachloride, hydrogen chloride is evolved and the acid chloride, *benzenesulphonyl chloride*, is formed.

$$C_6H_5.SO_2OH + PCl_5 \rightarrow C_6H_5.SO_2Cl + POCl_3 + HCl$$

The sulphonyl chlorides are intermediates in the manufacture of *sulphonamides*. These include the 'sulpha' drugs, which are used in treating dysentery, scarlet fever, pneumonia, etc. Sulphonamides are obtained by shaking the sulphonyl chlorides with concentrated aqueous ammonia.

$$C_6H_5.SO_2Cl + 2NH_3 \rightarrow C_6H_5.SO_2NH_2 + NH_4Cl$$
benzenesulphonamide

Chemical Test for Benzenesulphonic Acid or Sodium(I) Benzenesulphonate

Fusion with Sodium(I) Hydroxide. Mix together in a nickel crucible I g of the compound, half a dozen pellets of sodium(I) hydroxide, and 4–5 drops of water. Place the crucible on a pipe-clay triangle, and guard against 'spitting' of the mixture by interposing a glass screen (a clock-glass held over the crucible is satisfactory). Warm the mixture gently at first, and then heat it with a small flame for ten minutes so that the mixture is kept just molten. Allow the crucible to cool, and dissolve the residue by warming it with a little water. Transfer the solution to a boiling-tube and carefully add hydrochloric acid of medium concentration in small amounts. The acid first neutralizes the remaining alkali and then liberates sulphur dioxide from the sodium(I) sulphate(IV) (sodium sulphite) formed in the reaction. The gas can be recognized by its smell and by turning potassium(I) chromate(VI) paper green. Transfer the remaining solution to an evaporating-dish, and boil it. A strong smell of phenol is given off.

EXERCISE 22

Nitro-derivatives

1. (a) What is the formula of the nitryl cation? (b) Give the equation for the reaction by which nitryl cations are formed in nitrating mixture.

2. Indicate the steps in the mechanism postulated for nitration of benzene by nitryl cations.

3. Which of the following are correct for nitration of benzene to nitrobenzene?

(a) An excess of nitrating mixture should be used.

(b) The reaction is exothermic.

(c) The reaction is reversible.

(d) The yield of nitrobenzene is greater at higher temperatures.

(e) The sulphuric(VI) acid acts both as a catalyst and dehydrating agent.

4. Give the names of the organic compounds formed when the following are reduced with tin and hydrochloric acid: (a) nitro-benzene, (b) 1,3-dinitrobenzene, (c) methyl-4-nitrobenzene.

Sulphonic-derivatives

5. Write the conventional structures of (a) the hydrogen-sulphate(IV) (bisulphite) ion (HSO_3^-), (b) the sulphonic group ($-SO_3H$), (c) the benzenesulphonate ion ($C_6H_5SO_3^-$).

6. Which of the following are true of benzenesulphonic acid: (a) it is an oily liquid, (b) it is soluble in water, (c) it is a strong acid, (d) it forms a precipitate with barium(II) chloride solution, (e) the $-SO_3H$ group is not removed by boiling aqueous sodium(I) hydroxide?

7. Give the names and formulæ of the chief organic compounds formed in the reactions between (a) methylbenzene and concentrated sulphuric(VI) acid, (b) sodium(I) benzenesulphonate and excess of sodium(I) hydroxide, (c) benzenesulphonic acid and phosphorus pentachloride.

23 Aromatic Amines—Diazonium Compounds

AROMATIC AMINES

Aromatic amines, like aliphatic amines, are classified into primary, secondary, and tertiary amines. **A primary aromatic amine** *is a compound derived from an aromatic hydrocarbon by substituting a hydrogen atom of the latter by an amino ($—NH_2$) group.* Thus, by substituting a hydrogen atom of benzene or methylbenzene by an amino group, the following primary aromatic amines are obtained (the names given in brackets below the systematic names are current trivial names):

Primary Aromatic Amines

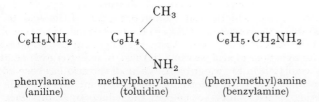

phenylamine (aniline)	methylphenylamine (toluidine)	(phenylmethyl)amine (benzylamine)

Alternatively, these amines can be regarded as derived from ammonia by replacing a hydrogen atom of the ammonia molecule by the aromatic radicals phenyl, $C_6H_5—$, methylphenyl, $C_6H_4(CH_3)—$, and phenylmethyl, $C_6H_5.CH_2—$. Further substitution in the ammonia molecule by either alkyl or aromatic radicals produces secondary or tertiary aromatic amines. In this way secondary and tertiary aromatic amines which are partially aliphatic or wholly aromatic in character are obtained.

Secondary Aromatic Amines	Tertiary Aromatic Amines
$C_6H_5.NH.CH_3$ *N*-methylphenylamine	$C_6H_5.N(CH_3)_2$ *N,N*-dimethylphenylamine
$(C_6H_5)_2NH$ diphenylamine	$(C_6H_5)_3N$ triphenylamine

There are important differences between primary amines which have the functional $NH_2—$ group attached to the nucleus and those in which the group is present in a side-chain. For this reason the two types of primary aromatic amine are studied separately.

NUCLEAR AMINES

The nuclear amines, of which phenylamine is by far the most important, are colourless liquids or solids when pure. The commercial compounds usually have a yellowish colour and, on keeping, turn brown and then black. This is probably due to oxidation of traces of sulphur compounds, which are present as impurities. The liquids have a faint odour, which is neither ammoniacal nor fishy (contrast aliphatic amines). The compounds are only slightly soluble in water, the effect of the hydrophobic nucleus outweighing that of the hydrophilic $—NH_2$ group. They dissolve readily, however, in organic solvents like ethoxyethane and benzene. Both the liquids and the vapours are toxic.

293

PHENYLAMINE, $C_6H_5NH_2$

Boiling point: 184°C   *Density:* 1·02 g cm^{-3}

Laboratory Preparation of Phenylamine

Primary aromatic amines are usually prepared by reduction of the corresponding nitro-compounds. Thus phenylamine (aniline) is obtained by reducing nitrobenzene with tin and concentrated hydrochloric acid. (Tin is preferred to zinc because the latter gives rise to wasteful side-reactions).

$$C_6H_5NO_2 + 6H \rightarrow C_6H_5NH_2 + 2H_2O$$

The tin is first converted to tin(II) chloride, but this also reduces nitrobenzene to phenylamine, so that ultimately the tin is oxidized to tin(IV) chloride. The latter combines with the phenylamine and hydrochloric acid to form a complex salt called *phenylammonium hexachlorostannate(IV)*.

$$2C_6H_5NH_2 + 2HCl + SnCl_4 \rightleftharpoons (C_6H_5NH_3)_2SnCl_6$$

When sodium(I) hydroxide solution is added the complex salt decomposes and liberates phenylamine. This is separated by steam distillation. The preparation can be divided into three stages as now described.

Experiment

(*i*) *Reduction of Nitrobenzene.* Put into a 50 cm³ pear-shaped flask 5 cm³ of nitrobenzene and 12 g of granulated tin. Partially immerse the flask in a beaker of cold water standing on a tripod and gauze,

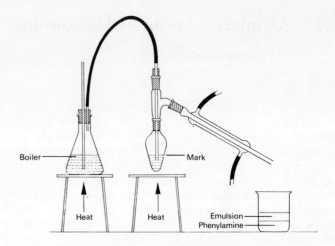

Fig. 23-1. Steam distillation of phenylamine

and attach a reflux air-condenser. Measure out 25 cm³ of concentrated hydrochloric acid. Add about 5 cm³ of the acid to the flask down the condenser, and swirl the flask round. Heat is evolved and the flask becomes warm. At intervals of about two minutes add further portions of acid, swirling the flask round after each addition and cooling it in the water-bath. When all the acid has been added heat the water in the water-bath to boiling point, and maintain the heating for ten minutes to complete the reduction. At this stage the phenylamine is present as the complex salt.

(*ii*) *Liberation of Phenylamine and Steam Distillation.* Allow the apparatus to cool, and remove the water-bath. Prepare a solution of 15 g of sodium(I) hydroxide pellets in 15 cm³ of water in a conical flask, and cool the latter under the tap. Detach the pear-shaped flask from the condenser and add the sodium(I) hydroxide solution in portions of 2–3 cm³, swirling the flask round and cooling it under the tap. Divide the liquid into two roughly equal parts, keeping one part in the flask. Add a little powdered pumice, and connect the flask to an apparatus for steam distillation (Fig. 23-1).

Pass steam into the flask until the liquid coming over no longer looks milky. During the distillation adjust the flames under the boiler and flask so as to keep the level of the liquid in the flask about the same (mark the original level with chalk or a strip of paper). The volume of the distillate should be about 30 cm³. Repeat the steam distillation with the second portion of the alkaline liquid and combine the two distillates.

(*iii*) *Ethoxyethane Extraction, Drying, and Fractional Distillation.* The distillate separates into two layers—a lower oily layer of phenylamine and an upper milky-looking layer which consists partly of a solution, and partly of an emulsion, of phenylamine in water. Phenylamine has only a small solubility in water, but an appreciable amount can be recovered from the aqueous solution by adding common salt, phenylamine being insoluble in a concentrated solution of this compound. Dissolve about 10 g of common salt in the distillate in the beaker, and then transfer the distillate to a separating-funnel. Having first extinguished any flames in the vicinity, add 10 cm³ of ethoxyethane to the funnel and insert the stopper. Invert the funnel and shake the liquids together vigorously, opening the tap from time to time to relieve the pressure of the ethoxyethane vapour. Run off the lower aqueous layer into a beaker, and pour the ethoxyethane extract of phenylamine into a dry conical flask. Put the aqueous liquid back into the separating-funnel, and repeat the extraction with another 10 cm³ of ethoxyethane. Combine the two extracts, add some pellets of potassium(I) hydroxide, and leave the liquid in the stoppered flask for two or three hours to dry. (Anhydrous calcium(II) chloride cannot be used to dry amines because it combines with them.)

Decant the dried solution into the dry pear-shaped flask and add a little powdered pumice. Connect the flask through an adapter, fitted with a 250°C thermometer, to a condenser and cooled receiver, the latter being arranged as in the preparation of ethoxyethane (Fig. 8-3). Distil off the ethoxyethane on a water-bath previously heated to 70–80°C. When no more ethoxyethane is obtained remove the water-bath, and change the receiver. Also, run the water out of the condenser, so that it now acts as an air-condenser. Distil the phenylamine over a small direct flame, collecting separately the fraction which passes over at 180°–185°C. The yield is about 4 g (4 cm³).

Principle of Steam Distillation

Steam distillation is used to isolate an organic liquid (*i*) when the liquid tends to decompose at temperatures approaching its normal boiling point; or (*ii*) when the liquid is mixed with several other substances and cannot be separated conveniently by other methods (this is the case in the preparation of phenylamine). The liquid must be immiscible, or nearly immiscible, with water. Phenylamine and water dissolve in each other to a small extent, but for theoretical purposes they can be regarded as immiscible liquids. A liquid, or a mixture of immiscible liquids, boils when its vapour pressure is equal to the pressure of the atmosphere above it. If a mixture of two immiscible liquids is distilled each liquid contributes to the total vapour pressure, and therefore the mixture boils at a temperature which is lower than the boiling point of either liquid separately. The relative masses, m_1 and m_2, of the two liquids in the distillate are given by

$$\frac{m_1}{m_2} = \frac{p_1 M_r}{p_2 M_r{}'}$$

where p_1, p_2, M_r and $M_r{}'$ are the respective vapour-pressures and relative molecular masses of the liquids. The temperature at which a mixture of phenylamine and water distils is just below 100°C at standard atmospheric pressure (101 325 N m^{-2}). At this temperature the vapour-pressure of phenylamine is small compared with that of water, but its relative molecular mass is large. Hence a fairly large proportion of the distillate consists of phenylamine.

It was found by Young that at an atmospheric pressure of 101 900 N m^{-2} a mixture of phenylamine and water distilled at 98·75°C. At the latter temperature the individual vapour pressures of phenylamine and water are 5 813 N m^{-2} and 96 900 N m^{-2} respectively. (The sum of these is slightly greater than the atmospheric pressure given, but the discrepancy is due to the fact that phenylamine and water are not perfectly immiscible.) The relative molecular masses of phenylamine and water are 93 and 18. Inserting the values of the vapour pressures and relative molecular masses in the expression given previously produces the following results:

$$\frac{\text{Mass of phenylamine}}{\text{Mass of water}} = \frac{5\ 813 \times 93}{96\ 900 \times 18} = \tfrac{1}{3} \text{ (approximately)}$$

Manufacture and Uses of Phenylamine

Phenylamine is made on a large scale by catalysed reduction of nitrobenzene with hydrogen gas. The reduction may be carried

$$C_6H_5NO_2 + 3H_2 \rightarrow C_6H_5NH_2 + 2H_2O$$

out either in the liquid phase or gaseous phase, different catalysts being employed in the two cases. In one process nitrobenzene is diluted with a relatively large volume of phenylamine itself (to diminish side-reactions), and reduction takes place at 160°–170°C with hydrogen and a nickel catalyst. In another process a mixture of nitrobenzene vapour and hydrogen is passed through a bed of silica and finely divided copper (the catalyst) at 270°C.

Phenylamine is chiefly important as an intermediate in the production of other compounds. Over half of the industrial output is absorbed in the manufacture of antioxidants and vulcanization accelerators for rubber. It is also used to make plastics, dyes, drugs, and photographic chemicals.

(A) Reactions of the Amino Group in Phenylamine

In general the properties of an amino group attached to the nucleus are similar to those of the same group in aliphatic amines. In both nuclear amines and aliphatic amines the presence of a lone pair of electrons on the nitrogen atom gives them nucleophilic character, so that they react with electrophiles like acids, alkyl halides, and acid chlorides. Nuclear amines differ from aliphatic amines, however, in their special reaction with nitric(III) acid (nitrous acid) under suitable conditions.

1. **Basic Character.** Whereas primary aliphatic amines are stronger bases than ammonia, corresponding nuclear amines are weaker. An aqueous solution of phenylamine does not affect litmus, although the solution contains a few phenylammonium (anilinium) ions and hydroxyl ions. These are formed by the nitrogen atom donating a share in its lone pair of electrons to a proton from a water molecule.

$$C_6H_5NH_2 + HOH \rightleftharpoons C_6H_5NH_3{}^+ + OH^-$$

Phenylamine combines with acids to form substituted ammonium salts. The salts most frequently encountered are the following:

$C_6H_5NH_3{}^+Cl^-$
phenylammonium chloride

$(C_6H_5NH_3{}^+)_2SO_4{}^{2-}$
phenylammonium sulphate(VI)

$C_6H_5NH_3{}^+HSO_4{}^-$
phenylammonium hydrogensulphate(VI)

Phenylammonium hydrogensulphate(VI) is formed by phenylamine in the presence of excess of sulphuric(VI) acid.

Formerly the salts were regarded as addition compounds of the base and acid. Thus phenylammonium chloride was formulated as $C_6H_5NH_2.HCl$, and called aniline hydrochloride. Both the name and formula are still met with occasionally.

The salts of phenylamine are colourless crystalline substances which darken on keeping. They are mostly soluble in water, although the sulphate(VI) and hydrogensulphate(VI) are only sparingly soluble. They are partially hydrolysed by water, giving acidic solutions. This is in marked contrast to the salts of primary aliphatic amines, which form neutral solutions. Phenylammonium salts are decomposed in the cold by alkalies with liberation of phenylamine.

$$C_6H_5NH_3{}^+ + OH^- \rightarrow C_6H_5NH_2 + H_2O$$

2. The Isocyano Reaction. Primary aromatic amines resemble primary aliphatic amines in giving sickly-smelling isocyano-derivatives when warmed with trichloromethane and an ethanolic solution of potassium(I) hydroxide. With phenylamine the product is isocyanobenzene (phenyl isocyanide).

$$C_6H_5NH_2 + CHCl_3 + 3KOH \rightarrow C_6H_5NC + 3KCl + 3H_2O$$

The reaction is carried out as described under trichloromethane in Chapter 9.

3. Reactions with Nitric(III) Acid (Nitrous Acid). A primary nuclear amine yields different products with nitric(III) acid (acidified sodium(I) nitrate(III) solution) according to the temperature. At 0°–10°C a reaction called *diazotization* takes place,

and a *diazonium* compound is formed in solution. Thus, when phenylamine is dissolved in dilute hydrochloric acid, the solution cooled, and a similarly cooled solution of sodium(I) nitrate(III) is added, benzenediazonium chloride is produced. This is an ionic compound.

$$C_6H_5NH_2 + HCl + HNO_2 \rightarrow C_6H_5N_2{}^+Cl^- + 2H_2O$$
benzenediazonium chloride

Diazonium compounds (described later in this chapter) are important as intermediates in the preparation of dyes from nuclear amines.

The benzenediazonium ion is unstable in aqueous solution at higher temperatures. If the solution is warmed, nitrogen is evolved, and phenol is obtained.

$$C_6H_5N_2{}^+ + H_2O \rightarrow C_6H_5OH + N_2 + H^+$$

By adding together the last two equations and omitting particles common to both sides, the following equation, representing the overall reaction at higher temperatures, is obtained.

$$C_6H_5NH_2 + HNO_2 \rightarrow C_6H_5OH + N_2 + H_2O$$

This equation is similar to the one for the reaction between a primary aliphatic amine and nitric(III) acid (Chapter 15).

Experiment

Dissolve two or three drops of phenylamine in 5 cm³ of dilute hydrochloric acid in a test-tube, and cool the tube in a beaker containing iced water (or water made cold by dissolving ammonium nitrate(V) crystals). Similarly cool some dilute sodium(I) nitrate(III) solution.

Add 1–2 cm³ of the latter solution to the phenylammonium chloride solution, and allow the mixture to stand in the cooling-bath for two minutes. Then divide the resulting solution of benzenediazonium chloride into three parts in three separate test-tubes. Leave two of these in the cooling bath until required. Carry out the following tests:

(i) Heat the first portion of the liquid until it boils. Effervescence occurs owing to the liberation of nitrogen, and the smell of phenol can be detected at the mouth of the tube.

(ii) Dissolve 0·5 g of phenol in 5 cm³ of dilute sodium(I) hydroxide solution. To this solution add the second portion of benzenediazonium chloride solution drop by drop. A yellow dye is precipitated.

(iii) Repeat the last experiment, using naphthalen-2-ol (β-naphthol) in place of phenol. A brilliant red dye is precipitated.

The names of the dyes obtained in (ii) and (iii) and the equations for their formation are given later in this chapter.

4. Ethanoylation and Benzenecarbonylation.

As seen in Chapter 12, ethanoylation, or acetylation, refers to the substitution of a hydrogen atom in a molecule by the ethanoyl, or acetyl, group ($CH_3.CO—$). Phenylamine undergoes ethanoylation with ethanoyl chloride in the cold. A hydrogen atom of the amino group is substituted and N-phenylethanamide, or N-phenyl-acetamide, is formed. (The product is also well known under the trivial name of *acetanilide*.)

$$C_6H_5NH_2 + CH_3.COCl \rightarrow C_6H_5NH.CO.CH_3 + HCl$$

Experiment

To a few drops of phenylamine in a dry test-tube add a few drops of ethanoyl chloride *in single drops*. A vigorous reaction occurs, and when the remaining liquid is poured into cold-water a white precipitate of N-phenylethanamide is obtained.

In the usual laboratory method of preparing N-phenylethanamide a mixture of glacial ethanoic acid and ethanoic anhydride is used as the ethanoylating agent. (With the acid alone refluxing must be continued for about eight hours.)

$$C_6H_5NH_2 + CH_3COOH \rightarrow C_6H_5NH.COCH_3 + H_2O$$

$$C_6H_5NH_2 + (CH_3CO)_2O \rightarrow C_6H_5NH.COCH_3 + CH_3COOH$$

Experiment

Put into a dry 50 cm³ pear-shaped flask 8 cm³ of phenylamine, 8 cm³ of glacial ethanoic acid, and 8 cm³ of ethanoic anhydride. Attach a reflux condenser to the flask, and boil the mixture gently for half an hour over a small flame. Pour the remaining liquid into 100 cm³ of cold water in a beaker. Filter the precipitate of N-phenylethanamide at the pump, and recrystallize the crude product from a mixture of equal volumes of glacial ethanoic acid and water.

N-Phenylethanamide forms colourless crystals (m.p. 114°C). It is only slightly soluble in water, but dissolves readily in ethanol. It is used (as described presently) to 'protect' the amino group in preparing nuclear substitution derivatives of phenylamine. When N-phenylethanamide is refluxed with medium sulphuric(VI) acid it is hydrolysed and phenylamine is re-formed.

$$C_6H_5NH.COCH_3 + H_2O \rightleftharpoons C_6H_5NH_2 + CH_3COOH$$

Benzenecarbonylation, or *benzoylation*, of phenylamine corresponds to ethanoylation, and consists of substituting a hydrogen atom of the amino group by a benzenecarbonyl, or benzoyl, group ($C_6H_5.CO$—). When benzenecarbonyl, or benzoyl, chloride is added to phenylamine heat is evolved and a white precipitate of *N*-phenylbenzenecarboxamide, or *N*-phenylbenzamide, is formed. (The product is also known by the trivial name *benzanilide*.)

$$C_6H_5NH_2 + C_6H_5COCl \rightarrow C_6H_5NH.COC_6H_5 + HCl$$
N-phenylbenzenecarboxamide

The benzenecarbonylation of phenylamine is discussed further in Chapter 25.

5. Reaction with Alkyl Halides. If an ethanolic solution of phenylamine is heated in a sealed vessel with iodomethane, the hydrogen atoms of the amino group are replaced in turn by methyl groups. *N*-Methylphenylamine and *N,N*-dimethylphenylamine are obtained partly as free amines and partly in the form of substituted ammonium salts. The reactions are analogons to those which take place between aliphatic amines and iodomethane (Chapter 15).

$$C_6H_5NH_2 + CH_3I \rightarrow C_6H_5NHCH_3 + HI$$
$$\rightleftharpoons C_6H_5NH_2CH_3{}^+I^-$$
methylphenylammonium iodide

$$C_6H_5NHCH_3 + CH_3I \rightarrow C_6H_5N(CH_3)_2 + HI$$
$$\rightleftharpoons C_6H_5NH(CH_3)_2{}^+I^-$$
dimethylphenylammonium iodide

If sufficient alkyl halide is present and the reaction is continued, the final product is a quaternary ammonium compound.

$$C_6H_5N(CH_3)_2 + CH_3I \rightarrow C_6H_5N(CH_3)_3{}^+I^-$$
trimethylphenylammonium iodide

As in the case of aliphatic amines a mixture of products is always formed, the proportion of each depending on the proportions of the reactants and the duration of the reaction.

(B) Nuclear Substitution Reactions

Phenylamine is a resonance hybrid with five contributing structures corresponding to those of phenol (Chapter 20). In both compounds the $+M$ effect of the substituting group is much stronger than the $-I$ effect, and as a result the nucleus is highly activated, particularly in the 2-, 6-, and 4- positions. This leads to modifications in some of the usual substitutions reactions.

1. Halogenation. The carbon atoms in the 2-, 6-, and 4- positions in the phenylamine molecule are extremely sensitive to attack by chlorine or bromine. Thus when phenylamine is treated with chlorine water or bromine water the three hydrogen atoms attached to these carbon atoms are simultaneously replaced. The reaction takes place in the cold, and no halogen carrier is required.

2,4,6-tribromophenylamine

Experiment

Dissolve two drops of phenylamine in 5 cm³ of dilute hydrochloric acid. To the solution add a few drops of chlorine water or bromine water. A white precipitate of 2,4,6-trichlorophenylamine or 2,4,6-tribromophenylamine is formed at once.

To obtain the monochloro- and monobromo-derivatives of phenylamine the activating effect of the amino group on the nucleus is reduced by substituting one of its hydrogen atoms by the electron-attracting ethanoyl, or acetyl, group. Thus phenylamine is first converted into N-phenylethanamide (acetanilide), which is then treated with bromine. The monobromo-derivative of N-phenylethanamide is obtained (almost wholly as the 1,4-isomer). When the latter is refluxed with medium sulphuric(VI) acid the ethanoyl group is removed and the amino group restored.

$$C_6H_5NH.COCH_3 \xrightarrow{Br_2} C_6H_4\begin{smallmatrix}NH.COCH_3\\[4pt]Br\end{smallmatrix} \xrightarrow{H_2O} C_6H_4\begin{smallmatrix}NH_2\\[4pt]Br\end{smallmatrix}$$

| N-phenylethan-amide | bromo-N-phenyl-ethanamide | bromophenyl-amine |

Monochlorophenylamine can be prepared in the same way, but in this case an appreciable amount of the 1,2-isomer is formed in addition to the 1,4-isomer. The isomers can be separated by making use of the volatility in steam of the 1,2- compound.

2. Nitration. If phenylamine is treated with nitrating mixture a vigorous reaction takes place, but this consists mostly of oxidation of the phenylamine. To introduce the nitro group into the phenylamine nucleus the amino group must be 'protected.'

This is again done by introducing the ethanoyl group into the amino group. Nitration of N-phenylethanamide with nitrating mixture yields the 1,2- and 1,4-isomers of nitro-N-phenylethanamide. The product is treated with ethanol to remove the 1,2-compound, which is much more soluble, and the remaining 1,4-compound is then hydrolysed to 4-nitrophenylamine.

$$C_6H_5NH.COCH_3 \xrightarrow[H_2SO_4]{HNO_3} C_6H_4\begin{smallmatrix}NH.COCH_3\\[4pt]NO_2\end{smallmatrix} \xrightarrow{H_2O} C_6H_4\begin{smallmatrix}NH_2\\[4pt]NO_2\end{smallmatrix}$$

2-Nitrophenylamine can be prepared from the 2-nitro-N-phenylethanamide. The 1,3-isomer is obtained from 1,3-dinitrobenzene by reduction of one of the nitro groups as described in Chapter 22.

3. Sulphonation. As seen earlier, at ordinary temperatures phenylamine and sulphuric(VI) acid yield the salts phenylammonium sulphate(VI) and phenylammonium hydrogensulphate(VI) according to the proportions of base and acid used. When phenylamine is heated with excess of concentrated sulphuric(VI) acid at 180°–190° 4-*aminobenzenesulphonic acid* (sulphanilic acid) is produced. Only traces of the 1,2-isomer are obtained. The overall reaction can be expressed simply as follows:

With concentrated sulphuric(VI) acid the time required for sulphonation is about five hours, but this can be reduced to one hour by using equal volumes of the concentrated acid and fuming sulphuric(VI) acid. The 4-aminobenzenesulphonic acid is precipitated when the remaining liquid is poured into cold water. The product is used in the manufacture of dyestuffs.

4. 'Coupling' with Diazonium Compounds. Phenylamine, like other compounds containing the amino group attached to the nucleus, reacts with diazonium compounds such as benzenediazonium chloride to form dyes. This type of reaction is described presently.

(C) Oxidation of Phenylamine

Phenylamine readily undergoes oxidation to give a wide variety of complex products, the nature of which depends upon the oxidizing agent and the conditions used. Many of the products are highly coloured, and some of them are dyes. The first coal-tar dye to be discovered ('mauveine') was prepared by William Perkin in 1856 by oxidizing phenylamine with potassium(I) dichromate(VI) and sulphuric(VI) acid. Perkin's dye has long been obsolete, but *phenylamine black*, another oxidation product, is still in use. The experiments now described illustrate some of the ways in which phenylamine can be oxidized.

Experiment

(*i*) *Oxidation by Acidified Potassium(I) Dichromate(VI)*. To two or three drops of phenylamine in a test-tube add gradually 2 cm³ of concentrated sulphuric(VI) acid. A precipitate of phenylammonium hydrogensulphate(VI) is first formed, but this soon redissolves. Add the solution to 2 cm³ of dilute potassium(I) dichromate(VI) solution, and boil the mixture gently for three or four minutes. The colour gradually darkens. Cool and filter the liquid. A residue of phenylamine black (better known as 'aniline black') is left in the filter-paper.

(*ii*) *Oxidation by Sodium(I) Chlorate(I) (Sodium Hypochlorite)*. Add one drop of phenylamine to a dilute solution of sodium(I) chlorate(I). A violet colour, which slowly darkens, is produced.

(*iii*) *Oxidation by Iron(III) Chloride*. Dissolve one or two drops of phenylamine in dilute hydrochloric acid, and add a few drops of iron(III) chloride solution. A bright green solution is formed.

Chemical Tests for Phenylamine

The following reactions, all of which have been described, are commonly used as tests for phenylamine:

(*i*) *The Isocyano Reaction*. This reaction is also given by other primary amines, such as the methylphenylamines or (phenylmethyl)amine, but not by secondary or tertiary amines (e.g. N-methylphenylamine or N,N-dimethylphenylamine.

(*ii*) *Diazotization*. A solution of benzenediazonium chloride, obtained by diazotization of phenylamine yields a red dye with a solution of naphthalen-2-ol (β-naphthol) in sodium(I) hydroxide solution. A similar reaction is given by the methylphenylamines, but not by (phenylmethyl)amine nor by secondary or tertiary amines.

(*iii*) *Reaction with Sodium(I) Chlorate(I)*. This is a very sensitive test for phenylamine. It is used in detecting traces of the compound. No violet colour is obtained with the methylphenylamines or the secondary and tertiary amines, providing these compounds are free from traces of phenylamine.

METHYLPHENYLAMINES, $C_6H_4\genfrac{}{}{0pt}{}{CH_3}{NH_2}$

Methylphenylamine (toluidine) exists as the 1,2-, 1,3-, and 1,4-isomers. These are prepared by reduction of the corresponding methylnitrobenzenes (nitrotoluenes) by means of tin, or iron, and hydrochloric acid.

$$C_6H_4\genfrac{}{}{0pt}{}{CH_3}{NO_2} + 6H \rightarrow C_6H_4\genfrac{}{}{0pt}{}{CH_3}{NH_2} + 2H_2O$$

Since nitration of methylbenzene yields the 1,2- and 1,4-isomers of methylnitrobenzene, the methylphenylamines usually encountered are the 1,2- and 1,4-compounds. 2-Methylphenylamine is an oily liquid which boils at 197°C, while 4-methylphenylamine is a colourless crystalline solid which melts at 45°C and boils at 198°C. 3-Methylphenylamine is a liquid which boils at 200°C.

The chemical properties of the methylphenylamines closely resemble those of phenylamine. All three isomers give the isocyano reaction. At 0°C–10°C the methylphenylamines undergo diazotization with nitric(III) acid (nitrous acid) and form the corresponding methylbenzenediazonium salts. Methylbenzenediazonium salts, like benzenediazonium salts, are used in the manufacture of dyes.

$$C_6H_4\genfrac{}{}{0pt}{}{CH_3}{NH_2} + HCl + HNO_2 \rightarrow \left[C_6H_4\genfrac{}{}{0pt}{}{CH_3}{N_2}\right]^+ Cl^- + 2H_2O$$

methylbenzenediazonium chloride

At higher temperatures nitric(III) acid converts the methylphenylamines into the corresponding hydroxy-compounds, which are known as methylphenols (cresols).

$$C_6H_4\genfrac{}{}{0pt}{}{CH_3}{NH_2} + HNO_2 \rightarrow C_6H_4\genfrac{}{}{0pt}{}{CH_3}{OH} + N_2 + H_2O$$

methylphenol

N-ALKYLPHENYLAMINES

N-Methylphenylamine and **N,N-Dimethylphenylamine.** The letters prefixed to these names indicate that the alkyl radicals are attached to the nitrogen atom, and not to the nucleus. The first compound is a secondary aromatic amine, while the second is a tertiary aromatic amine.

$C_6H_5N\genfrac{}{}{0pt}{}{H}{CH_3}$ $C_6H_5N\genfrac{}{}{0pt}{}{CH_3}{CH_3}$

N-methylphenylamine   N,N-dimethylphenylamine

N-Methylphenylamine is isomeric with the three methylphenyl-amines (CH_3—C_6H_4—NH_2) and with (phenylmethyl)amine ($C_6H_5.CH_2NH_2$).

When pure, N-methylphenylamine and N,N-dimethyl-phenylamine are colourless liquids which boil at 190°C and 192°C respectively. The commercial compounds invariably darken on keeping. They are seldom prepared in the laboratory, although both are produced when phenylamine is heated with an ethanolic solution of iodomethane in a sealed vessel. The N-alkylphenyl-amines are made on a large scale by heating phenylamine with methanol and concentrated sulphuric(VI) acid under pressure. They are used for making dyes.

Like phenylamine, N-alkylphenylamines are weak bases. They are only slightly soluble in water, but dissolve readily in dilute mineral acids to form salts such as methylphenylammonium chloride ($C_6H_5NH_2CH_3{}^+Cl^-$).

Distinction Between Primary, Secondary, and Tertiary Aromatic Amines

Phenylamine, the methylphenylamines, N-methylphenylamine, and N,N-dimethylphenylamine can be distinguished by means of their different reactions with nitric(III) acid.

(i) *Phenylamine and the Methylphenylamines.* If a solution of phenylamine (or one of the methylphenylamines) in dilute hydrochloric acid is cooled to 0°–10°C, and a cooled solution of sodium(I) nitrate(III) (sodium nitrite) is added, the correspond-ing diazonium compound is formed in solution. When the latter is added to an alkaline solution of naphthalen-2-ol (β-naphthol) a red dye is precipitated. The reaction is carried out as described earlier.

(ii) *N-Methylphenylamine.* This secondary amine reacts with nitric(III) acid at ordinary temperatures in the same way as an aliphatic secondary amine. It yields a nitrosoamine in the form of a yellow, or brown, oil. The nitroso group (—NO) is derived from nitric(III) acid in the same way that the nitro group (—NO_2) is derived from nitric(V) acid.

$$\begin{matrix} C_6H_5 \\ \diagdown \\ \quad NH + HO.NO \to \\ \diagup \\ CH_3 \end{matrix} \qquad \begin{matrix} C_6H_5 \\ \diagdown \\ \quad N.NO + H_2O \\ \diagup \\ CH_3 \end{matrix}$$

N-methyl-N-nitrosophenylamine

Experiment

Dissolve 1 cm³ of N-methylphenylamine in 5 cm³ of dilute hydro-chloric acid. Cool the solution under the tap, and add 5 cm³ of 10 per cent sodium(I) nitrate(III) solution. Shake the tube well for two or three minutes and then allow it to stand. A yellow, or brown, oil forms as a separate layer at the bottom of the tube.

(iii) *N,N-Dimethylphenylamine.* Tertiary aromatic amines react readily with nitric(III) acid. In this case, however, it is the benzene nucleus, and not the nitrogen-containing group, which is attacked. The hydrogen atom in the 4-position is replaced by the nitroso group. If excess of sodium(I) hydroxide solution is added after the reaction a bright-green precipitate of the nitroso compound is formed. This is called 4-nitroso-N,N-dimethyl-phenylamine.

$$(CH_3)_2N\hexagon + HO.NO \to (CH_3)_2N\hexagon NO + H_2O$$

Experiment

Dissolve two or three drops of N,N-dimethylphenylamine in 5 cm³ of dilute hydrochloric acid, and cool the solution under the tap. Add drop by drop about 2 cm³ of dilute sodium(I) nitrate(III) solution. During the addition an orange-coloured precipitate of a salt (formed from the acid and the nitroso-base) may be produced. Allow the mixture to stand for two minutes, and then add excess of 'bench' sodium(I) hydroxide solution. A green precipitate of the nitroso-compound is obtained.

SIDE-CHAIN AROMATIC AMINES

Aromatic amines containing the amino group in a side-chain are relatively unimportant. (Phenylmethyl)amine (benzylamine) may be regarded as a typical side-chain aromatic amine.

(PHENYLMETHYL)AMINE, $C_6H_5.CH_2NH_2$

Boiling point: 184°C   *Density:* 0·98 g cm⁻³

The chief methods of preparing (phenylmethyl)amine are given below. These should be compared with the methods used for obtaining primary aliphatic amines.

(*i*) *From Phenylethanamide.* (Phenylmethyl)amine can be prepared by treating phenylethanamide, or phenylacetamide, with bromine and sodium(I) hydroxide solution. The reactions are analogous to those which occur in the preparation of methylamine from ethanamide, or acetamide (Chapter 15). The net result of the reactions is that the carbonyl group is eliminated from the phenylethanamide molecule.

$$C_6H_5.CH_2.CONH_2 \rightarrow C_6H_5.CH_2NH_2$$

phenylethanamide   (phenylmethyl)-amine

(*ii*) *From (Chloromethyl)benzene.* (Phenylmethyl)amine is formed (chiefly as its salt with hydrochloric acid) when (chloromethyl)-benzene (benzyl chloride) is heated with an ethanolic solution of ammonia in a sealed vessel at 100°C.

$$C_6H_5.CH_2Cl + NH_3 \rightarrow C_6H_5.CH_2NH_2 + HCl$$

(*iii*) *From Benzenecarbonitrile, or Benzonitrile.* This compound is reduced to (phenylmethyl)amine when it is dissolved in absolute ethanol and sodium is added to the boiling solution.

$$C_6H_5CN \xrightarrow{4H} C_6H_5.CH_2NH_2$$

(Phenylmethyl)amine is a colourless liquid, which darkens on keeping. It is more closely related to aliphatic amines than nuclear amines. Thus it is readily miscible with water, and the solution gives an *alkaline* reaction with litmus. The amine forms salts with acids.

(Phenylmethyl)amine resembles phenylamine in giving the isocyano reaction, but differs in its behaviour with nitric(III) acid at 0°–10°C. Compounds containing the amino group in the side-chain cannot be diazotized. They are converted into alcohols by nitric(III) acid both at ordinary, and at lower, temperatures. (Phenylmethyl)amine thus yields phenylmethanol(benzyl alcohol).

$$C_6H_5.CH_2NH_2 + HNO_2 \rightarrow C_6H_5.CH_2OH + N_2 + H_2O$$

DIAZONIUM COMPOUNDS

As already noted, phenylamine reacts with nitric(III) acid in the presence of hydrochloric acid at 0°–10°C to give the ions of a salt, benzenediazonium chloride. All primary aryl amines undergo 'diazotization' in this way. The products are called *diazonium* salts.

$$ArNH_2 + HNO_2 + HCl \rightarrow ArN_2{}^+ + Cl^- + 2H_2O$$

'Azo' is derived from 'azote,' an old English name for nitrogen. The ending-*ium* indicates an analogy with the ammonium ion and phenylammonium (anilinium) ion. Indeed, the benzene-diazonium ion can be regarded as derived theoretically from the phenylammonium ion by substituting the three hydrogen atoms of the latter by a nitrogen atom.

$$\left[\begin{array}{c} H \\ | \\ H-N-H \\ | \\ H \end{array}\right]^+ \quad \left[\begin{array}{c} H \\ | \\ C_6H_5-N-H \\ | \\ H \end{array}\right]^+ \quad \left[\begin{array}{c} C_6H_5-N \\ ||| \\ N \end{array}\right]^+$$

ammonium ion　　　phenylammonium ion　　benzenediazonium ion

Diazonium salts have the general formula $ArN_2{}^+X^-$, where X can be Cl, NO_3, HSO_4, etc. They resemble ammonium salts in being crystalline solids soluble in water, but most of them differ from ammonium salts in being explosive when dry. This is not a practical drawback, however, because they are normally used in aqueous solution.

Preparation of Benzenediazonium Chloride in Solution

The preparation of a solution of a diazonium salt is a matter requiring considerable care in regard to the particulars now mentioned.

(*i*) *Temperature.* To prevent decomposition of the diazonium salt by water the temperature must be kept between 0°C and 10°C. At 0°C diazotization tends to be slow, and in practice a temperature of about 5°C is preferred. The reaction is exothermic, so that external cooling is necessary to avoid a rise in temperature.

(*ii*) *Proportions of the Reactants.* Excess of phenylamine and a large excess of sodium(I) nitrate(III) must be avoided. The former reacts with the diazonium salt produced (see later), and the latter causes side-reactions to occur. Sufficient hydrochloric acid is used to convert all the phenylamine to phenylammonium chloride and to provide a small excess of nitric(III) acid. Theoretically the preparation of benzenediazonium chloride requires for each mole of phenylamine one mole of sodium(I) nitrate(III) and two moles of hydrochloric acid (one to react with the sodium(I) nitrate(III) and one for the reaction with the phenylamine). In practice the substances are used in the following proportions:

$$1.0 \ C_6H_5NH_2 : 1.1 \ NaNO_2 : 2.2 \ HCl$$
$$(93 \ g) \qquad (69 \ g) \qquad (36.5 \ g)$$

From the molar quantities given it can be seen that for every 10 cm³ (10.2 g) of phenylamine about 8.5 g of sodium(I) nitrate(III) and 9 g of hydrochloric acid are required. This amount of hydrochloric acid is contained in about 25 cm³ of the concentrated acid, which has a concentration of about 10 mol dm⁻³.

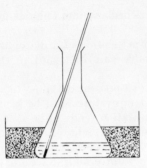

Fig. 23-2. Diazotization of phenylamine

Experiment

Mix 25 cm³ of concentrated hydrochloric acid with 25 cm³ of water in a 200 cm³ conical flask, and dissolve 10 cm³ of phenylamine in the acid. Put a thermometer into the flask, and stand the latter in water containing lumps of ice in a small glass trough (Fig. 23-2). While the temperature of the liquid is falling to 5°C dissolve 8·5 g of sodium(I) nitrate(III) in 20 cm³ of water. When the liquid in the flask has reached 5°C run in the sodium(I) nitrate(III) solution 2-3 cm³ at a time, allowing two or three minutes between each addition, so that the heat liberated in the reaction is dissipated. Keep the mixture in the flask stirred, and do not allow the temperature to rise above 10°C. After the final addition of the sodium(I) nitrate(III) solution the liquid should give an *immediate* blue-black colour when tested with starch-potassium(I) iodide paper (indicating an excess of nitric(III) acid).

The solution of benzenediozonium chloride should be left standing in the cooling-bath until required. Portions of the liquid can be used for the reactions described in the following sections. *Any solution left over should be thrown away as the solid formed on evaporation is highly explosive.*

Reactions of Diazonium Salts

The reactions of a diazonium salt in aqueous solution are essentially those of the diazonium ion. As the latter is positively charged it is electrophilic in character, and reacts with nucleophilic reagents. Diazonium salts have two types of application:

(*i*) they are used to prepare simple derivatives of aromatic hydrocarbons.

(*ii*) they are intermediates in the preparation of dyestuffs.

Only the more important reactions are given here.

1. Reaction with Water. As seen earlier, when an aqueous solution of benzenediazonium chloride is heated phenol is produced.

$$C_6H_5N_2^+ + H_2O \rightarrow C_6H_5OH + N_2 + H^+$$

2- or 4-Methylbenzenediazonium chloride (toluenediazonium chloride) yields the corresponding methylphenol (cresol).

2. Sandmeyer's Reactions. These are a series of reactions in which substitution of the —N_2^+ group occurs under the catalytic influence of a copper(I) salt. Thus chlorobenzene, bromobenzene, or benzenecarbonitrile are respectively obtained when a solution of benzenediazonium chloride is heated with a solution of copper(I) chloride in concentrated hydrochloric acid, copper(I) bromide in concentrated hydrobromic acid, or copper(I) cyanide in potassium(I) cyanide solution.

$$C_6H_5N_2^+ + Cl^- \xrightarrow{\text{CuCl}} C_6H_5Cl + N_2$$

$$C_6H_5N_2^+ + Br^- \xrightarrow{\text{CuBr}} C_6H_5Br + N_2$$

$$C_6H_5N_2^+ + CN^- \xrightarrow{\text{CuCN}} C_6H_5CN + N_2$$

The usefulness of Sandmeyer's reactions lies in the means they afford for obtaining derivatives which are not readily prepared directly from the hydrocarbons. Thus, as was shown in Chapter 21, chlorination or bromination of methylbenzene yields a mixture of 1,2- and 1,4-derivatives which are difficult to separate owing to the closeness of their boiling points. The individual compounds can be made indirectly by nitrating methylbenzene and separating the 1,2- and 1,4-nitro-derivatives. These are reduced to the corresponding methylphenylamines, which are then diazotized and treated by Sandmeyer's method.

The reaction of a diazonium salt with copper(I) cyanide is important because the —CN group is readily hydrolysed to a —COOH group. Thus the reaction can be used to introduce a carboxyl group into the benzene nucleus. Benzocarbonitrile, or benzonitrile (C_6H_5CN), yields benzenecarboxylic, or benzoic, acid ($C_6H_5 . COOH$).

3. Reaction with Potassium(I) Iodide Solution. Replacement of the diazonium group by an iodine atom does not require the use of a catalyst. It can be brought about simply by treating the solution of diazonium salt with potassium(I) iodide solution. Benzenediazonium chloride solution gives iodobenzene.

$$C_6H_5N_2^+ + I^- \rightarrow C_6H_5I + N_2$$

This is the usual method of preparing iodobenzene because iodine does not normally react with benzene.

4. Reduction. A solution of benzenediazonium chloride is reduced to phenylhydrazine ($C_6H_5NH . NH_2$) by a solution of tin(II) chloride in concentrated hydrochloric acid. When a cooled solution of the reagent is added to a cooled solution of the diazonium compound a white precipitate of the salt $C_6H_5NH . NH_3^+Cl^-$ is formed.

$$C_6H_5N_2^+ + 2Sn^{2+} + 4H^+ \rightarrow C_6H_5NH . NH_3^+ + 2Sn^{4+}$$

To obtain phenylhydrazine itself the precipitate is treated with sodium(I) hydroxide solution. Phenylhydrazine is liberated as a reddish oil. When purified it is a colourless liquid (b.p. 241°C).

5. Coupling Reactions. Since the $C_6H_5N_2^+$ ion is an electrophile it might be expected to bring about substitution in the benzene nucleus as in the case of NO_2^+. Diazonium ions, however, are only weakly electrophilic (they are stabilized by mesomerism), and substitute in the nucleus only when a strongly activating atom or group is attached to the latter.

(*i*) *Coupling with Aryl Hydroxy-compounds in Alkaline Solution.* The formation of a yellow precipitate when an alkaline solution of phenol is added to benzenediazonium chloride solution has been described earlier in this chapter. Phenol itself reacts only slowly. Alkali converts phenol to the phenoxide ion ($C_6H_5O^-$), which is more reactive because the negatively charged oxygen atom has a $+I$ effect, whereas the —OH group has a $-I$ effect. The reaction is represented by the equation shown.

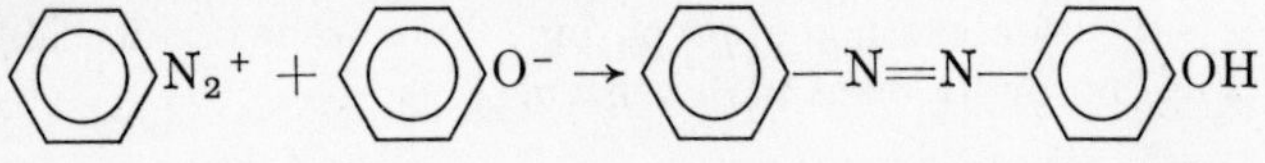

$$\langle\!\bigcirc\!\rangle N_2{}^+ + \langle\!\bigcirc\!\rangle O^- \rightarrow \langle\!\bigcirc\!\rangle\!-\!N\!=\!N\!-\!\langle\!\bigcirc\!\rangle OH$$

(4-hydroxyphenyl)azobenzene

In the above reaction two benzene nuclei become 'coupled' together by means of the *azo* group (—N=N—). All compounds containing the azo group are highly coloured, and many are used as dyes. Coupling takes place almost exclusively in the 4- position with respect to the hydroxyl group of phenol. This is usual. If the 4- position is already occupied, the 2- position is attacked instead.

When a solution of naphthalen-2-ol (β-naphthol) in aqueous sodium(I) hydroxide is added to benzenediazonium chloride solution a red dye is precipitated. This reaction has been used earlier in testing for a primary aryl amine.

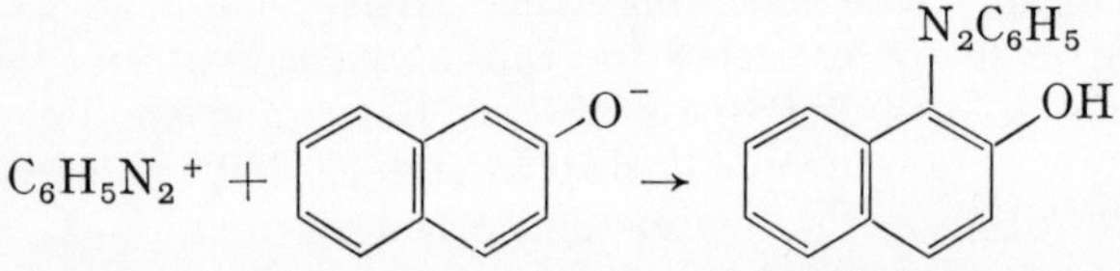

$$C_6H_5N_2{}^+ + \; [\text{naphthalen-2-olate}] \rightarrow [\text{azo dye structure}]$$

red dye

(It should be explained that when a molecule contains two benzene nuclei 'fused' together, as in the case of naphthalen-2-ol, the nuclei are represented by the older type of symbol, and not

by plain hexagons with inscribed circles. This is because in naphthalen-2-ol and similar molecules the common orbital of the π electrons extends over *both* nuclei.)

(*ii*) *Coupling with Nuclear Amines.* In the preparation of benzenediazonium chloride from phenylamine an excess of phenylamine must be avoided because the excess reacts with the diazonium salt and produces a yellow precipitate. The latter is not formed by the benzenediazonium ion attacking the phenylamine nucleus, but by the ion reacting with the nitrogen atom through its lone pair of electrons. At the same time one of the hydrogen atoms of the amino group is expelled as a proton.

$$C_6H_5N_2{}^+ + \; :\!\underset{\underset{\displaystyle H}{|}}{\overset{\overset{\displaystyle H}{|}}{N}}\!-\!C_6H_5 \rightarrow C_6H_5\!-\!N\!=\!N\!-\!\underset{\underset{\displaystyle H}{|}}{N}\!-\!C_6H_5 + H^+$$

N-(phenylazo)phenylamine

If the precipitate of *N*-(phenylazo)phenylamine is left standing in the mother-liquor, it slowly undergoes an isomeric change, in which the —NH— group 'migrates' to the 4- position. This gives the compound which would have been obtained if the diazonium ion had attacked the nucleus directly.

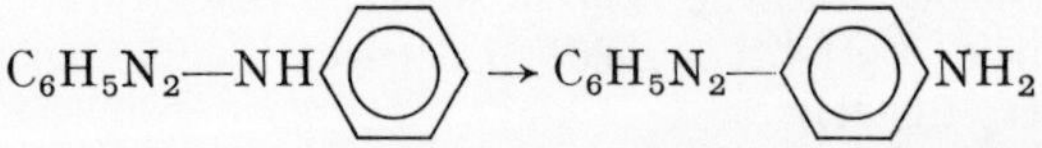

$$C_6H_5N_2\!-\!NH\langle\!\bigcirc\!\rangle \rightarrow C_6H_5N_2\!-\!\langle\!\bigcirc\!\rangle NH_2$$

4-(phenylazo)phenylamine

The two-stage reaction of phenylamine with benzenediazonium chloride solution is typical of the behaviour of primary and secondary nuclear amines. Tertiary nuclear amines do not have a replaceable hydrogen atom attached to their nitrogen atom, and undergo substitution in the nucleus directly. This is illustrated by the reaction which gives the well known indicator *methyl orange*. The latter is obtained by coupling N,N-dimethylphenylamine with diazotized 4-aminobenzenesulphonic acid (sulphanilic acid).

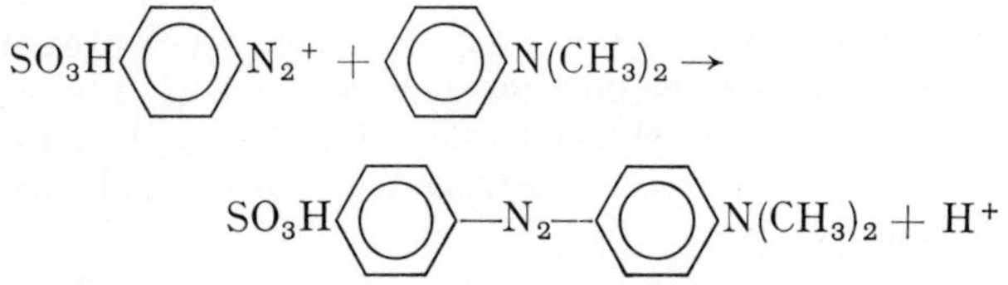

$$SO_3H\langle\bigcirc\rangle N_2{}^+ + \langle\bigcirc\rangle N(CH_3)_2 \rightarrow$$

$$SO_3H\langle\bigcirc\rangle{-}N_2{-}\langle\bigcirc\rangle N(CH_3)_2 + H^+$$

methyl orange

EXERCISE 23

1. In the laboratory preparation of phenylamine from nitrobenzene, tin, and hydrochloric acid state:

(*a*) Whether the reaction is endothermic or exothermic;

(*b*) The name of the complex salt $(C_6H_5NH_3)_2$ Sn Cl_6 first formed;

(*c*) The reagent used to liberate phenylamine from the complex salt;

(*d*) The solvent used to extract phenylamine from the distillate obtained by steam distillation;

(*e*) The reagent used to dry the extract.

2. A liquid X, which is immiscible with water, has a relative molecular mass of 90. If X is steam-distilled at standard atmospheric pressure, state:

(*a*) Whether X distils over, above, or below 100°C;

(*b*) Whether the composition of the distillate alters or remains constant;

(*c*) The relative masses of X and water in the distillate if the ratio of their vapour-pressures at the distillation temperature is 1 : 20.

3. Give the names and formulæ of the organic compounds formed in the reactions between phenylamine and (*a*) hydrochloric acid, (*b*) nitric(III) acid (nitrous acid) in the presence of hydrochloric acid at 5°C, (*c*) a mixture of ethanoic acid and ethanoic anhydride, (*d*) bromine water, (*e*) concentrated sulphuric(VI) acid (on refluxing).

4. State what you *see* in the reactions between the following:

(*a*) Phenylamine and dilute sulphuric(VI) acid, (*b*) an aqueous solution of phenylammonium chloride and aqueous sodium(I) hydroxide, (*c*) phenylamine and benzenecarbonyl or benzoyl, chloride, (*d*) phenylamine and sodium(I) chlorate(I) solution, (*e*) benzenediazonium chloride solution and a solution of naphthalen-2-ol in sodium(I) hydroxide solution.

5. The conversion of phenylamine to a mixture of 2-nitrophenylamine and 4-nitrophenylamine can be represented as follows:

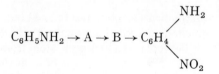

What are the names and formulæ of A and B?

6. Give the systematic names of the compounds which have the molecular formula C_7H_9N.

7. State which of the following react with 2-methylphenylamine only, with (phenylmethyl)amine only, with both, or with neither: (*a*) trichloromethane and ethanolic potassium(I) hydroxide, (*b*) ethanoyl, or acetyl, chloride, (*c*) an ethanolic solution of iodomethane, (*d*) aqueous sodium(I) hydroxide, (*e*) nitric(III) acid.

8. Suggest a method, involving not more than three stages, by which phenylammonium chloride could be converted into (phenylmethyl)amine (indicate the reagents used).

24 Phenols and Aromatic Alcohols

PHENOLS

Phenols *are derivatives of aromatic hydrocarbons in which one or more hydrogen atoms of the benzene nucleus have been replaced by the corresponding number of hydroxyl groups.*

Phenols are described as monohydric, dihydric, etc., according to the number of hydroxyl groups attached to the nucleus. Only monohydric phenols are considered here. The most important monohydric phenol is phenol itself, which can be regarded as derived theoretically from benzene by replacing a hydrogen atom of the latter by a hydroxyl group.

phenol

Methylbenzene gives rise to two kinds of hydroxy-compounds according to whether the hydroxyl group has replaced a hydrogen atom from the nucleus or from the side-chain.

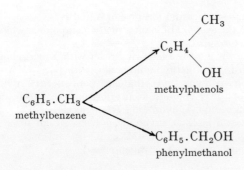

The three isomeric methylphenols (cresols) belong to the class of phenols, while phenylmethanol (benzyl alcohol) is a member of the *aromatic alcohols*.

PHENOL, C_6H_5OH

Melting point: 43°C   *Boiling point:* 181°C

In the laboratory there are two general methods of introducing an —OH group into the nucleus of benzene or one of its homologues. One method involves nitration, the other sulphonation, of the nucleus. In the case of benzene the two methods may be summarized as follows:

(*i*) Benzene → nitrobenzene → phenylamine → benzenediazonium chloride (or sulphate(VI)) → phenol.

(*ii*) Benzene → benzenesulphonic acid → sodium(I) benzenesulphonate → sodium(I) phenoxide → phenol.

Usually the actual starting-point in the two methods is either phenylamine or sodium(I) benzenesulphonate.

Laboratory Preparation of Phenol

(*i*) *From Phenylamine.* Phenol is produced when phenylamine is diazotized by means of nitric(III) acid (nitrous acid) at 0°–10°C and the resulting aqueous solution of benzenediazonium salt is heated.

$$C_6H_5NH_2 + HNO_2 + HCl \rightarrow C_6H_5N_2{}^+ + Cl^- + 2H_2O$$

$$C_6H_5N_2{}^+ + HOH \rightarrow C_6H_5OH + N_2 + H^+$$

Practical details for carrying out the diazotization were given in the last chapter. After the diazotization the flask containing the reaction mixture is heated in a water-bath at $50°$–$55°C$ for half an hour. Nitrogen is evolved, and phenol is formed. The remaining liquid is distilled in steam. The phenol in the distillate is extracted with ethoxyethane, and the extract is dried with anhydrous sodium(I) sulphate(VI). The ethoxyethane is removed by distillation on a hot water-bath, and finally the phenol is distilled over, using an air-condenser. The phenol forms colourless crystals on cooling.

(ii) From Sodium(I) Benzenesulphonate. Phenol is prepared from sodium(I) benzenesulphonate by fusing the latter with excess of sodium(I) hydroxide and a little water. Sodium(I) phenoxide and sodium(I) sulphate(IV) (sodium sulphite) are obtained.

$$C_6H_5SO_3Na + NaOH \rightarrow C_6H_5OH + Na_2SO_3$$

$$C_6H_5OH + NaOH \rightarrow C_6H_5ONa + H_2O$$
$$\text{sodium phenoxide}$$

The fusion is carried out in a nickel crucible or basin at about $250°C$ as described in Chapter 22, but the heating is continued for an hour. The brown pasty residue is cooled, dissolved in a small amount of water, and acidified by careful addition of a mixture of equal volumes of concentrated hydrochloric acid and water. Phenol separates as an oil.

$$C_6H_5ONa + HCl \rightarrow C_6H_5OH + NaCl$$

Manufacture and Uses of Phenol

A certain amount of phenol is still extracted from the 'middle oil' fraction obtained in the distillation of coal-tar (Chapter 19). It is mostly manufactured, however, from benzene by different methods. The *sulphonation process* is essentially the method outlined in the last section. In the *chlorobenzene process* (chiefly used in the U.S.A.) benzene is firstly converted to chlorobenzene. The latter is then heated with aqueous sodium(I) hydroxide under conditions which make the chlorine atom 'active,' so that hydrolysis occurs (see Chapter 21).

The most important method of manufacturing phenol is now the *cumene process*. Benzene is made to combine with propene in the vapour phase at $250°C$ and 25 atmospheres pressure, anhydrous aluminium(III) chloride or phosphoric(V) acid being used as a catalyst. The combination results in (1-methylethyl)-benzene (cumene).

$$C_6H_6 + CH_3\!-\!CH\!=\!CH_2 \xrightarrow{AlCl_3} C_6H_5.\underset{\underset{CH_3}{|}}{\overset{\overset{CH_3}{|}}{CH}}$$
$$\text{(1-methylethyl)benzene}$$

The vapour obtained is condensed, and the liquid is made into an emulsion with very dilute sodium(I) carbonate solution. The (1-methylethyl)benzene is then oxidized to a hydroperoxide by blowing in air at $120°C$ and a few atmospheres pressure. The oxidation product is then decomposed catalytically by dilute sulphuric(VI) acid at $100°C$ to phenol and propanone, the latter being a valuable by-product.

$$\underset{\underset{CH_3}{|}}{C_6H_5.\overset{\overset{H}{|}}{C}-CH_3} \xrightarrow{O_2} \underset{\underset{CH_3}{|}}{C_6H_5.\overset{\overset{O-O-H}{|}}{C}-CH_3} \xrightarrow{H_2SO_4} C_6H_5OH + (CH_3)_2CO$$

Over 50 per cent of industrial phenol is used in the production of phenolic resins and plastics (Chapter 26). It is also an intermediate in making nylon, aspirin, weed-killers, and dyes. Phenol is a powerful germicide. It was the first substance used (by Joseph Lister in 1867) for producing aseptic conditions in major surgical operations. It is no longer employed in this way because of its destructive effect on the tissues, but it is still used to a minor extent in 'carbolic' soap.

Properties of Phenol

Phenol forms colourless hygroscopic crystals with a characteristic sweet smell. The crystals turn red on keeping owing to oxidation of traces of impurities. Phenol is extremely poisonous, and causes painful blisters if allowed to come into contact with the skin. *It must on no account be handled with the fingers.* At ordinary temperatures phenol has only a small solubility in water, but dissolves readily in organic solvents. If phenol is mixed with water, two liquid layers are formed. The upper layer is a saturated solution of phenol in water, the lower layer a saturated solution of water in phenol. Above 68°C phenol and water are completely miscible.

The chemical reactions of phenol can be divided into those involving only the hydroxyl group and those in which the nucleus takes part. Each of these modifies the character of the other because of resonance in the phenol molecule. As explained in Chapter 20, there are five contributing structures to the normal state of the molecule.

(A) Reactions of the Hydroxyl Group

The —OH group in phenol resembles the same group in aliphatic alcohols in the way it reacts with sodium(I) hydroxide, metallic sodium, phosphorus pentachloride (partial resemblance), acid chlorides, and acid anhydrides. It differs from the —OH group in aliphatic alcohols in the strength of its acidic character and in its behaviour towards acids, oxidizing agents, and dehydrating agents. Phenol gives rise to ethers and esters, but these can only be obtained in satisfactory yield by indirect methods.

1. **Acidic Character of Phenol.** The acidic character of phenol is reflected in the name 'carbolic acid.' In aqueous solution phenol is slightly dissociated into phenoxide (phenate) ions and oxonium ions.

$$C_6H_5OH + H_2O \rightleftharpoons C_6H_5O^- + H_3O^+$$

The corresponding reaction between ethanol and water (Chapter 8) occurs to an even smaller extent.

Phenol is an extremely weak acid, and its aqueous solution does not affect litmus. It is weaker than carbonic acid, and does not liberate carbon dioxide from carbonates or hydrogencarbonates. On the contrary, carbonic acid or carbon dioxide decomposes an aqueous solution of a phenoxide and liberates phenol.

$$2C_6H_5ONa + H_2CO_3 \rightarrow 2C_6H_5OH + Na_2CO_3$$

The failure of phenol to liberate carbon dioxide from a solution of sodium(I) carbonate or hydrogencarbonate can be used to distinguish phenol from a carboxylic acid such as benzene-carboxylic, or benzoic, acid.

As might be expected, phenol dissolves readily in aqueous sodium(I) hydroxide, giving a solution of sodium(I) phenoxide. The reaction is reversible, sodium(I) phenoxide being partially hydrolysed by water.

$$C_6H_5OH + OH^- \rightleftharpoons C_6H_5O^- + H_2O$$

2. Reaction with Sodium. When a solution of phenol in ethoxy-ethane is treated in the cold with metallic sodium the hydrogen of the hydroxyl group is displaced and sodium(I) phenoxide is formed. This can be obtained by evaporating the solution.

Sodium(I) phenoxide is used for the preparation of aromatic ethers by reaction with alkyl halides (compare Williamson's synthesis of ethers in Chapter 8). Thus if sodium(I) phenoxide is heated with iodomethane in ethanolic solution methoxybenzene (methylphenyl ether) is produced. This is a pleasant-smelling liquid (b.p. 154°C).

$$C_6H_5ONa + CH_3I \rightarrow C_6H_5\text{---}O\text{---}CH_3 + NaI$$
$$\text{methoxybenzene}$$

3. Reaction with Phosphorus Pentachloride. Phosphorus pentachloride reacts in the cold with phenol, and hydrogen chloride is evolved as with alcohols. The reaction, however, takes place less readily with phenol, and is complicated by side-reactions. Only a little chlorobenzene is formed.

4. Ethanoylation and Benzenecarbonylation. Esters of phenol with carboxylic acids can be obtained by warming phenol with the acid chlorides or anhydrides. Phenol and ethanoyl, or acetyl, chloride yield phenyl ethanoate, or acetate.

$$CH_3.COCl + C_6H_5OH \rightarrow CH_3.COOC_6H_5 + HCl$$
$$\text{phenyl ethanoate}$$

In practice the reaction is usually carried out in alkaline solution, which converts phenol into the more reactive phenoxide ions. With the latter the reactions take place readily in the cold and the yields are better.

$$CH_3.COCl + C_6H_5O^- \rightarrow CH_3.COOC_6H_5 + Cl^-$$

Phenol undergoes a parallel reaction with benzenecarbonyl, or benzoyl, chloride ($C_6H_5.COCl$). The product is phenyl benzenecarboxylate, or phenyl benzoate ($C_6H_5.COOC_6H_5$).

Unlike alcohols, phenols do not react with hydrohalogen acids (HCl, HBr, and HI) and they do not yield simple aldehydes or ketones with oxidizing agents. Also, they do not produce alkenes when warmed with concentrated sulphuric(VI) acid.

(B) Reactions of the Nucleus

As was seen in Chapter 20, introduction of the —OH group into the benzene nucleus strongly activates the latter, so that nuclear substitution by electrophilic reagents occurs much more readily than with benzene. A similar effect was noted for the introduction of the —NH$_2$ group. One drawback of the strong activation of the nucleus is that it becomes very sensitive to attack by oxidizing agents (compare the nucleus in aniline).

1. Halogenation. Different products are obtained by chlorination or bromination of phenol according to the conditions used.

(*i*) A mixture of 2- and 4-chlorophenol in about equal proportions is formed when chlorine is passed into melted phenol at 50°–60°C until the theoretical increase in mass has taken place. No chlorine carrier is required. By continuing the chlorination more highly substituted derivatives can be produced.

$$C_6H_5OH + Cl_2 \rightarrow C_6H_4\begin{matrix} OH \\ \\ Cl \end{matrix} + HCl$$

chlorophenol

Corresponding bromo-derivatives can be prepared by heating bromine with melted phenol, the substances being used in the theoretical proportions.

(*ii*) Addition of chlorine water, or bromine water, to an aqueous solution of phenol results in an immediate white precipitate of 2,4,6-trichlorophenol, or 2,4,6-tribromophenol.

The reaction with bromine water is used for estimating phenol in aqueous solution. Excess of standard bromine solution is added to a known volume of the phenol solution. The excess of bromine is found by liberation of iodine from potassium(I) iodide solution, the liberated iodine being titrated with standard sodium(I) thiosulphate(VI) solution.

2. Nitration. Phenol undergoes nitration so readily that dilute nitric(V) acid alone will bring it about (slowly). Nitric(V) acid of medium concentration is usually employed. The temperature must be kept below 20°C if formation of dinitrophenol is to be avoided. Mononitration results in a mixture of 2- and 4-nitrophenols.

$$C_6H_5OH + HNO_3 \rightarrow C_6H_4\begin{matrix} OH \\ \\ NO_2 \end{matrix} + H_2O$$

nitrophenol

The preparation and properties of the nitrophenols are described shortly.

If phenol is treated with concentrated nitric(V) acid 2,4-dinitrophenol is obtained, while ordinary nitrating mixture (concentrated nitric(V) acid and concentrated sulphuric(VI) acid) yields symmetrical, or 2,4,6-, trinitrophenol. In both cases considerable oxidation of the phenol takes place with formation of resinous products, and the yields of the nitro-derivatives are low.

3. Sulphonation. Phenol can be sulphonated merely by dissolving phenol in concentrated sulphuric(VI) acid at ordinary temperatures. A mixture of 2- and 4-hydroxybenzenesulphonic acids is formed. These are much more soluble in water than phenol itself because of the extra hydrophilic character introduced by the sulphonic group.

$$C_6H_5OH + H_2SO_4 \rightarrow C_6H_4 \underset{SO_3H}{\overset{OH}{\diagup}} + H_2O$$

hydroxybenzenesulphonic acid

4. Coupling with Diazonium Salts. Phenol, like other compounds containing a hydroxyl group attached to the nucleus, reacts with diazonium salts in aqueous solution to produce dyes.

The reaction is actually carried out with an alkaline solution of phenol (that is, with phenoxide ions), as described in Chapter 23.

5. Condensation with Aldehydes. Condensation reactions occur between phenol and aldehydes (aliphatic and aromatic) under the catalytic influence of acids or alkalies. The condensation of phenol with methanal (formaldehyde) is of great industrial importance because it is the basis of the manufacture of phenolic resins and plastics. The reaction is described in Chapter 26.

6. Oxidation. Phenol is readily oxidized by the common oxidizing agents such as an acidified, or alkaline, solution of potassium(I) manganate(VII), nitric(V) acid, etc. The reactions which occur are complex and depend on the oxidizing agent used. In some cases resinous products of unknown constitution are formed. When phenol is boiled with an acidified solution of potassium(I) manganate(VII) the nucleus is broken down and several degradation products, including ethanedioic acid and carbon dioxide, are obtained.

7. Reduction. When hydrogen is blown through a suspension of finely divided nickel in melted phenol at about 160°C the phenol is reduced to cyclohexanol.

$$C_6H_5OH + 3H_2 \rightarrow C_6H_{11}OH$$

Cyclohexanol is a colourless liquid (b.p. 161°C) used in the manufacture of nylon. It is not a phenol, but a cyclic aliphatic alcohol. Thus the ring of six carbon atoms in the cyclohexanol molecule is not planar, as in aromatic molecules, but has a puckered structure.

cyclohexanol

Chemical Tests for Phenol

(i) *Reaction with Iron(III) Chloride.* To an aqueous solution of phenol add one drop of iron(III) chloride solution. A violet colour is produced. This is a general test for compounds having a hydroxyl group attached to the nucleus, although the colour obtained is not always violet. When other groups are also present in the molecule the colour may be red or blue.

(ii) *Reaction with Bromine Water.* Add bromine water drop by drop to an aqueous solution of phenol. A white precipitate of 2,4,6-tribromophenol is formed.

(iii) *Coupling with Diazonium Ions.* Add drop by drop a cold (5°C) solution of benzenediazonium chloride (Chapter 23) to a cold solution of phenol in sodium(I) hydroxide solution. A yellow precipitate of (4-hydroxyphenyl)azobenzene is obtained.

(iv) *Liebermann's Nitroso-reaction.* Warm a very small amount (0·25 g) of sodium(I) nitrate(III) (sodium nitrite) with 0·5 g of phenol in a dry test-tube for about twenty seconds. Add to the tube 1 cm³ of concentrated sulphuric(VI) acid drop by drop. Allow the tube to stand for a minute, and then pour the contents into 10 cm³ of cold water. A red solution is produced, and this turns green or blue when excess of sodium(I) hydroxide solution is added.

Nitrophenols

The mononitro-derivatives (1,2- and 1,4-) of phenol are best prepared by melting phenol with a little water and dropping the mixture slowly into a solution of sodium(I) nitrate(V) in medium sulphuric(VI) acid contained in a flask. The flask must be cooled in iced water, as heat is evolved and appreciable amounts of dinitrophenol are produced if the temperature is allowed to rise above 20°C. The resulting 2- and 4-nitrophenols are separated by steam distillation, only the 2-compound being volatile in steam. Crude 4-nitrophenol is deposited when the residue left in the flask is cooled. It is purified by extracting the crystals with boiling dilute hydrochloric acid, the compound crystallizing out again as the hot solution cools.

3-Nitrophenol cannot be prepared directly from phenol. It is obtained by first reducing one of the nitro groups of 1,3-dinitrobenzene to give 3-nitrophenylamine (see Chapter 22). The latter is then diazotized and the diazonium salt hydrolysed by heating the aqueous solution.

$$C_6H_4 {NH_2 \atop NO_2} \xrightarrow[5°C]{HNO_2} C_6H_4 {N_2Cl \atop NO_2} \xrightarrow[50°C]{H_2O} C_6H_4 {OH \atop NO_2}$$

The nitrophenols are all stronger acids than phenol itself. They are only sparingly soluble in water, but the solutions turn blue litmus red and liberate carbon dioxide from sodium(I) carbonate solution. When the nitrophenols are reduced with tin and hydrochloric acid they are converted into the corresponding aminophenols ($HO.C_6H_4.NH_2$). These are used in the manufacture of dyes.

A point of interest concerning the nitrophenols is the large difference between the melting point of the 1,2-compound on the one hand and those of the 1,3- and 1,4-compounds on the other. (2-Nitrophenol melts at 46°C, 3-nitrophenol at 97°C, and 4-nitrophenol at 112°C). The difference is explained by different kinds of hydrogen-bonding. In the 1,2-compound *intramolecular* hydrogen-bonding occurs. The hydrogen atom of the hydroxyl group forms a hydrogen bond with one of the oxygen atoms of the adjacent nitro group, thus forming a six-membered ring of atoms as shown.

Similar hydrogen-bonding cannot take place in 3- or 4-nitrophenol because the hydroxyl group and nitro group are not

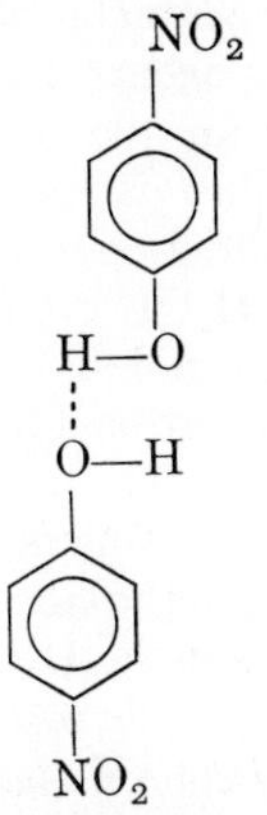

NO_2

H—O

O—H

NO_2

adjacent. In these compounds, however, *intermolecular* hydrogen bonding occurs, the —OH group of one molecule being joined to the —OH group of another molecule. The compounds thus exist in an associated condition and, since additional energy is required to break hydrogen bonds between molecules, the melting points of the compounds are considerably higher. For the same reason the volatilities are much lower, only the 1,2-isomer being appreciably volatile in steam.

METHYLPHENOLS, $C_6H_4 \begin{smallmatrix} CH_3 \\ OH \end{smallmatrix}$

The methylphenols (cresols) are the immediate homologues of phenol. They exist as the 1,2-, 1,3-, and 1,4-isomers.

2-methylphenol   3-methylphenol   4-methylphenol

All three compounds occur in the 'carbolic oil' fraction of coal-tar, and are obtained industrially from this source. They can be made in the laboratory from the corresponding methylphenylamines in the same way that phenol is prepared from phenylamine, that is, by diazotization followed by hydrolysis of the methylbenzenediazonium salt with water.

2-Methylphenol is a solid (m.p. 31°C), 3-methylphenol is a liquid (b.p. 203°C), and 4-methylphenol again is a solid (m.p. 35°C). Like phenol, the methylphenols are used as disinfectants and in the manufacture of resins and plastics. The higher phenols and their chloro-derivatives are more powerful germicides than phenol itself. They are also less harmful to the body tissues. Chlorodimethylphenol is 256 times more powerful than phenol as a germicide. It is used in preparations of the 'Dettol' type.

AROMATIC ALCOHOLS

PHENYLMETHANOL, $C_6H_5.CH_2OH$

Boiling point: 205°C   *Density*: 1·05 g cm^{-3}

Preparation

Phenylmethanol (benzyl alcohol) is prepared both in the laboratory and in industry from (chloromethyl)benzene (benzyl chloride) obtained by chlorination of methylbenzene. Since (chloromethyl)benzene is an ester of phenylmethanol, it yields the latter on hydrolysis with aqueous alkalies.

$$C_6H_5.CH_2Cl + OH^- \rightarrow C_6H_5.CH_2OH + Cl^-$$

In practice aqueous sodium(I) carbonate is used for the hydrolysis in preference to aqueous sodium(I) hydroxide, which produces wasteful side-reactions. In the laboratory preparation (chloromethyl)benzene is refluxed with an excess of aqueous sodium(I) carbonate until the smell of the chloro-compound has disappeared. After cooling, the phenylmethanol is extracted from the remaining liquid with ethoxyethane. The extract is dried with anhydrous potassium(I) carbonate, the ethoxyethane is distilled off on a water-bath, and then the alcohol is distilled over.

Phenylmethanol is also obtained from benzenecarbaldehyde, or benzaldehyde, by Cannizzaro's reaction (see Chapter 25).

Phenylmethanol is employed as a plasticizer and solvent in the manufacture of nitrocellulose lacquers. It is also used in making esters, such as phenylmethyl ethanoate (benzyl acetate), which are used in perfumery.

Properties of Phenylmethanol

Phenylmethanol is a colourless oily liquid with a smell resembling that of orange peel. It is only sparingly soluble in water, but dissolves readily in ethanol and ethoxyethane.

In general the reactions of the —CH_2OH group in aromatic alcohols correspond with those of the same group in aliphatic alcohols. Thus phenylmethanol reacts in the same way as ethanol

with sodium, phosphorus pentachloride, acids, acid chlorides and anhydrides, and oxidizing agents. One difference is that phenylmethanol does not form an alkene when heated with concentrated sulphuric(VI). This is because the phenylmethanol molecule does not have a hydrogen atom attached to the carbon atom adjacent to the —CH_2OH group (see Chapter 8).

Phenylmethanol and phenol show both chemical similarities and differences. These are outlined below.

Resemblances to Phenol. (*i*) The hydrogen atom of the hydroxyl group can be displaced by treating phenylmethanol with metallic sodium. Sodium phenylmethoxide ($C_6H_5.CH_2ONa$), corresponding to sodium phenoxide (C_6H_5ONa), is formed in solution.

(*ii*) Phosphorus pentachloride reacts with the hydroxyl group in the cold, and hydrogen chloride is evolved.

$$C_6H_5.CH_2OH + PCl_5 \rightarrow C_6H_5.CH_2Cl + POCl_3 + HCl$$

(*iii*) Phenylmethanol yields esters with acid chlorides (in the cold) and acid anhydrides (on heating). Thus with ethanoyl, or acetyl, chloride the ester phenylmethyl ethanoate, or acetate is obtained.

$$C_6H_5.CH_2OH + CH_3COCl \rightarrow CH_3COOCH_2.C_6H_5 + HCl$$
phenylmethyl ethanoate

(*iv*) Phenylmethanol shows its aromatic character in undergoing nitration and sulphonation in the nucleus, but the reactions are accompanied by others due to the reactive —CH_2OH group, and the yields of nuclear derivatives are small. These are best prepared from (chloromethyl)benzene (benzyl chloride), the chlorine atom being subsequently replaced by hydrolysis. The nuclear derivatives are of little importance.

Differences From Phenol. (*i*) Phenylmethanol does not react with aqueous alkalies to form salts.

(*ii*) Phenylmethanol undergoes esterification with acids under suitable conditions. (Chloromethyl)benzene is obtained in good yield by heating phenylmethanol with concentrated hydrochloric acid.

$$C_6H_5.CH_2OH + HCl \rightleftharpoons C_6H_5.CH_2Cl + H_2O$$

Phenylmethyl esters of organic acids can be prepared by refluxing the alcohol and acid together in the presence of moderately concentrated sulphuric(VI) acid (hot concentrated sulphuric(VI) acid resinifies phenylmethanol). Under these conditions phenylmethanol and ethanoic acid yield phenylmethyl ethanoate.

(*iii*) When phenylmethanol is heated with moderately concentrated sulphuric(VI) acid it forms an ether, (phenylmethoxymethyl)benzene (dibenzyl ether).

$$2C_6H_5.CH_2OH \rightarrow (C_6H_5.CH_2)_2O + H_2O$$

(*iv*) With oxidizing agents, such as dilute nitric(V) acid or acidified potassium(I) manganate(VII), phenylmethanol is oxidized firstly to benzenecarbaldehyde, or benzaldehyde ($C_6H_5.CHO$), and then to benzenecarboxylic, or benzoic, acid ($C_6H_5.COOH$). Phenol also reacts with oxidizing agents, but in this case the reactions are more complex and usually result in breakdown of the nucleus.

(*v*) Reduction of phenylmethanol with hydrogen (using a palladium catalyst) converts the —CH_2OH group to —CH_3.

$$C_6H_5.CH_2OH + H_2 \rightarrow C_6H_5.CH_3 + H_2O$$
methylbenzene

In reduction of phenol by hydrogen (with a nickel catalyst) the hydrogen is added to the nucleus to give cyclohexanol.

(*vi*) Phenylmethanol differs from phenol in not coupling with diazonium salts.

Chemical Tests for Phenylmethanol

Phenylmethanol is most easily recognized by means of its oxidation products benzenecarbaldehyde, or benzaldehyde, and benzenecarboxylic, or benzoic, acid.

(*i*) Boil 1 cm³ of phenylmethanol with 3–4 cm³ of dilute nitric(V) acid in a boiling-tube, using a very small flame. In a few minutes the almond-like odour of benzenecarbaldehyde can be detected.

(*ii*) Repeat the experiment, but use an acidified solution of potassium(I) manganate(VII) instead of dilute nitric(V) acid. After three or four minutes' boiling cool the tube under the tap. A white crystalline precipitate of benzenecarboxylic acid is formed.

EXERCISE 24

1. Indicate by chemical formulæ the stages involved in converting benzene to phenol (*a*) *via* benzenediazonium chloride, (*b*) *via* benzenesulphonic acid.

2. In the 'cumene' process for manufacturing phenol from benzene:
(*a*) Name the hydrocarbon which combines with benzene;
(*b*) Name the catalyst used;
(*c*) Give the chemical formula of 'cumene';
(*d*) Give the chemical formula of 'cumene' hydroperoxide;
(*e*) Name the important by-product obtained in the process.

3. With which of the following reagents does phenol behave in the same way as ethanol: (*a*) dilute aqueous sodium(I) hydroxide, (*b*) hydrobromic acid, (*c*) ethanoic anhydride, (*d*) concentrated sulphuric(VI) acid, (*e*) benzenecarbonyl, or benzoyl, chloride?

4. Name the organic compounds formed in the reactions (if any) between phenol and the reagents given in question 3.

5. State what you *see* (if anything) when the following are added together: (*a*) blue litmus paper and aqueous phenol, (*b*) carbon dioxide and sodium(I) phenoxide solution, (*c*) bromine water and aqueous phenol, (*d*) iron(III) chloride solution and aqueous phenol, (*e*) benzenediazonium chloride solution and sodium(I) phenoxide solution.

6. With which of the following does phenylmethanol react in the same way as phenol: (*a*) ethanoyl chloride, (*b*) dilute nitric(V) acid, (*c*) hydrogen (with a suitable catalyst), (*d*) hydrochloric acid, (*e*) sodium?

7. Name the organic compounds formed in the reactions between phenylmethanol and the reagents in question 6.

8. Name the five isomers which have the molecular formula C_7H_8O.

9. Indicate the stages by which the following conversions could be brought about: (*a*) benzene to phenylmethanol, (*b*) phenylmethanol to benzene.

25 Aromatic Aldehydes, Ketones, and Carboxylic Acids

AROMATIC ALDEHYDES

Under the heading Aromatic Aldehydes only those compounds in which the functional aldehydic group is directly attached to the benzene nucleus will be discussed. Aldehydes which have the —CHO group linked to a carbon atom in a ring system are named systematically by adding the suffix-*carbaldehyde* to the name of the ring system. Typical examples are benzenecarbaldehyde and methylbenzenecarbaldehyde. (For the first compound the trivial name benzaldehyde is at present a 'recommended' alternative name.)

$$C_6H_5-\overset{\displaystyle H}{\underset{\displaystyle O}{C}}$$

$$C_6H_4\overset{\displaystyle CH_3}{\underset{\displaystyle CHO}{<}}$$

benzenecarbaldehyde

methylbenzene-
carbaldehyde

Compounds like phenylethanal ($C_6H_5.CH_2.CHO$), in which the aldehydic group is present in a side-chain, are regarded as phenyl-substitution derivatives of aliphatic aldehydes. In these compounds the properties of the —CHO group are similar to those of the same group in aliphatic aldehydes. When the —CHO group is directly attached to the nucleus its behaviour differs in certain ways from that of the —CHO group in aliphatic aldehydes.

Benzenecarbaldehyde is the most important aromatic aldehyde, and may be taken as representative of the series.

BENZENECARBALDEHYDE, $C_6H_5.CHO$

Boiling point: 179°C   *Density:* 1·05 g cm^{-3}

Benzenecarbaldehyde is a colourless liquid with a strong smell of almonds (nitrobenzene has a similar smell). It is only sparingly soluble in water, but dissolves readily in organic solvents.

Laboratory Preparation

(*i*) *By Oxidation of Phenylmethanol.* Mild oxidizing conditions must be used because benzenecarbaldehyde readily oxidizes to benzenecarboxylic acid (benzoic acid). A suitable oxidizing agent is aqueous copper(II) nitrate(V) or lead(II) nitrate(V). Instead of using phenylmethanol as the starting-point, (chloromethyl)-benzene (benzyl chloride) may be used. When (chloromethyl)-benzene is refluxed with a 10 per cent solution of copper(II) nitrate(V) for about eight hours it is first hydrolysed to phenyl-methanol, which is then oxidized to benzenecarbaldehyde.

$$C_6H_5.CH_2Cl + H_2O \rightleftharpoons C_6H_5.CH_2OH + HCl$$

$$C_6H_5.CH_2OH + O \rightarrow C_6H_5.CHO + H_2O$$

During the reaction a slow current of carbon dioxide is passed through the reaction vessel to remove oxides of nitrogen which are produced. These tend to oxidize the aldehyde further. The remaining aldehyde is purified through its addition compound with sodium(I) hydrogensulphate(IV) (sodium bisulphite), as described later.

(ii) *By Synthesis from Benzene.* The aldehyde group can be introduced directly into the benzene nucleus. This is done by passing a mixture of carbon monoxide and hydrogen chloride into a solution of benzene in ethoxyethane, the latter containing (as catalyst) anhydrous aluminium(III) chloride and a trace of copper(I) chloride. The reaction is a modified form of the Friedel-Crafts reaction (Chapter 20). It probably depends on formation of unstable methanoyl, or formyl, chloride as an intermediate compound. The method is chiefly used for making higher aromatic aldehydes.

$$CO + HCl \rightleftharpoons HCOCl$$
methanoyl chloride

$$C_6H_6 + HCOCl \rightarrow C_6H_5.CHO + HCl$$

Manufacture and Uses of Benzenecarbaldehyde

On a large scale benzenecarbaldehyde is made directly from methylbenzene by oxidation. A mixture of methylbenzene vapour and air is passed over a catalyst (a vanadium compound) at 500°C, additional nitrogen being added to minimize further oxidation of the benzenecarbaldehyde.

$$C_6H_5.CH_3 + O_2 \rightarrow C_6H_5.CHO + H_2O$$

Benzenecarbaldehyde is used in the cheaper kinds of perfumery and as a flavouring agent ('essence of almonds') in confectionery. It is often used as an ingredient of scented soap. The aldehyde is also an intermediate in the production of dyes.

Reactions of Benzenecarbaldehyde

Many of the reactions of the —CHO group in benzenecarbaldehyde are similar to those of the same group in aliphatic aldehydes. Generally the reactions take place less readily with benzenecarbaldehyde because of the $-M$ and $-I$ effects of the aldehydic group (the five contributing structures of the resonance hybrid have been given in Chapter 20). Withdrawal of electrons from the nucleus into the side-chain results in the carbonyl carbon atom having a smaller partial positive charge and hence diminished reactivity with nucleophilic reagents. At the same time deactivation of the nucleus renders the latter less susceptible to attack by electrophilic reagents.

(A) Resemblances to Aliphatic Aldehydes.

1. **Oxidation.** As mentioned earlier, benzenecarbaldehyde is very readily oxidized to benzenecarboxylic acid ($C_6H_5.COOH$). Air alone brings about this oxidation, and crystals of benzenecarboxylic acid are often formed round the neck of a bottle of benzenecarbaldehyde. If the aldehyde is shaken with a warm acidified solution of potassium(I) manganate(VII) the solution is decolorized. The reduction of Tollens's reagent (ammoniacal silver(I) oxide) takes place more slowly than with aliphatic aldehydes, and it is necessary to heat the substances in a water bath and shake them well. Benzenecarbaldehyde does not reduce Fehling's solution.

2. Addition with Nucleophilic Reagents.

Benzenecarbaldehyde undergoes addition with hydrogen cyanide (in the presence of a trace of alkali as catalyst) and with sodium(I) hydrogensulphate(IV) (sodium bisulphite). The addition compounds, which consist of a colourless oil and a white crystalline solid respectively, have the systematic names and formulæ shown.

$$
\begin{array}{cc}
\overset{\displaystyle OH}{\underset{\displaystyle H}{C_6H_5-\overset{|}{\underset{|}{C}}-CN}} & \overset{\displaystyle OH}{\underset{\displaystyle H}{C_6H_5-\overset{|}{\underset{|}{C}}-SO_3^- \ Na^+}} \\
\text{2-hydroxy-2-phenyl-} & \text{sodium(I) hydroxyphenyl-} \\
\text{ethanonitrile} & \text{methylsulphate(IV)}
\end{array}
$$

The trivial names of the two addition compounds are benzaldehyde-cyanohydrin and sodium benzaldehyde-bisulphite.

The second addition compound is used in the laboratory purification of benzenecarbaldehyde. The impure aldehyde is shaken with a saturated solution of sodium(I) hydrogensulphate(IV), and the precipitate of the addition compound is filtered and washed with ethanol. The crystals are then treated with dilute sulphuric(VI) acid, which liberates benzenecarbaldehyde again. The latter is distilled in steam, and extracted from the distillate with ethoxyethane. The extract is dried with anhydrous calcium(II) sulphate(VI), and the benzenecarbaldehyde obtained by distillation.

3. Reduction.

Benzenecarbaldehyde undergoes reduction in the same way as aliphatic aldehydes. It can be reduced to phenylmethanol by means of sodium and ethanol, or by zinc and hydrochloric acid.

$$C_6H_5.CHO + 2H \rightarrow C_6H_5.CH_2OH$$

By using Clemmensen's method of reduction (zinc amalgam and concentrated hydrochloric acid) the reduction can be carried a stage further to methylbenzene.

$$C_6H_5.CHO + 4H \rightarrow C_6H_5.CH_3 + H_2O$$

4. Condensation with Ammonia and 'Substituted Ammonias.'

When benzenecarbaldehyde is left in contact with concentrated ammonia solution for a few days condensation occurs and a white crystalline substance (hydrobenzamide) is deposited.

$$3C_6H_5.CHO + 2NH_3 \rightarrow (C_6H_5.CH)_3N_2 + 3H_2O$$

The reaction may be compared with the condensation of methanal (formaldehyde) with ammonia to give hexamine, $(CH_2)_6N_4$.

Benzenecarbaldehyde resembles aliphatic aldehydes in the way it condenses with the 'substituted ammonias' phenylhydrazine and hydroxylamine to give crystalline derivatives. With phenylhydrazine yellow crystals of *benzenecarbaldehyde phenylhydrazone* are obtained.

$$C_6H_5.CHO + H_2N.NH.C_6H_5$$
$$\rightarrow C_6H_5.CH:N.NH.C_6H_5 + H_2O$$

Experiment

To 1 cm³ of glacial ethanoic acid in a test-tube add two or three drops of phenylhydrazine, and dilute the resulting solution of phenylhydrazine ethanoate with 2-3 cm³ of water. Add two or three drops of benzenecarbaldehyde and shake the tube. A yellow crystalline precipitate of the phenylhydrazone (m.p. 157°C) is produced.

Benzenecarbaldehyde similarly gives a yellow crystalline precipitate with 2,4-dinitrophenylhydrazine.

Benzenecarbaldehyde and hydroxylamine condense together to form *benzenecarbaldehyde oxime* (benzaldoxime), which consist of colourless needle-shaped crystals.

$$C_6H_5.CHO + H_2NOH \rightarrow C_6H_5.CH:NOH + H_2O$$
benzenecarbaldehyde oxime

(B) Differences from Aliphatic Aldehydes

In addition to its failure to reduce Fehling's solution, benzenecarbaldehyde differs from the majority of aliphatic aldehydes in the ways now described.

1. Reaction with Alkalies. Benzenecarbaldehyde does not polymerize with strong alkalies in the manner of ethanal, which forms firstly a dimer ('aldol') and eventually a brown resin. Instead *disproportionation* occurs, part of the aldehyde being oxidized to benzenecarboxylic acid, and part reduced to phenylmethanol (*Cannizzaro's reaction*).

$$2C_6H_5.CHO + KOH \rightarrow C_6H_5.CH_2OH + C_6H_5.COOK$$

The difference in behaviour of benzenecarbaldehyde is caused by the absence from the molecule of a hydrogen atom joined to a carbon atom in the 2- position. As was seen in Chapter 10, methanal, HCHO, and trimethylethanal, $C(CH_3)_3.CHO$, which similarly lack a suitably placed hydrogen atom, also undergo Cannizzaro's reaction with strong alkalies.

2. Polymerization. Benzenecarbaldehyde does not form cyclic polymers like ethanal trimer, $(CH_3.CHO)_3$, and ethanal tetramer, $(CH_3.CHO)_4$, under the catalytic influence of acids. It does, however, yield a dimer, commonly called *benzoin*, when it is heated with potassium(I) cyanide solution, the CN^- ions acting as a catalyst. The structure and systematic name of the dimer are given below.

$$C_6H_5-\overset{\overset{\displaystyle H}{|}}{\underset{\underset{\displaystyle O}{||}}{C}} + \overset{\overset{\displaystyle O}{||}}{\underset{\underset{\displaystyle H}{|}}{C}}-C_6H_5 \rightarrow C_6H_5-\overset{\overset{\displaystyle H}{|}}{\underset{\underset{\displaystyle OH}{|}}{C}}-\overset{\overset{\displaystyle O}{||}}{C}-C_6H_5$$

2-hydroxy-1,2-diphenyl-
ethanone

The dimer is a colourless crystalline compound (m.p. 137°C).

3. Nuclear Substitution. Benzenecarbaldehyde shows typical aromatic character in undergoing halogenation, nitration, and sulphonation in the nucleus. As already mentioned, the reactions occur more slowly than with benzene owing to deactivation of the nucleus. Owing to the tendency of some of the reagents to react with the aldehydic group, the yields of the nuclear derivatives are generally low.

Chlorination takes place at ordinary temperatures when chlorine is passed into the aldehyde in the presence of a chlorine carrier such as finely divided iron. The product is 3-chlorobenzene-carbaldehyde.

CHO

3-chlorobenzenecarb-
aldehyde

In view of the fact that dilute nitric(V) acid oxidizes benzene-carbaldehyde to benzenecarboxylic acid it is, perhaps, surprising to find that a mixture of concentrated nitric(V) acid and concentrated sulphuric(VI) acid nitrates the aldehyde to 3-nitrobenzenecarbaldehyde. A certain amount of oxidation occurs at the same time.

$$C_6H_4 \overset{CHO}{\underset{NO_2}{\diagup}}$$

nitrobenzenecarb-
aldehyde

$$C_6H_4 \overset{CHO}{\underset{SO_3H}{\diagup}}$$

sulphobenzenecarb-
aldehyde

Sulphonation of benzenecarbaldehyde takes place when the latter is refluxed with concentrated sulphuric(VI) acid. The product is 3-sulphobenzenecarbaldehyde.

Chemical Tests for Benzenecarbaldehyde

(*i*) *Schiff's reagent*. To 2 cm³ of Schiff's reagent add one drop of benzenecarbaldehyde, and shake the tube. A violet colour is obtained, but this develops more slowly than with ethanal.

(*ii*) *Cannizzaro's Reaction* (Small Scale). Dissolve three pellets of sodium(I) hydroxide in 2 cm³ of water in a test-tube. Add two drops of benzenecarbaldehyde, and warm the tube gently for five minutes, shaking it well. Then cool the tube under the tap, and pour the remaining liquid drop by drop into about 5 cm³ of a mixture of equal volumes of concentrated hydrochloric acid and water. A white precipitate of benzenecarboxylic acid is formed in the aqueous layer below the remaining aldehyde.

(*iii*) *Sodium(I) Hydrogensulphate(IV)* (*Sodium Bisulphite*). Add 2–3 cm³ of a concentrated solution of this salt to 1 cm³ of benzene-carbaldehyde and shake the tube. Heat is evolved, and when the tube is cooled a white precipitate of the addition compound of the aldehyde and salt is formed.

(*iv*) *Phenylhydrazine or 2,4-Dinitrophenylhydrazine*. Test benzene-carbaldehyde with either of these reagents as described earlier in this chapter.

AROMATIC KETONES

There are two classes of aromatic ketones. In one class (alkyl-aryl ketones) the carbonyl group is attached to an alkyl radical and an aromatic nucleus. In the other class (diaryl ketones) the carbonyl group is linked to two aromatic nuclei. The most important members of the two classes are the compounds shown.

$$\underset{\text{phenylethanone}}{CH_3-\overset{O}{\overset{\|}{C}}-C_6H_5}$$

$$\underset{\text{diphenylmethanone}}{C_6H_5-\overset{O}{\overset{\|}{C}}-C_6H_5}$$

The systematic names of aromatic ketones are based on the straight carbon chain (including the carbon atom of the carbonyl group) which is linked to one or both benzene nuclei. The names are derived from those of the alkanes corresponding to the straight carbon chain, the benzene nuclei are specified as phenyl substituents, and the final -e in the name of the alkane is replaced by -one. Thus in phenylethanone the straight carbon chain is two atoms long, the alkane is ethane, and there is one phenyl substituent. (Traditional names for phenylethanone and diphenylmethanone are acetophenone and benzophenone respectively.)

Preparation of Aromatic Ketones

The best way of preparing aromatic ketones in the laboratory is by means of a Friedel-Crafts reaction, consisting of acylation of an aromatic hydrocarbon under the catalytic influence of anhydrous aluminium(III) chloride. As shown earlier, acylation of aromatic hydrocarbons with 'methanoyl chloride,' or 'formyl chloride' (a mixture of carbon monoxide and hydrogen chloride), yields aldehydes. With other acyl chlorides ketones are obtained. Thus benzene and ethanoyl, or acetyl, chloride produce phenylethanone, while benzene and benzenecarbonyl chloride give diphenylmethanone. The reactions take place at ordinary temperature.

$$C_6H_6 + CH_3.COCl \xrightarrow{AlCl_3} CH_3.CO.C_6H_5 + HCl$$

$$C_6H_6 + C_6H_5.COCl \xrightarrow{AlCl_3} C_6H_5.CO.C_6H_5 + HCl$$

Phenylethanone is obtained commercially as a by-product in the manufacture of phenol by the cumene process (Chapter 24). This process yields not only phenol and propanone, but also a small proportion (about 1 per cent) of the ketone. It is separated from the other products and used as a soporific ('hypnone'), as an ingredient of perfumes, and in the manufacture of other compounds.

Properties of Aromatic Ketones

Aromatic ketones are either colourless oily liquids or crystalline solids. Phenylethanone melts at 20·5°C and boils at 202°C. Diphenylmethanone exists in two crystalline forms, a stable rhombic form melting at 49°C and a metastable monoclinic form melting at 27°C. It boils at 306°C. Both ketones are insoluble in water, but dissolve in organic solvents.

Many of the reactions of the carbonyl group in aromatic ketones resemble those of the carbonyl group in aliphatic ketones, but there are some differences. The carbonyl group is less reactive in aromatic ketones owing to stabilization of the molecules by resonance. Thus no addition compounds are formed with sodium(I) hydrogensulphate(IV) (sodium bisulphite), and only a little combination occurs with hydrogen cyanide. Addition with Grignard reagents, however, takes place normally. Condensation reactions with phenylhydrazine and hydroxylamine are similar to those given by aliphatic ketones.

Phenylethanone is gradually oxidized to benzenecarboxylic acid when it is refluxed with acidified potassium(I) manganate(VII).

$$C_6H_5.CO.CH_3 + 4O \rightarrow C_6H_5.COOH + H_2O + CO_2$$

Diphenylmethanone is scarcely affected by oxidizing agents.

The carbonyl group of an aromatic ketone is reduced to a secondary alcohol group by sodium and ethanol. Phenylethanone yields

1-phenylethanol, the number being explained by the phenyl radical being attached to the first carbon atom of the carbon chain joined to the nucleus. By using Clemmensen's method of reduction (zinc amalgam and concentrated hydrochloric acid) reduction can be carried a stage further and hydrocarbons obtained.

$$C_6H_5.CO.CH_3 + 2H \rightarrow C_6H_5.CHOH.CH_3$$
1-phenylethanol

$$C_6H_5.CO.CH_3 + 4H \rightarrow C_6H_5.CH_2.CH_3 + H_2O$$
ethylbenzene

AROMATIC CARBOXYLIC ACIDS

Aromatic acids in which a carboxylic group is directly attached to the benzene nucleus are named systematically by adding *-carboxylic acid* to the name of the parent ring system. The most important acids in this group are benzenecarboxylic acid, 2-hydroxybenzenecarboxylic acid (salicylic acid), and the three benzenedicarboxylic acids (phthalic acids). These have the following formulæ:

benzenecarboxylic acid   2-hydroxybenzene-carboxylic acid   benzenedicarboxylic acids

In the case of benzenecarboxylic acid the trivial name benzoic acid may be used as an alternative name.

BENZENECARBOXYLIC ACID, $C_6H_5.COOH$

Melting point: 121·5°C   *Boiling point:* 249°C

Laboratory Preparation

Benzenecarboxylic acid can be prepared by oxidizing phenylmethanol or benzenecarbaldehyde with a warm acidified solution of potassium(I) dichromate(VI) or potassium(I) manganate(VII).

$$C_6H_5.CH_2OH + O \rightarrow C_6H_5.CHO + H_2O$$

$$C_6H_5.CHO + O \rightarrow C_6H_5.COOH$$

When the remaining solution is cooled a white precipitate of benzenecarboxylic acid is formed. The acid is purified by recrystallization from hot water.

A common method of preparation is to reflux (chloromethyl)-benzene (benzyl chloride) with an *alkaline* solution of potassium(I) manganate(VII) for about one and a half hours. The halogen compound is hydrolysed to phenylmethanol, which is then oxidized to the acid. During the reaction a brown precipitate of hydrated manganese(IV) oxide is formed. This is removed at the end of the reaction by shaking with a concentrated solution of sodium(I) sulphate(IV) (sodium sulphite) or sulphur dioxide $(MnO_2 + SO_2 \rightarrow MnSO_4)$. A white crystalline mass of benzenecarboxylic acid remains.

A general method of introducing a —COOH group into the nucleus of an aromatic hydrocarbon has been described in Chapter 23. The method involves the use of one of the Sandmeyer reactions.

Manufacture and Uses of Benzenecarboxylic Acid

Like benzenecarbaldehyde the acid is made on a large scale by catalysed oxidation of methylbenzene with air. The oxidation is carried out either in the gas phase at 500°C, using a vanadium compound as a catalyst, or in the liquid phase with a manganese or cobalt compound as catalyst.

$$2C_6H_5.CH_3 + 3O_2 \rightarrow 2C_6H_5.COOH + 2H_2O$$

Benzenecarboxylic acid is a powerful antiseptic, and is chiefly used as a preservative for pharmaceutical preparations and food products (e.g. pickles, sauces, and fruit cordials). For this purpose it is employed in the form of its sodium(I) salt. Ethyl benzenecarboxylate, or benzoate, and other esters are used in perfumery. The acid is also an intermediate in the manufacture of other compounds.

Reactions of Benzenecarboxylic Acid

In general the —COOH group in aromatic carboxylic acids behaves in the same way as in alkanoic acids. The presence of the benzene nucleus decreases the acidic character of the group, as shown by the fact that benzenecarboxylic acid (C_6H_5—COOH) is a weaker acid than methanoic acid (H—COOH). Benzenecarboxylic acid, however, is stronger than ethanoic acid and the other alkanoic acids.

Derivatives of benzenecarboxylic acid include esters, an acyl chloride, an acid amide, and a nitrile. These are prepared as described shortly. The acid undergoes the usual halogenation, nitration, and sulphonation reactions in the nucleus, the product in each case consisting mainly of the 1,3-isomer. The reactions occur more slowly than with benzene owing to deactivation of the nucleus by the carboxyl group. The nuclear substitution, derivatives are of little importance.

Chemical Tests for Benzenecarboxylic Acid

(i) *Sodium(I) Carbonate Solution.* Warm a little of the acid with sodium(I) carbonate solution. Effervescence occurs and carbon dioxide is given off.

(ii) *Iron(III) Chloride Solution.* To a solution of benzenecarboxylic acid or its sodium(I) salt add a few drops of iron(III) chloride solution. A buff-coloured precipitate of iron(III) benzenecarboxylate is obtained.

$$3C_6H_5.COONa + FeCl_3 \rightarrow (C_6H_5.COO)_3Fe \downarrow + 3NaCl$$

(iii) *Esterification.* Warm together in a test-tube for a few minutes about 1 g of benzenecarboxylic acid, 2 cm³ of ethanol, and a few drops of concentrated sulphuric(VI) acid. Ethyl benzenecarboxylate is formed and can be detected by its characteristic odour of peppermint.

(iv) *Decarboxylation.* If a mixture of benzenecarboxylic acid and soda-lime is heated in a dry tube, benzene vapour is evolved and can be ignited at the mouth of the tube. This experiment has been described in Chapter 19.

Esters of Benzenecarboxylic Acid

Ethyl benzenecarboxylate ($C_6H_5.COOC_2H_5$) is a colourless liquid which boils at 213°C and has a density of 1·04 g cm^{-3}. It resembles ethyl ethanoate, or acetate, in most of its reactions. It is formed when ethanol is refluxed with benzenecarboxylic acid in the presence of a strong mineral acid as a catalyst. If concentrated sulphuric(VI) acid is used as the catalyst, however, considerable charring takes place and the yield of ester is small. In the *Fischer-Speier* method of esterification (see Chapter 13) the difficulty is avoided by employing dry hydrogen chloride as the catalyst.

$$C_6H_5.COOH + C_2H_5OH \xrightleftharpoons{HCl} C_6H_5.COOC_2H_5 + H_2O$$
ethyl benzenecarboxylate

In practice a nearly saturated solution of hydrogen chloride in absolute ethanol is refluxed with benzenecarboxylic acid for about one and a half hours. The remaining ethanol is then distilled off, leaving most of the ester because of its much higher boiling point. The ester is purified by dissolving it in tetrachloromethane and shaking the mixture in turn with water and sodium(I) carbonate solution. The tetrachloromethane, a heavy liquid, is used because the ester has almost the same density as water and separation from the aqueous layers would be difficult. After drying with anhydrous calcium(II) chloride, the ester and the tetrachloromethane are separated by fractional distillation.

Phenylmethyl benzenecarboxylate (benzyl benzoate), $C_6H_5.COOCH_2C_6H_5$, can be prepared by esterifying phenylmethanol with benzenecarboxylic acid, using the method which has been described for ethyl benzenecarboxylate. The phenylmethyl ester is a solid (m.p. 21°C), but it supercools readily and is often encountered as an oily liquid. It boils at 323°C. It is used in perfumery and in medicine.

Phenyl benzenecarboxylate ($C_6H_5.COOC_6H_5$) cannot be obtained by treating phenol with benzenecarboxylic acid. It is prepared as described in the next section. It is a white solid, which melts at 69°C and boils at 314°C.

Benzenecarbonyl Chloride

The acyl chloride of benzenecarboxylic acid is prepared by warming the acid with phosphorus pentachloride or sulphur dichloride oxide (thionyl chloride) in a flask attached to a reflux condenser. The warming is conducted on a water-bath. After the reaction the benzenecarbonyl chloride is separated by fractional distillation.

$$C_6H_5.COOH + PCl_5 \rightarrow C_6H_5.COCl + POCl_3 + HCl$$

$$C_6H_5.COOH + SOCl_2 \rightarrow C_6H_5.COCl + SO_2 + HCl$$

Benzenecarbonyl, or benzoyl, chloride is a colourless oily liquid (b.p. 220°C) with an irritating smell. The —COCl group is much less reactive than in ethanoyl, or acetyl, chloride. Thus benzenecarbonyl chloride is hydrolysed only slowly by water or aqueous alkalies at ordinary temperatures. This is often an advantage because it enables the acyl chloride to be used in an aqueous medium or in the presence of alkalies.

The chief use of benzenecarbonyl chloride is in *benzenecarbonylation*, or *benzoylation*, that is, the introduction of the benzenecarbonyl, or benzoyl, group ($C_6H_5.CO$—) into a molecule in place of a hydrogen atom. This process is applied chiefly to hydroxy-compounds and amines (primary and secondary); e.g.

$$C_2H_5OH + C_6H_5.COCl \rightarrow C_6H_5.COOC_2H_5 + HCl$$
ethyl benzenecarboxylate

$$C_6H_5NH_2 + C_6H_5.COCl \rightarrow C_6H_5NH.COC_6H_5 + HCl$$
N-phenylbenzenecarboxamide

Benzenecarbonylation is frequently used to assist in the identification of hydroxy-compounds and amines by means of the melting points of the derivatives. For this purpose it is often more convenient to use benzenecarbonyl derivatives than ethanoyl derivatives, because the former can be prepared in the presence of water, and, as they are insoluble, they are precipitated when the reaction is carried out in an aqueous medium.

A reagent often preferred to benzenecarbonyl chloride itself is the 3,5-dinitro-derivative.

COCl

O_2N ⬡ NO_2

3,5-dinitrobenzene-
carbonyl chloride

As mentioned in Chapter 8, this compound is particularly useful for identifying the lower aliphatic alcohols. Under suitable conditions these yield precipitates of the 3,5-dinitrobenzenecarboxylate esters, even when the alcohols are present in dilute aqueous solution. This situation arises when the alcohol component of an ester is being identified by alkaline hydrolysis of the ester, followed by distillation of the alcohol (Chapter 13).

Benzenecarbonylation is usually performed by the *Schotten-Baumann* method. This consists of shaking the hydroxy-compound or amine in the cold with benzenecarbonyl chloride and an excess of aqueous sodium(I) hydroxide until the smell of the acyl chloride has disappeared. The reason for the use of the alkali is different in the two cases. Hydroxy-compounds (particularly phenols) react only slowly with benzenecarbonyl chloride alone, but reaction is rapid in the presence of sodium(I) hydroxide. The latter converts phenols into their more reactive anions.

$$C_6H_5.COCl + C_6H_5O^- \rightarrow C_6H_5.COOC_6H_5 + Cl^-$$
phenyl benzenecarboxylate

In the case of an amine, such as phenylamine, benzene-carbonylation occurs readily with the acyl chloride alone, but in the absence of alkali only half of the amine is substituted. The other half combines with the hydrochloric acid formed in the reaction.

$$2C_6H_5NH_2 + C_6H_5COCl \rightarrow$$
$$C_6H_5NH.COC_6H_5 + C_6H_5NH_3^+Cl^-$$
N-phenylbenzene-   phenylammonium
carboxamide   chloride

The alkali makes all the amine available for benzenecarbonylation by combining itself with the hydrochloric acid.

When benzenecarbonyl chloride is distilled with sodium(I) benzenecarboxylate it yields *benzenecarboxylic anhydride* (compare the preparation of ethanoic anhydride in Chapter 12).

$$C_6H_5.COCl + C_6H_5COONa \rightarrow (C_6H_5.CO)_2O + NaCl$$

Benzenecarboxylic, or benzoic, anhydride is a white solid (m.p. 42°C), which changes into the acid only slowly with water. It can be used for benzenecarbonylation, but does not react as rapidly as the acyl chloride.

Benzenecarboxamide

Benzenecarboxamide, or benzamide, can be obtained by the same general methods as alkanoic acid amides (Chapter 14). It is most conveniently prepared by adding benzenecarbonyl chloride drop by drop to concentrated aqueous ammonia cooled in iced water. The acid amide precipitated is recrystallized from hot water.

$$C_6H_5.COCl + 2NH_3 \rightarrow C_6H_5.CONH_2 + NH_4Cl$$
benzenecarboxamide

Benzenecarboxamide is a white crystalline solid (m.p. 129°C). It is only slightly soluble in water, but dissolves in organic solvents. The reactions of the amido group ($—CONH_2$) in this compound are similar to those of the group in ethanamide. Thus when benzenecarboxamide is heated with mineral acids or aqueous alkalies it is hydrolysed to benzenecarboxylic acid.

$$C_6H_5.CONH_2 + H_2O \rightarrow C_6H_5.COOH + NH_3$$

If benzenecarboxamide is distilled with phosphorus(V) oxide,

it is dehydrated to *benzenecarbonitrile*, or benzonitrile. This is a colourless oil (b.p. 191°C).

$$C_6H_5.CONH_2 - H_2O \rightarrow C_6H_5CN$$
benzenecarbonitrile

Homologues of Benzenecarboxylic Acid

The immediate homologues of benzenecarboxylic acid are the three methylbenzenecarboxylic acids, $C_6H_4(CH_3)COOH$. These are isomeric with phenylethanoic acid ($C_6H_5CH_2.COOH$). The methylbenzenecarboxylic acids are colourless crystalline compounds, which are obtained by hydrolysis of the corresponding nitriles. These in turn are derived from the corresponding methylphenylamines through diazotization and Sandmeyer's reaction (Chapter 23). The three acids, which closely resemble benzenecarboxylic acid, are little used.

2-HYDROXYBENZENECARBOXYLIC ACID,

Melting point: 159°C   *Boiling point:* (decomposes)

Of the three isomeric hydroxybenzenecarboxylic acids only the isomer specified above is important. This compound is well known under its trivial name *salicylic acid*. (For convenience we shall use the trivial name.) The acid occurs naturally as its methyl ester in *oil of wintergreen*, a fragrant oil derived from certain evergreen shrubs. The acid itself is found in willow-bark, and to this it owes its trivial name (Latin *salix* = a willow).

Salicylic acid forms colourless crystals, which resemble those of benzenecarboxylic acid in being only slightly soluble in cold water, but fairly soluble in hot water. The crystals dissolve readily in ethanol and ethoxyethane.

Reactions of Salicylic Acid

Salicylic acid combines the reactions of a phenol and a carboxylic acid. With aqueous sodium(I) hydroxide it yields two salts according to the amount of alkali used. The first product is sodium(I) salicylate, $C_6H_4(OH).COONa$, but with more alkali disodium(I) salicylate, $C_6H_4(ONa).COONa$, is formed. With sodium *carbonate* solution only the first salt is produced. If salicylic acid or sodium(I) salicylate is heated with soda-lime decarboxylation occurs and phenol vapour is evolved.

$$C_6H_4 \begin{matrix} OH \\ \\ COONa \end{matrix} + NaOH \rightarrow C_6H_5OH + Na_2CO_3$$

Chemical Tests for Salicylic Acid

Salicylic acid resembles phenol in giving a violet colour with iron(III) chloride solution. It can be distinguished from phenol by warming with methanol and concentrated sulphuric(VI) acid. The ester methyl salicylate is formed and this has the characteristic smell of oil of wintergreen.

Another method of distinguishing salicylic acid from phenol is to add the acid to sodium(I) carbonate solution. Carbon dioxide is evolved with salicylic acid, but not with phenol.

Manufacture and Uses

Salicylic acid is made on a large-scale by the *Kolbe-Schmitt synthesis*, which consists of heating solid sodium(I) phenoxide with carbon dioxide under pressure at 150°C. The reaction, which is complicated, can be represented simply by the equation shown.

$$C_6H_5ONa + CO_2 \rightarrow C_6H_4 \begin{matrix} OH \\ \\ COONa \end{matrix}$$

The free acid is obtained by acidifying an aqueous solution of the remaining sodium(I) salicylate.

Salicylic acid and its chief derivatives are powerful germicides. The acid is used in treating skin diseases and rheumatism, and forms the basis of many corn-cures. Methyl salicylate is an ingredient of tooth-pastes and liniments. Ethanoylsalicylic acid, or acetylsalicylic acid, is the well-known drug *aspirin*, which is used for its antipyretic (fever-reducing) and analgesic (pain-relieving) properties. It is made by ethanoylating salicylic acid with ethanoic anhydride.

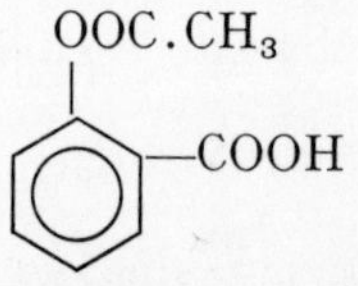

aspirin

BENZENEDICARBOXYLIC ACIDS, $C_6H_4(COOH)_2$

The three isomeric benzenedicarboxylic acids (phthalic acids) are distinguished by individual trivial names as follows:

phthalic acid   isophthalic acid   terephthalic acid
(1,2-)   (1,3-)   (1,4-)

The different benzenedicarboxylic acids can be prepared in the laboratory by refluxing the corresponding dimethylbenzenes for 24 hours with dilute nitric(V) acid or sodium(I) dichromate(VI) and 50 per cent sulphuric(VI) acid.

$$C_6H_4\begin{matrix}CH_3\\ \\CH_3\end{matrix} + 6O \rightarrow C_6H_4\begin{matrix}COOH\\ \\COOH\end{matrix} + 2H_2O$$

The benzenedicarboxylic acids are colourless crystalline compounds, which are only slightly soluble in cold water, but fairly soluble in hot water or ethanol. The 1,2- and 1,3- isomers melt at 195°C and 348°C respectively, but the 1,4-compound sublines without melting at about 300°C.

1,2-Benzenedicarboxylic acid is little used but its anhydride (phthalic anhydride) is an important compound. The anhydride is formed when the acid is heated at about 200°C. It is used to make esters like dibutyl 1,2-benzenedicarboxylate (dibutyl phthalate), which are employed in the manufacture of plastics.

1,2-benzenedicarboxylic
anhydride

1,4-Benzenedicarboxylic acid is one of the basic materials used in the manufacture of Terylene (Chapter 26). It is produced on a large scale by oxidizing 1,4-dimethylbenzene with air, a cobalt compound being used as catalyst.

EXERCISE 25

Aromatic Aldehydes and Ketones

1. Mentioning any catalysts used, name the reagents used for (a) simultaneous hydrolysis and oxidation of (chloromethyl)benzene to benzenecarbaldehyde, (b) the laboratory purification of benzenecarbaldehyde, (c) introduction of the —CHO group directly into the benzene nucleus, (d) large-scale production of benzenecarbaldehyde directly from methylbenzene.

2. State which of the following react with benzenecarbaldehyde in the same way as with ethanal: (a) sodium(I) hydrogensulphate(IV), (b) aqueous potassium(I) hydroxide, (c) zinc amalgam and concentrated hydrochloric acid, (d) Fehling's solution, (e) hydroxylamine.

3. Give the names and formulæ of the organic compounds formed in the reactions (if any) between benzenecarbaldehyde and the reagents in question 2.

4. Give the systematic names and formulæ of the organic compounds formed in the reactions between (a) benzene and ethanoyl chloride (AlCl$_3$ catalyst), (b) diphenylmethanone and phenylhydrazine, (c) an ethanolic solution of phenylethanone and sodium, (d) phenylethanone and acidified potassium(I) manganate(VII).

Aromatic Acids and Derivatives

5. Name the organic compounds formed in the reactions between benzenecarboxylic acid and (a) soda-lime, (b) iron(III) chloride, (c) sulphur dichloride oxide, (d) chlorine (iron catalyst).

6. Give the formulæ of the organic compounds formed in the reactions between benzenecarbonyl chloride and (a) propan-1-ol, (b) phenol dissolved in aqueous sodium(I) hydroxide, (c) sodium(I) benzenecarboxylate, (d) concentrated aqueous ammonia.

7. Which of the following statements are true?

(a) A white precipitate is formed when medium hydrochloric acid is added to a concentrated solution of sodium(I) benzenecarboxylate.

(b) Benzenecarboxylic acid undergoes nitration more rapidly than benzene.

(c) Benzenecarboxylic acid is a weaker acid than methanoic acid.

(d) Phenyl benzenecarboxylate cannot be made from phenol and benzenecarboxylic acid.

8. Give the systematic names of the following:

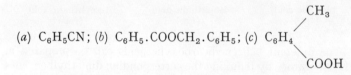

(a) C_6H_5CN; (b) $C_6H_5.COOCH_2.C_6H_5$; (c) $C_6H_4\begin{smallmatrix}CH_3\\COOH\end{smallmatrix}$

(d) $C_6H_4\begin{smallmatrix}OH\\COOH\end{smallmatrix}$; (e) $C_6H_4\begin{smallmatrix}COOH\\COOH\end{smallmatrix}$

part four

26 Synthetic High Polymers

GENERAL FEATURES

High polymers are composed of 'giant' molecules (macro-molecules), their relative molecular masses ranging from a few thousand to over a million. Naturally occurring compounds of this class include starch, cellulose, and proteins, which have been studied earlier. The key to the synthesis of high polymers was discovered in the 1920s, when X-ray analysis was applied to materials like cotton, wool, silk, and rubber. It was found that the molecules making up these materials have a chain-like structure, in which the same structural unit is repeated over and over again. This discovery prompted chemists to attempt the synthesis of high polymers by joining together small molecules of suitable types. The success of their efforts led to the founding of those branches of chemical industry associated with the manufacture of resins and plastics, synthetic rubber, and synthetic fibres. In the last fifty years these man-made materials have played an increasingly important part in our daily lives, so that names like polythene, Perspex, nylon, and Terylene are now familiar household terms.

There are two basic methods of synthesizing high polymers, and these result in molecular structures of various types, as now described.

1. Addition Polymerization. This method consists of linking together simple unsaturated molecules, which usually contain an alkene double bond. The 'monomers' may be alkenes, like ethene and propene, or substituted ethenes of the type $CH_2=CHX$ containing the ethenyl (vinyl) group $CH_2=CH—$. Thus under suitable conditions ethene molecules ($CH_2=CH_2$) combine together to form poly(ethene), ($—CH_2—CH_2—)_n$, in which the molecular chains may contain a thousand or more monomer units. Addition polymerization of chloroethene (vinyl chloride), $CH_2=CHCl$, yields the well known plastic poly(chloroethene) (polyvinyl chloride, or PVC),

$$(—CH_2—CHCl—)_n$$

In both of the above cases the molecular chains are terminated, not by carbon atoms with free valencies, but by 'end-groups,' the nature of which depends on several factors, including the method of manufacture. Thus, if benzoyl peroxide is used to 'initiate' the polymerization (see later), the phenyl radical $C_6H_5—$ may form one of the end-groups. In view of the length of the molecular chains the end-groups may be ignored. Poly-(ethene) and poly(chloroethene) are examples of 'true' poly-merization, that is, polymerization in the limited sense of the term used earlier in the book. The polymer is formed by addition of identical molecules, and its relative molecular mass is an integral multiple of that of the monomer.

In poly(ethene) and poly(chloroethene) the molecular chains consist of similar units arranged end to end. This arrangement is characteristic of *linear* high polymers. Although the majority of the chains are unbranched, a certain amount of branching does occur. Thus in a linear polymer there may be molecular chains of the types shown.

—A—A—A—A—A—
unbranched linear polymer

branched linear polymer

High polymers are often made by addition polymerization of a mixture of two different monomers. The product is then known as a *copolymer*. It consists of molecular chains in which the different monomer units are joined haphazardly, and it is there-fore described as a *random linear copolymer*. Another type of copolymer is obtained by first growing molecular chains from one monomer and then adding a second monomer to continue the growth of the chains. In this case the product is described as a *linear block copolymer*. In both cases some branching of the molecular chains may take place.

—A—A—B—A—B—B—
random linear copolymer

—A—A—A—B—B—B—B—
linear block copolymer

In copolymerization the term 'polymerization' is used in a broader sense than previously. In this case the relative molecular mass of the product is *not* an integral multiple of the relative molecular mass of either monomer.

Addition polymerization is a 'chain reaction' in a double sense. It not only produces molecules which are chain-like in form, but it proceeds by the three steps characteristic of chain reactions—initiation, propagation, and termination. These steps may be brought about either by a free radical mechanism or an ionic mechanism. More advanced books must be consulted for details of the mechanisms.

In the free radical mechanism initiation involves breaking, or 'opening out,' the π bond of the alkene double bond. This is brought about by adding to the system a small amount of an 'initiator.' The initiator is a compound which on heating yields free radicals. Two commonly used initiators are potassium(I) peroxidesulphate(VI) (potassium persulphate), $K_2S_2O_8$, and di(benzoyl)peroxide. The latter decomposes at about 70°C, giving firstly free benzoate, or benzenecarboxylate, radicals and then free phenyl radicals.

$$C_6H_5COO\!-\!OOCC_6H_5$$
di(benzoyl)peroxide

$$\downarrow$$

$$2C_6H_5\overset{.}{C}OO$$

$$\downarrow$$

$$2C_6H_5 \ + 2CO_2$$

An important advance in addition polymerization came with the introduction in 1953 of a new type of catalyst (thought to work by an ionic mechanism). The new catalysts were called *Ziegler catalysts* after their German discoverer. A Ziegler catalyst is actually a combination of two substances. One is a compound of a transition metal (usually titanium(IV) chloride, $TiCl_4$), while the other is a metal alkyl such as triethylaluminium(III), $Al(C_2H_5)_3$. The new catalysts were able to bring about polymerizations which had previously been difficult or impossible. Thus propene was first polymerized with the aid of a Ziegler catalyst in 1954 by Natta, an Italian chemist.

One of the most important features of Ziegler catalysts is their ability to impart a regular structure to the polymers obtained. When an ethenyl monomer $CH_2\!=\!CHX$ is polymerized the X atoms or groups in the molecular chain can be distributed in three ways relative to each other and the chain axis. In one arrangement, described as *atactic* (Greek a = not, *taktos* = ordered), they have a random, or disorderly, distribution. In the other two arrangements the distribution is orderly. One of these is shown in the diagram below.

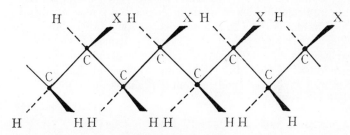

isotactic ethenyl polymer

In the above diagram the carbon atoms are supposed to be in the plane of the paper, and the bonds between them are represented by ordinary lines (all the bond angles are approximately

337

109·5°). Broken lines stand for bonds directed below the plane of the paper, and the wedge-shaped symbols for bonds directed above. It will be seen that all the X atoms or groups lie on the same side of the dividing plane. For this reason the polymer is described as *isotactic*. (Commercial poly(propene) consists mainly of this form.) In the second type of orderly structure the X atoms or groups are situated alternately on opposite sides of the dividing plane, and the polymer is said to be *syndiotactic*.

As will be explained shortly, the properties of an ethenyl polymer depend on whether it is atactic or whether it has one of the two ordered structures. Atactic polymers are formed when ethenyl monomers are polymerized by free radical mechanisms. By the use of appropriate Ziegler catalysts ethenyl polymers can be made in one or other of the ordered forms.

2. Condensation Polymerization. This process also consists of repeated combination of small molecules, but is accompanied by the elimination of water, ammonia, or some other simple substance (again the product is not a 'true' polymer). High polymers can be formed by condensation of molecules of the same monomer, providing these are of the right kind. Thus cellulose is derived in Nature by repeated condensation of β-glucose molecules.

$$n C_6H_{12}O_6 \rightarrow (C_6H_{10}O_5)_n + n H_2O$$
cellulose

In industry condensation polymers are made from two different monomers, so that the products are actually copolymers.

(*i*) *Polyesters*. The reactions by which polyesters are obtained are basically similar to the condensation of ethanol with ethanoic, or acetic, acid to give the ester ethyl ethanoate, or acetate. In this reaction each of the reacting molecules contains only one functional group, and no further combination can take place.

$$CH_3.COOH + HOC_2H_5 \rightarrow CH_3.COOC_2H_5 + H_2O$$
ethyl ethanoate

The situation is different when both of the reacting molecules contain *two* functional groups, that is, when they are both *bifunctional*. Thus, if HOOC—X—COOH represents a dicarboxylic acid and HO—Y—OH a dihydric alcohol, combination of one molecule of the acid with one molecule of the alcohol will yield a monoester analogous to ethyl ethanoate.

$$HOOC—X—COOH + HO—Y—OH$$
$$\rightarrow HOOC—X—COO—Y—OH + H_2O$$
monoester

The monoester, having a carboxyl group at one end of the molecule and a hydroxyl group at the other, is also bifunctional. It can therefore undergo further reaction as follows:

(*a*) The —COOH group of the monoester can react with an —OH group of another alcohol molecule, or the —OH group of the monoester can react with the —COOH group of another acid molecule, e.g.

$$HOOC—X—COO—Y—OH + HOOC—X—COOH$$
$$\rightarrow HO(OC—X—COO—Y—O—)OC—X—COOH + H_2O$$
diester

(*b*) Two molecules of monoester can react together through their terminal —COOH and —OH groups.

$$HOOC—X—COO—Y—OH + HOOC—X—COO—Y—OH$$
$$\rightarrow HO(OC—X—COO—Y—O—)(OC—X—COO—Y—O—)H$$
$$+ H_2O$$

triester

The above reactions represent early stages in the building up of a long molecular chain containing the repeating structural unit (—OC—X—COO—Y—O—). This unit has been placed in brackets in the formulæ of the diester and triester. Since the diester and triester are still bifunctional, combination can continue in the ways illustrated above, yielding eventually a polyester with the following general formula:

$$HO(—OC—X—COO—Y—O—)_nH$$

The best known polyester is Terylene, in which X is C_6H_4 and Y is C_2H_4. Terylene is manufactured (as described shortly) from benzene-1,4-dicarboxylic acid (terephthalic acid),

$$HOOC—C_6H_4—COOH,$$

and ethane-1,2-diol (ethylene glycol), $HO—C_2H_4—OH$.

(*ii*) *Polyamides*. Instead of condensing a dicarboxylic acid with a dihydric alcohol the acid can be condensed with a diamine of the type $H_2N—Y—NH_2$. In this case the carboxyl and amino groups react with elimination of water.

$$—COOH + H_2N— \rightarrow —CO.NH— + H_2O$$

The resulting high polymer is known as a *polyamide*. If the monomers are $HOOC—X—COOH$ and $H_2N—Y—NH_2$, the polyamide has the formula:

$$HO(—OC—X—CO—NH—Y—NH—)_nH$$

The various forms of nylon are polyamides. The most common form is manufactured by condensation polymerization of hexanedioic acid (adipic acid), $HOOC(CH_2)_4COOH$, and hexane-1,6-diamine (hexamethylenediamine), $H_2N(CH_2)_6NH_2$. The method of manufacture is described later.

(*iii*) *Three-dimensional High Polymers*. If one of the monomers taking part in condensation polymerization is *trifunctional*, while the other is bifunctional, there is greater scope for the manner in which the molecules can combine. Condensation may now yield not only linear molecules (branched and unbranched), but also a three-dimensional 'cross-linked' structure. In this the molecular chains are joined sideways in different directions, forming a network in space, so that the polymer is actually a single giant molecule.

The formation of three-dimensional high polymers is illustrated by the reaction used to manufacture Bakelite from methanal (formaldehyde) and phe.iol. The methanal molecule acts as a bifunctional molecule because its oxygen atom combines with two hydrogen atoms derived from different molecules of phenol. The phenol molecule is trifunctional because it can supply hydrogen atoms from either the 1,2- or 1,6- positions or from the 1,4- position. Thus simple condensation of one molecule of methanal with two molecules of phenol is represented by the following equation.

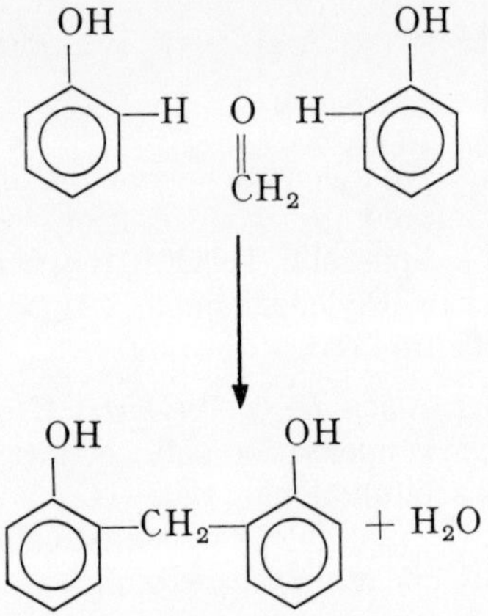

Since the resulting molecule still has reactive hydrogen atoms in 2- and 4- positions relative to the —OH groups further reaction can occur with methanal molecules, giving straight-chain, branched-chain, or cross-linked polymers according to how far the reaction proceeds. The final product is the three-dimensional cross-linked copolymer known as Bakelite, the structure of which may be represented (in two dimensions) as follows:

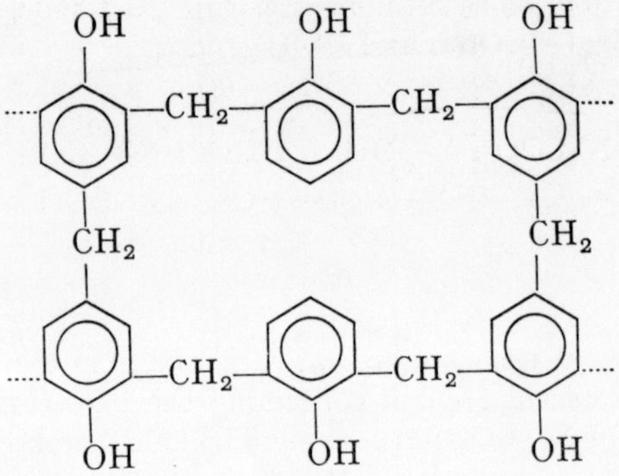

Bakelite

General Properties of High Polymers

The commercial uses of high polymers depend very largely on their physical properties. These are determined by the size, structure, and shape of the molecules and how they are assembled in the bulk materials. The properties vary with temperature, and can be modified by mixing the polymers with certain other substances, e.g. plasticizers (see later).

Molecules of high polymers differ from simple molecules in not having a fixed size, or a definite relative molecular mass. In a linear polymer the molecular chains are of varying lengths, but in a given sample these do not differ greatly from the general average. The average length of the chains can be controlled by arresting the polymerization at the desired stage. In free radical reactions this is done by adding special chemicals called 'stoppers' or 'inhibitors' while in condensation polymerizations polymerization is arrested by chilling the reaction mixture.

The average length of the molecular chains in a linear polymer is indicated by the average relative molecular mass. This may have a wide range of values, depending on the use for which the polymer is required. The normal ranges for some common plastics are as follows (for special applications the values are sometimes much higher): poly(ethene) 15 000–50 000, poly-(chloroethene) 50 000–80 000, poly(phenylethene) (polystyrene) 50 000–300 000. The average relative molecular mass is usually determined by making a solution of known concentration in a suitable organic solvent and measuring the viscosity of the solution. Other methods include measurement of the osmotic pressure of the solution or of its rate of sedimentation in an ultracentrifuge. As already seen, a three-dimensional cross-linked polymer of the Bakelite type consists of a single giant molecule. In this case the term 'relative molecular mass' has no meaning.

Linear polymers may be amorphous or 'crystalline.' In an amorphous polymer, such as natural rubber, the molecular chains are twisted and tangled together like a number of separate lengths of string (Fig. 26-1a). In a 'crystalline' polymer there are regions where the molecular chains run parallel to each other for all, or part of their length, giving an orderly arrangement of particles characteristic of a crystal (Fig. 26-1b). These orderly regions, which are called *crystallites*, are separated by others in which the chains have a disorderly distribution.

The ability of the molecular chains to form crystallites depends on how easily they can fit together, and thus on their shape. Crystallization is greatly favoured by the molecules having a regular structure. Thus poly(propene), which consists mostly of

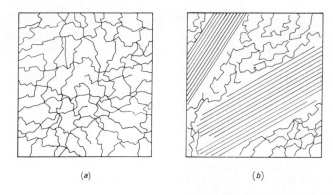

Fig. 26-1. Distribution of molecular chains in (a) an amorphous linear polymer, (b) a 'crystalline' linear polymer

the isotactic variety, is very largely crystalline. The degree of crystallinity influences many of the physical properties of the material.

Density. High polymers are relatively light materials. This is partly because they are composed of relatively light atoms and partly because the molecules are not tightly packed together. If a polymer has both an amorphous form and a crystalline form, as in the case of poly(ethene), the crystalline form has the higher density owing to closer packing of the molecules.

Action of Heat. Many linear polymers can exist in either a rigid 'glassy' state or a soft 'rubbery' state according to the temperature. In the first state the molecular chains have fixed positions and are capable only of vibrational movement. Most linear polymers are in this state at ordinary temperatures, and are glass-like in appearance (Perspex and poly(phenylethene) are

341

conspicuous examples). If the glassy polymer is heated, it changes over a range of temperature into a 'rubbery' state, in which it is soft and plastic. At still higher temperatures it melts and forms a viscous liquid.

The change from the glassy state to the rubbery state is due to the molecular chains acquiring sufficient kinetic energy to overcome the van der Waals force of attraction between them. The chains can then move relative to each other like snakes in a basket. Rubber exists in the 'rubbery' state at ordinary temperatures and changes into the glassy state only below $-70°C$. A well known experiment which illustrates this change is to cool a rubber ball in liquid air and show that it breaks into fragments when struck by a hammer.

The transition between the glassy state and the rubbery state is reversible providing the molecular chains retain their linear form on heating. Thus, if a transparent poly(phenylethene) (polystyrene) ball-point pen is placed in boiling water it softens, but becomes hard again when it is removed from the water. This can be repeated indefinitely. Polymers which behave in this way are described as *thermosoftening*, or *thermoplastic*.

Some linear polymers, however, are converted by heat into hard, three-dimensional, cross-linked products, which cannot be softened by heating to still higher temperatures. In this case the linear polymers are said to be *thermohardening*, or *thermosetting*. Thus, condensation of phenol with methanal first yields a linear polymer, but when this is heated (preferably with a catalyst) it undergoes cross-linking to give Bakelite.

Action of Solvents. Most linear polymers are soluble in appropriate organic solvents (propanone, benzene, etc.). Dissolving usually takes place very slowly, and the solutions are colloidal in character because of the large size of the polymer molecules.

They are also viscous owing to entanglement of the molecular chains. Three-dimensional cross-linked polymers like Bakelite are insoluble in all solvents. This is to be expected since dissolving would involve the breaking of strong C—C bonds.

Mechanical Strength. Linear polymers derive their strength from the van der Waals force of attraction between the molecular chains. Since the attraction is much larger when the chains are aligned and thus close together, a crystalline polymer is stronger than an amorphous one. Even in a crystalline polymer, however, there are weaker regions of molecular disorder, and these can act as starting-points for cracks when the material is under stress. Thus linear polymers tend to be brittle, and are unsuitable for bearing heavy loads. Brittleness can be reduced by incorporating a 'plasticizer' such as dibutyl benzene-1,2-dicarboxylate (dibutyl phthalate) or calcium(II) ethanoate. Plasticizers act as internal lubricants, and allow the molecular chains to 'give,' or slide easily over each other.

Three-dimensional polymers might be expected to have high mechanical strength because the bonds between the atoms are strong covalent bonds. However, cross-linking is never 100 per cent complete, and where it is absent points of weakness occur, causing brittleness. This can be largely overcome by adding a *filler*, which serves as a reinforcement. Fillers are usually fibrous materials such as wood flour, asbestos, and fibre glass.

Electrical Properties. Since the bonds between the atoms are normal covalent bonds, high polymers are non-conductors of electricity both in the glassy state and the rubbery state. Some high polymers (e.g. poly(ethene), poly(chloroethene), and Bakelite) are used as insulators in electrical apparatus.

Chemical and Biological Properties. Although, as mentioned earlier, the uses of synthetic high polymers are largely determined by their physical properties, chemical and biological properties are also important. A feature of most high polymers is their resistance to attack by air, water, acids, alkalies, bacteria and moulds. This inertness is advantageous when high polymers are in use, but, since they do not corrode or decay naturally, their subsequent disposal as waste matter is a serious economic problem. One factor which limits the usefulness of polymers based on a backbone of carbon atoms is that the polymers begin to decompose at relatively low temperatures (200°–300°C).

Synthetic polymers fall into three groups according to their properties and uses—(*i*) rubbers, (*ii*) resins and plastics, and (*iii*) fibres. There are no sharp dividing lines between the groups, and some high polymers belong to more than one group. Thus poly(propene) and nylon are used both as plastics and as fibres.

NATURAL AND SYNTHETIC RUBBERS

The characteristic property of a 'rubber' is that of elasticity, and for this reason rubber and rubber-like materials are known scientifically as *elastomers*. They consist of linear high polymers, and are normally amorphous. They are elastic because their molecular chains, although twisted and coiled, are extremely flexible. If an elastomer is stretched, the chains straighten and become aligned by rotation about the C—C bonds in the carbon backbone. When the stretching force is removed the chains spring back to their former configurations, and the elastomer regains its previous length. The structural difference between an unstretched elastomer and a stretched one is clearly shown by their different X-ray diffraction patterns. The pattern given by the unstretched polymer is diffuse and characteristic of an amorphous substance; that from the stretched polymer consists of a regular pattern of dots and is typical of a crystalline material.

Natural Rubber

Raw rubber is a polymer of the hydrocarbon methylbuta-1,3-diene (isoprene), and has the formula $(C_5H_8)_n$, the average value of n being about 11 000. As the systematic name of the monomer shows, the latter is a methyl derivative of buta-1,3-diene (Chapter 6).

Monomer

$$CH_2{=}\overset{\overset{\displaystyle CH_3}{|}}{C}{-}CH{=}CH_2$$

methylbuta-1,3-diene

Polymer

$$\left({-}CH_2{-}\overset{\overset{\displaystyle CH_3}{|}}{C}{=}CH{-}CH_2{-}\right)_n$$

natural rubber

It will be noticed that the opening out of the two double bonds in the monomer units is accompanied by the appearance of a new double bond in the units. The presence of double bonds in the carbon chain is very important. Before raw rubber can be made into tyres, hot-water bottles, etc., it has to undergo a number of treatments. The most important of these is *vulcanization*. The latter consists of heating the rubber with 1-3 per cent of

343

sulphur, together with small amounts of special chemicals ('activators' and 'accelerators') which assist the process. The sulphur atoms combine with the unsaturated rubber molecules by establishing cross-linkages at the double bonds. In this way they form 'bridges' between adjacent molecular chains.

The proportion of sulphur used in vulcanization is only small, and the degree of cross-linking is correspondingly low. It is sufficient, however, to reduce the excessive flexibility of the molecular chains, and thus increase the rigidity, toughness, and elasticity of the material. If the proportion of sulphur is increased to about 30 per cent the degree of cross-linking is much higher, and the hard inelastic substance called *ebonite* is obtained.

Other additives to rubber include fillers and anti-oxidants. Fillers increase hardness, tensile strength, and resistance to abrasion. The most important filler is carbon black, which may constitute nearly a third of the mass of a tyre tread. A serious defect of rubber is its tendency to 'perish' on keeping. This is a complex change in which the rubber loses its elasticity and disintegrates. 'Perishing' is mainly caused by atmospheric oxygen and exposure to sunlight. It can be retarded (but not prevented) by addition of suitable organic compounds.

Synthetic Rubber

'Synthetic rubber' is not identical with natural rubber. The term refers to man-made rubber-like materials, of which a dozen or more forms are in common use. Nearly all the forms are made from petrochemicals. Synthetic rubber is more costly than natural rubber, and is still used to a smaller extent than the latter.

The properties of synthetic rubber are generally similar to those of natural rubber, but differ in degree. Thus the synthetic material is often more resistant to solvents, is a better insulator, or has greater tensile strength. Frequently it is 'compounded' with natural rubber, giving a product superior to either in certain respects.

Neoprene. Neoprene is the trade name of one of the simplest forms of synthetic rubber. It is made by addition polymerization of 2-chlorobuta-1,3-diene (chloroprene). An initiator is used to start the reaction, which therefore has a free-radical mechanism. An unusual feature is that zinc(II) oxide is used for vulcanization instead of the usual sulphur.

Monomer	Polymer
$CH_2\!\!=\!\!CH\!\!-\!\!CCl\!\!=\!\!CH_2$	$(-CH_2\!\!-\!\!CH\!\!=\!\!CCl\!\!-\!\!CH_2\!\!-\!\!)_n$
2-chlorobuta-1,3-diene	Neoprene

Neoprene has high resistance to solvents, chemicals, and heat. It is used for making hoses and conveyor belts, and as a lining for reaction vessels.

Phenylethene-butadiene Copolymer. Phenylethene (styrene) $(C_6H_5\!\!-\!\!CH\!\!=\!\!CH_2)$ is a colourless liquid made by catalytic dehydrogenation of ethylbenzene (Chapter 20). Buta-1,3-diene is a colourless gas, which liquefies readily under a small pressure. It is obtained by catalytic dehydrogenation of but-1-ene or but-2-ene (Chapter 6).

A mixture of the two monomers is made into an emulsion with water, the proportions being varied according to the properties required in the product. The mixture is polymerized at $5°C$, using an initiator, and the copolymer is vulcanized with sulphur. The copolymer is represented by the following formula:

$$\left(\begin{array}{c} C_6H_5 \\ | \\ -CH-CH_2 \end{array} \right)_m -(CH_2-CH\!\!=\!\!CH-CH_2-)_n$$

The formula given for the copolymer merely shows that m phenylethene units have combined with n units of buta-1,3-diene. It does not show the order in which the monomer units are joined. The values of m and n depend on the proportions of the two monomers used.

Phenylethene-butadiene copolymers are used more than any other form of synthetic rubber. They have a particularly good resistance to abrasion, and are used (with a carbon filler) to make tyre treads and soles for boots and shoes.

SYNTHETIC RESINS AND PLASTICS

There is no clear distinction between a 'resin' and a 'plastic.' The first term is usually applied to the glassy-looking substances formed when momomers are polymerized, while the second is used for the moulded products, which often contain other ingredients (such as plasticizers and fillers) in addition to the polymers. Essentially a plastic is a material which is soft enough at some stage of its manufacture to be pressed into any desired shape, and which can then be hardened so that it keeps that shape. (Clay is a naturally occurring plastic).

Synthetic resins and plastics are divided into two classes— those which are thermosoftening, or thermoplastic, and those which are thermohardening, or thermosetting. In this discussion attention will be confined to the simpler members of each class.

Thermosoftening Resins and Plastics

Ethenyl Polymers. As seen earlier, these are made from monomers containing the ethenyl (vinyl) group ($CH_2=CH—$). Ethenyl polymers form the most important group of synthetic resins.

More than half the total output of synthetic resins consists of polymers of ethene, propene, phenylethene, and chloroethene. According to the uses for which they are intended the resins are made into sheets, extruded through slits or holes as filaments or, pipes, or granulated to give moulding powders, pellets, etc.

Poly(ethene), ($—CH_2—CH_2—$)$_n$. Poly(ethene), which is commonly called *polythene*, is the most widely used synthetic resin. It is manufactured in two forms—a lighter amorphous form known as 'low-density' poly(ethene) and a heavier crystalline form called 'high-density' poly(ethene).

'Low-density' poly(ethene) is made by heating ethene at about 250°C at a pressure of 1500–3000 atmospheres, a trace of oxygen being added to act as an initiator. The product is a waxy-looking solid, which floats on water and softens on heating at 85°C. It is a tough general purpose resin which is suitable for moulding into houseware (bottles, bowls, buckets, etc.) as well as toys. A major application is transparent film for packaging garments and food. In addition "low-density" poly(ethene) is utilized as a wire and cable insulator.

'High-density' poly(ethene) is made with the help of a Ziegler catalyst (titanium(IV) chloride and an aluminium alkyl). In this case the ethene is dissolved in a petroleum oil, and much milder conditions are used (the temperature is only 75°C and the pressure about six atmospheres). The polymer obtained has a high degree of crystallinity. It can be heated to 125°C before softening. The high degree crystallinity imparts greater liquidity, and this makes 'high-density' poly(ethene) suitable for the production of crates and large boxes (e.g. for packaging fish).

'High-density' poly(ethene) is used extensively in the manufacture of bleach and detergent bottles. A recent development has been the production of film to replace grease proof paper and paper bags.

Poly(tetrafluoroethene) $(C_2F_4)_n$. This is not an ethenyl polymer but is included here because of its close relationship with poly(ethene), $(C_2H_4)_n$. The trade name of the British product is *Fluon*, while that of the American one is *Teflon*.

The monomer, tetrafluoroethene, is obtained by treating trichloromethane with hydrogen fluoride and heating the resulting chlorodifluoromethane.

$$CHCl_3 + 2HF \rightarrow CHF_2Cl + 2HCl$$

$$2CHF_2Cl \rightarrow CF_2 = CF_2 + 2HCl$$

The monomer, which is a gas, is polymerized under pressure with the help of an initiator.

Fluon has a high softening point (over 300°C) and excellent resistance to attack by solvents and corrosive chemicals. It is also a very good electrical insulator. Its best known use is as a lining for 'non-stick' frying-pans, but it is also used in chemical plant and as a sheath for underground electric cables. It is one of the most expensive synthetic resins.

Poly(propene). Polymerization of propene with a Ziegler catalyst is carried out in the same way as that of ethene. The polymer is known commercially as 'Propathene.'

Monomer	Polymer
CH_3	CH_3
$\|$	$\|$
$CH = CH_2$	$\left(-CH-CH_2-\right)_n$
propene	poly(propene)

The polymer resin is more crystalline than poly(ethene), and has a higher softening temperature (150°C). It is also lighter, harder and stronger. Many of its applications are similar to those of the two forms of poly(ethene). However, the higher softening-point of poly(propene) enables it to be used in applications where sterilization by steam is necessary (e.g. in hospital-ware). Poly(propene) can also be extruded as a fibre (see later under 'Man-made Fibres').

Poly(phenylethene). As stated earlier, phenylethene (styrene) is obtained as a colourless liquid by catalytic dehydrogenation of ethylbenzene. The liquid is polymerized by heating it with an initiator. The resin obtained looks like glass, but is much lighter, It is also more brittle. In one technique synthetic rubber is introduced during the polymerization stage, and this gives a tougher, and less brittle, product, which is, however, somewhat opaque.

Monomer	Polymer
C_6H_5	C_6H_5
$\|$	$\|$
$CH = CH_2$	$\left(-CH-CH_2-\right)_n$
phenylethene	poly(phenylethene)

Major applications of poly(phenylethene) (polystyrene) include the production of house-ware, toys, ball-point pens, and display-stands. The toughened resin is used to make transistor radio and television casings and the inner linings of refrigerators.

Poly(phenylethene) is also produced in an 'expanded' form. During polymerization a volatile hydrocarbon is added to the ingredients, and each individual bead, or 'pearl,' obtained at the end of polymerization contains a little of the vapour. When the beads are heated and softened the vapour escapes, causing the beads to expand. The beads are then moulded in steam chests under pressure to produce insulating board, ceiling tiles, and packings for fragile articles such as cameras and radios.

Poly(chloroethene). Chloroethene (vinyl chloride), $CH_2=CHCl$ is prepared as a colourless gas by the method described in Chapter 7. It is liquefied under pressure, made into an emulsion with water and polymerized by heating with an initiator. When dried the polymer, $(-CH_2-CHCl-)_n$, consists of a white powder, and is commonly called polyvinyl chloride (PVC).

Before the powder can be moulded or extruded other ingredients have to be incorporated. Two specific types of product are manufactured, these being described as (a) rigid, and (b) plasticized. The mixture for the rigid material is made by adding stabilizers and antioxidants to the poly(chloroethene) powder. The mixture is heated and extruded under pressure to give pipes, guttering, and corrugated sheet.

For the plasticized material the mixture contains about 53 per cent of a plasticizer such as dibutyl benzene-1,2-dicarboxylate (dibutyl phthalate) in addition to the ingredients already mentioned. The presence of the plasticizer enables flexible articles to be produced. In this way leather-cloth (for car seat covers), wall coverings, mackintoshes, etc., are obtained. Cable and wire insulants are also made from plasticized poly(chloroethene).

Propenoic Polymers. In this group the monomers are based on propenoic acid (acrylic acid), $CH_2=CH-COOH$. Only the most important member of the group will be described here. This is the polymer obtained from the methyl ester of 2-methylpropenoic acid, $CH_2=C(CH_3)-COOH$.

Poly(methyl 2-methylpropenoate) (*Perspex*). The monomer is made from propanone. The latter is combined with hydrogen cyanide, and the addition compound is then converted into 2-methylpropenoic acid by treatment with concentrated sulphuric(VI) acid. (The latter first dehydrates the addition compound and then hydrolyses the —CN group to —COOH) The 2-methylpropenoic acid is esterified by means of methanol and sulphuric(VI) acid. The changes can be summarized as follows:

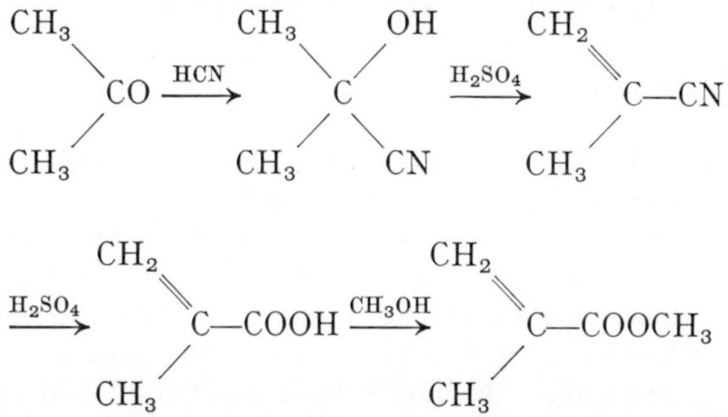

$$
\begin{array}{c}
CH_3 \\
\diagdown \\
\quad CO \xrightarrow{HCN} \\
\diagup \\
CH_3
\end{array}
\begin{array}{c}
CH_3 \quad OH \\
\diagdown \diagup \\
C \xrightarrow{H_2SO_4} \\
\diagup \diagdown \\
CH_3 \quad CN
\end{array}
\begin{array}{c}
CH_2 \\
\| \\
C-CN \\
\diagup \\
CH_3
\end{array}
$$

$$
\xrightarrow{H_2SO_4}
\begin{array}{c}
CH_2 \\
\| \\
C-COOH \\
\diagup \\
CH_3
\end{array}
\xrightarrow{CH_3OH}
\begin{array}{c}
CH_2 \\
\| \\
C-COOCH_3 \\
\diagup \\
CH_3
\end{array}
$$

Methyl 2-methylpropenoate, a colourless liquid is polymerized at about 70°C, using an initiator (e.g. di(benzoyl)peroxide). The structure of the polymer is shown below.

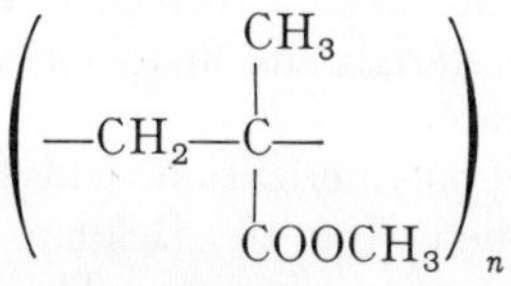

$$
\left(-CH_2-\underset{\underset{COOCH_3}{|}}{\overset{\overset{CH_3}{|}}{C}}-\right)_n
$$

Perspex

The outstanding property of Perspex is its high transparency. It is sometimes described as an 'organic glass'. It is used in aeroplanes as safety glass for the windows and as a coating for the wings, to which it gives a smooth, weather-resistant finish. Other applications are in making optical lenses, instrument panels, electric light fittings, dental plates, and decorative products. Its chief drawback is that it scratches easily. Perspex softens at 80°–100°C.

THERMOHARDENING RESINS AND PLASTICS

Phenolic Resins

Earlier in this chapter it was shown how phenol and methanal react by condensation polymerization to give firstly a linear polymer and then a three-dimensional cross-linked product. If polymerization is stopped at the first stage a colourless resin called *Novolak* is obtained. In practice the latter is made by heating together phenol, aqueous methanal, and a catalyst. This may be an alkali (sodium(I) hydroxide) or an acid (ethanedioic acid) according to the characteristics required for the resin.

Novolak resins are excellent insulators. They are employed as varnishes for armature coils and as cements for sealing electric-light bulbs into their metal holders. They are also used for bonding together sheets of paper, cardboard, and wood, and from these 'laminated' materials are made airscrews, gear-wheels, control panels, and table tops.

The second stage of polymerization yields the infusible product called Bakelite (named after the Belgian chemist Baekeland, who introduced it as a plastic in 1907). The powdered resin from the first stage is mixed with a filler (usually wood flour) and hexamine, $(CH_2)_6N_4$ (Chapter 10). The latter not only supplies methylene groups for cross-linking the phenol nuclei, but also ammonia, which acts as a catalyst. The mixture is hardened to the desired shape in a heated steel mould by means of a hydraulic press. Bakelite is used to fabricate a wide variety of products, particularly electrical fittings such as plugs, switches, telephone parts, etc.

Amino Resins

These are obtained by condensation polymerization of methanal with certain compounds containing amino groups. A typical example is the resin derived from methanal and carbamide, or urea.

Carbamide-methanal Resin. When an aqueous solution of carbamide is heated with methanal solution a vigorous exothermic reaction sets in just below 100°C (or at about 50°C if a little ammonia is added as a catalyst). The product is a creamy resin composed of linear molecules in which carbamide residues are linked by methylene bridges. The first stage of the condensation polymerization can be represented as follows:

$$H_2N—CO—NH_2 + \overset{\overset{\textstyle CH_2}{\|}}{O} + H_2N—CO—NH_2$$
$$\rightarrow H_2N—CO—NH—CH_2—NH—CO—NH_2 + H_2O$$

The resin is used both for making plastics and as a bonding agent for laminated materials. It can be hardened (cross-linked) by heat or by adding a weak acid such as citric acid. For the manufacture of moulded products the resin is dried, powdered, and mixed with a filler, a hardening agent, and a pigment. When the mixture is compressed in a heated mould the resin first softens and then changes into the hard cross-linked polymer. Articles made from the plastic include toys, egg-cups, beakers, cups, saucers, plates, and cutlery.

Instructions for making, and using carbamide-methanal resin are given at the end of the chapter.

MAN-MADE FIBRES

Besides the naturally occurring fibres like cotton and wool, modern textile fibres include two classes of artificial, or 'man-made,' fibres. These are:

(*i*) fibres made by chemical treatment of naturally occurring high polymers (usually cellulose), and (*ii*) fibres composed of synthetic high polymers.

Synthetic fibre-forming polymers are closely related to those which make up thermosoftening resins. Both consist of long molecular chains, and frequently the same polymer can be used either as a fibre or as a plastic. A fibre is distinguished by having its molecular chains arranged parallel with each other, not in the form of straight lines, but usually in a spiral, or helical, pattern. In cotton and wool this alignment occurs naturally. In man-made fibres it is achieved mechanically by stretching the polymer during processing. As mentioned earlier, alignment of the molecular chains greatly increases the van der Waals force of attraction between them and thereby enhances the tensile strength. The tensile strength is further increased by twisting the fibres round each other to form a thread, in which there are frictional forces between the individual fibres.

Man-made Fibres Based on Cellulose

The first commercially successful process for producing artificial fibres from cellulose was invented by two English chemists, Cross and Bevan, in 1892. This process is still in use and is described below. Artificial fibres based on cellulose are known collectively as *rayon*.

Viscose Rayon

There are two chief stages in making viscose rayon. One consists of getting cellulose into solution, and the other of precipitating it again in the form of a continuous filament. In the first stage, sheets of wood pulp are soaked in a strong solution of sodium(I) hydroxide to form sodium cellulose. Excess of alkali is squeezed from the sheets, which are then shredded into a fluffy mass known as 'white crumbs.' These are stored for three or four days at a controlled temperature to 'ripen' them. They are then immersed in carbon disulphide, which converts the sodium(I) cellulose into cellulose xanthate, an orange-coloured substance ('yellow crumbs'). Using the simplified formula RONa for sodium(I) cellulose the formation of cellulose xanthate may be represented by the following equation:

$$RONa + CS_2 \rightarrow RO\!-\!\underset{\underset{S}{\|}}{C}\!-\!SNa$$

The cellulose xanthate is dissolved in a dilute solution of sodium(I) hydroxide, giving a viscous honey-coloured liquid ('viscose').

In the second stage the viscose solution, after being filtered, is forced through 'spinnerets' (metal discs perforated with holes about 0·1 mm in diameter) into a bath of hot dilute sulphuric(VI) acid. Here the cellulose is precipitated as a number of continuous

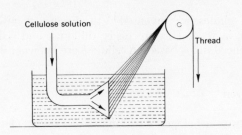

Cellulose solution

Thread

Fig. 26-2. Manufacture of viscose rayon

filaments. These pass round a series of revolving wheels, which stretch them and twist them into a single thread. The product is washed to remove acid, desulphurized with sodium(I) sulphide solution, bleached with sodium(I) chlorate(I) (sodium hypochlorite) solution, and finally dried.

Instead of being forced through holes, the viscose solution may be squeezed through a narrow slit into dilute sulphuric(VI) acid. In this case the cellulose is obtained as a thin sheet known as 'Cellophane.' This is used as a transparent, moisture-proof wrapping-material.

In the manufacture of viscose rayon, cellulose forms both the raw material and the final product. For this reason viscose rayon is often described as 'regenerated cellulose.'

Ethanoate, or Acetate, Rayon

The ester cellulose ethanoate, or acetate, is made by treating cotton or wood pulp with a mixture of glacial ethanoic acid, ethanoic anhydride, and concentrated sulphuric(VI) acid.

Ethanoylation, or acetylation, occurs and a product corresponding roughly to cellulose triethanoate, or triacetate, is formed in solution. This 'primary ethanoate' is precipitated by addition of water and allowed to stand in contact with the water for a time to bring about partial hydrolysis. The 'secondary ethanoate' thus obtained contains a lower proportion of combined ethanoic acid (it corresponds approximately to the diethanoate), but is soluble in propanone. A syrupy solution is made in this solvent and forced through spinnerets into a current of warm air. The propanone evaporates, leaving the cellulose ethanoate in the form of filaments. These are twisted together into threads, which are then stretched to align the molecules.

It will be noted that ethanoate rayon, unlike viscose rayon, is not regenerated cellulose. Ethanoate rayon has particularly good water resistance, and is suitable for making articles of clothing which are washed frequently.

Cellulose ethanoate is also used as a plastic. For this purpose it is mixed with a plasticizer and made into a dough with a little propanone. The dough is then rolled out into sheets. These are used as a packaging material and as a base for non-flammable photographic film.

Synthetic Fibres

Synthetic fibres account for about 40 per cent of the total output of man-made fibres in Britain. They are superior to natural fibres in many ways. Thus they are lighter and stronger, moth-proof and rot-proof, and they absorb moisture to a much smaller extent. Because of this they can be used to make 'drip-dry' fabrics, which require little or no ironing. At the same time they have certain disadvantages compared with natural fibres.

They are more expensive, often have a harsher feel, and, since they develop electrostatic charges more readily, they have a stronger attraction for dirt particles. These drawbacks can be lessened, however, by imparting special finishes to the fibres or by blending them with natural fibres.

Terylene

As seen earlier in this chapter, Terylene is a polyester formed by condensation polymerization of ethane-1,2-diol (ethylene glycol), $HO—CH_2.CH_2—OH$, with benzene-1,4-dicarboxylic acid (terephthalic acid), $HOOC—C_6H_4—COOH$. The systematic name of the polyester is poly(ethane-1,2-diyl benzene-1,4-dicarboxylate). It has the structure shown below.

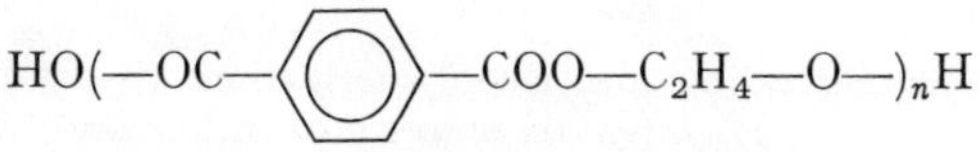

$$HO(—OC—\langle\bigcirc\rangle—COO—C_2H_4—O—)_n H$$

Terylene

In the industrial process the more reactive dimethyl ester of benzene-1,4-dicarboxylic acid is used instead of the acid. Since, however, methanol is eliminated (instead of water), the reaction may be regarded as taking place between the acid itself and the alcohol. Condensation is brought about by heating the reactants under reduced pressure. Methanol vapour is given off, and the polymer is left as a viscous liquid. Fibres are produced by 'melt spinning', the molten polymer being extruded through spinnerets to give filaments, which solidify on cooling. The filaments are twisted into a thread, stretched to three or four times their original length, and wound on to bobbins.

Terylene is used to make many kinds of wearing apparel. Blended with wool or cotton it is used in suits, socks, 'drip-dry' shirts, and lingerie. It is also employed as a filling in eiderdowns, sleeping-bags, and anoraks. 'Crimplene', a specially treated form of the fibre, is used to make 'uncrushable' fabrics.

Nylon

'Nylon' is the general name for polyamide fibres, of which there are several. As mentioned earlier, the common form of nylon is made from hexanedioic acid (adipic acid) and hexane-1,6-diamine (hexamethylenediamine), both of which are derived from phenol. This form is called 'nylon 66' because both monomers contain six carbon atoms in the molecule.

$$HOOC.(CH_2)_4.COOH \qquad\qquad H_2N.(CH_2)_6.NH_2$$

hexanedioic acid $\qquad\qquad\qquad$ hexane-1,6-diamine

The course of the condensation polymerization is not quite as simple as represented previously. Since one monomer is an acid and the other a base, mixing of the two (in methanol solution) first produces a salt by interaction of the carboxyl groups and amino groups. This 'nylon salt' has the following composition:

$$----\overset{-}{O}OC(CH_2)_4CO\overset{-}{O}\ \overset{+}{N}H_3(CH_2)_6\overset{+}{N}H_3----$$

nylon salt

When the salt is heated in an atmosphere of nitrogen at about

300°C water is eliminated, yielding a polyamide with the structure shown below.

$$HO[-OC(CH_2)_4CO-NH(CH_2)_6NH-]_nH$$
nylon 66

Fibres are obtained by 'melt spinning' as in the case of Terylene, and again stretching of the fibres is an important part of the processing.

Nylon is chiefly used in the manufacture of stockings. It is more suitable for this purpose than any other synthetic fibre because of its fineness, strength, elasticity, and cheapness. Nylon has a host of other applications, which include the production of clothing, carpets, tooth-brush bristles, combs, and fishing-lines.

Poly(propenonitrile)

Propenonitrile (acrylonitrile) can be made by direct combination of ethyne (acetylene) and hydrogen cyanide, but is obtained on a large scale from propene, ammonia, and air (see Chapter 6). Addition polymerization is carried out by heating an emulsion of the liquid monomer in water with an initiator.

Monomer

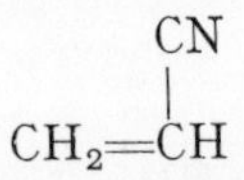

$$\underset{\text{propenonitrile}}{\overset{\displaystyle CN}{\underset{\displaystyle |}{CH_2=CH}}}$$

Polymer

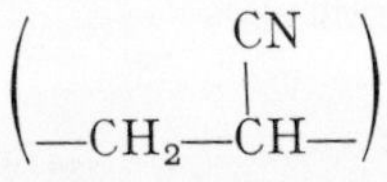

$$\underset{\text{poly(propenonitrile)}}{\left(\overset{\displaystyle CN}{\underset{\displaystyle |}{-CH_2-CH-}}\right)_n}$$

The resulting resin is insoluble in most solvents, but dissolves in N,N-dimethylmethanamide, $HCON(CH_3)_2$. The viscous solution is extruded into warm air, as in the case of cellulose ethanoate solution described earlier. Evaporation of the solvent leaves the polymer in the form of fine strands, which are twisted together and stretched in the usual manner.

Poly(propenonitrile) fabrics are produced under the trade names of Acrilan and Courtelle (Orlon in the U.S.A.). Their chief applications are as clothing and furnishing materials.

Poly(propene). Fibres are manufactured from poly(propene) resin (prepared as described earlier) by extrusion of the melted polymer, followed by spinning and stretching. Their main features are their lightness and strength, which make them suitable for use in light-weight blankets, tow-ropes, ship's hawsers, and climbing-ropes.

Experiments on Synthetic High Polymers

Addition Polymerization and Depolymerization. (*i*) *Preparation of Poly(phenylethene)*. Dissolve about 0·1 g of di(benzoyl) peroxide in 2-3 cm³ of phenylethene (styrene) in a dry test-tube, and stand the tube in a beaker of boiling water for about an hour. The polymer is formed as a white solid.

(*ii*) *Poly(propenamide) Resin*. Dissolve about 4 g of propenamide (acrylamide) in 20 cm³ of water in a boiling-tube, and heat the solution to about 80°C. Add a very small amount of potassium(I) peroxodisulphate(VI) (potassium persulphate) as an initiator. Shake the tube and stand it in a test-tube rack. Considerable heat is evolved, and the liquid becomes viscous as the polymer is formed.

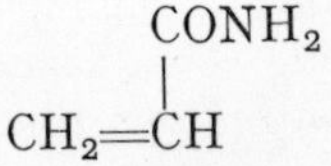

$$\underset{\text{propenamide}}{\overset{\displaystyle CONH_2}{\underset{\displaystyle |}{CH_2=CH}}}$$

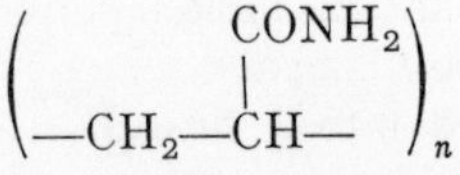

$$\underset{\text{poly(propenamide)}}{\left(\overset{\displaystyle CONH_2}{\underset{\displaystyle |}{-CH_2-CH-}}\right)_n}$$

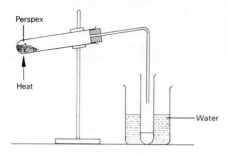

Fig. 26-3. Depolymerization of Perspex

(*iii*) *Depolymerization of Perspex.* (This experiment should be performed in the fume-cupboard as poisonous fumes are given off.) Depolymerization consists of breaking down the molecules of the polymer into those of the monomer.

Break some Perspex into small pieces and put a few grams into a hard-glass test-tube. Clamp the latter so that it slopes upwards. Fit the tube with a cork and delivery-tube leading to a test-tube cooled in water (Fig. 26-3), and heat the Perspex with a medium Bunsen flame. The polymer first melts and then gives off a vapour, which condenses in the receiver to a liquid (methyl 2-methylpropenoate). The distillate is usually coloured brown by traces of impurities.

The monomer obtained can be polymerized again by adding to it a small amount (0·5 g) of di(benzoyl)peroxide and standing the tube in boiling water for about an hour.

Condensation Polymerization

(*i*) *Preparation of Carbamide-methanal Resin and Plastic.* (The preparation of the resin should be carried out in the fume-cupboard.)

Dissolve 20 g of carbamide, or urea, crystals in 20 cm³ of 40 per cent formalin solution in a beaker in the cold. Heat the solution nearly to boiling point, when a vigorous exothermic reaction begins and the liquid turns milky. Remove the flame and allow the beaker to cool. A viscous white resin remains.

To illustrate the use of the resin in making laminated materials cut six strips of blotting paper 10 cm long and 2 cm wide. Soak the strips separately in the resin, lay them on top of each other, and squeeze out excess resin. Leave the compound strip for half-an-hour in a warm oven to dry out slowly. Then increase the temperature of the oven to 100°–120°C for 15 minutes. The heat causes cross-linking of the molecular chains and hardens the resin, giving a laminated strip. This will easily support the weight of half a brick.

To illustrate the use of a filler and hardening by an acid make a very dilute (1 per cent) solution of citric acid in water, and add three or four drops only to the remaining resin in the beaker. Stir into the mixture sufficient sawdust to make a stiff paste, and transfer this to a metal lid. Leave the lid in a warm place for a few hours to dry and harden. (The beaker should be cleaned as soon as possible with a dilute alkali.)

(*ii*) *Nylon Fibre.* Nylon 66 is most conveniently made in the laboratory from hexane-1,6-diamine and the acid chloride of hexanedioic acid. The first stage in the condensation polymerization is represented by the following equation:

$$H_2N(CH_2)_6NH_2 + ClOC(CH_2)_4COCl \rightarrow$$

hexane-1,6-diamine   hexanedioyl dichloride

$$H_2N(CH_2)_6NH.OC(CH_2)_4COCl + HCl$$

Prepare a 5 per cent solution of hexanedioyl dichloride in tetrachloromethane, and put about 5 cm³ of the solution into a 20 cm³

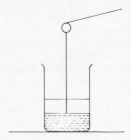

Fig. 26-4. Drawing a nylon fibre

beaker. Prepare also a 5 per cent solution of hexane-1,6-diamine in water, and carefully run 5 cm³ of the solution on to the top of the solution in the beaker. A polyamide film is produced at the interface of the two liquids. By inserting a wire loop or tweezers through the film a continuous filament of nylon 66 can be withdrawn from the beaker (Fig. 26-4).

EXERCISE 26

1. Give the names used to describe addition polymers of the following types:

(*a*) —A—A—A—A—   (*b*) —A—B—B—A—B—A—

(*c*) —A—A—A—B—B—B—   (*d*) —A—A—A—A
$\qquad\qquad\qquad\qquad\qquad\qquad$ |
$\qquad\qquad\qquad\qquad\qquad\qquad$ A
$\qquad\qquad\qquad\qquad\qquad\qquad$ |
$\qquad\qquad\qquad\qquad\qquad\qquad$ A
$\qquad\qquad\qquad\qquad\qquad\qquad$ |

2. Give the terms used for the following:
(*a*) A substance which starts an addition polymerization reaction by supplying free radicals.
(*b*) A polymerization catalyst consisting of titanium(IV) chloride and an aluminium alkyl.
(*c*) An ethenyl polymer (—CH_2—CHX—)$_n$, in which the X atoms or groups have the same orientation with respect to the carbon chain.
(*d*) The types of high polymer formed by condensing a dicarboxylic acid with a dihydric alcohol and a diamine respectively.
3. Which of the following are true of three-dimensional cross-linked high polymers: (*a*) their molecules have no definite size, (*b*) they are crystalline, (*c*) they are thermosoftening, (*d*) they dissolve in organic solvents, (*e*) they have low electrical conductivity?
4. Give the names applied to the following: (*a*) rubber and synthetic forms of rubber, (*b*) the monomer of natural rubber, (*c*) cross-linking of rubber molecules by heating with sulphur, (*d*) compounds added to counteract the 'perishing' of rubber, (*e*) the formula of synthetic rubber obtained by polymerizing the compound CH_2=CH—CCl=CH_2.

5. Poly(ethene) is manufactured in a 'low-density' form (A) and a high-density form (B). State which form has (a) the greater crystallinity, (b) the stronger attraction between the molecules, (c) the lower softening point, (d) the greater mechanical strength, (e) the higher transparency.

6. Deduce the probable nature of the plastics **A**, **B**, **C**, and **D** from the following:

(a) **A** is thermosoftening. Its monomer is made from two hydrocarbons. In one form it is highly transparent. In another it is used for making models in chemistry;

(b) **B** is thermosoftening, and contains a halogen. It is noted for its high resistance to corrosive chemicals. It is used in cooking-utensils;

(c) **C** is thermohardening. One of the compounds used to make it is an aromatic hydroxy-compound. It is widely used in electrical fittings;

(d) **D** is thermosoftening, and is a hydrocarbon. It is found in chemistry laboratories. One form is used as a wrapping material.

7. (a) Give the names and formulæ of the acid and alcohol used in the manufacture of Terylene. (b) What is the formula of the repeating structural unit in Terylene?

8. (a) Give the names and formulæ of the monomers used in the manufacture of nylon 66. (b) What is the formula of the repeating structural unit in nylon 66?

9. What is the chemical nature of (a) non-flammable photographic film base, (b) Acrilan, (c) Crimplene, (d) Propathene?

MISCELLANEOUS PROBLEMS

The letters in brackets after some of the questions indicate the examining bodies from whose G.C.E. papers questions have been taken. The abbreviations used are as follows: *J.M.B*—Joint Matriculation Board of the Universities of Manchester, Liverpool, Leeds, Sheffield, and Birmingham; *W.J.E.C.*—Welsh Joint Education Committee; *C.L.*—University of Cambridge Local Examinations Syndicate; *O. and C.*—Oxford and Cambridge Schools' Examination Board; *S.U.*—Southern Universities' Joint Board for School Examinations.

$$(H = 1, \ C = 12, \ N = 14, \ O = 16, \ S = 32,$$
$$Cl = 35 \cdot 5, \ Br = 80, \ Ag = 107 \cdot 9)$$

Advanced Level

1. The following data refer to oxidation of certain organic compounds containing carbon, hydrogen, and, possibly, oxygen. Calculate the percentages of these elements present in the compounds. (Give answers to one decimal place.)

	Mass of compound/g	Mass of CO_2/g	Mass of H_2O/g
(i)	0·039	0·132	0·027
(ii)	0·0253	0·0484	0·0297
(iii)	0·044	0·0645	0·0264

2. In an estimation of chlorine 0·225 g an organic compound gave 0·287 g of silver(I) chloride. Find the percentage of chlorine in the compound.

3. Find the empirical formulæ of the following compounds from the percentage compositions given (no percentage is given for oxygen):

	Per cent carbon	Per cent hydrogen	Per cent nitrogen	Per cent chlorine
(A)	75·0	25·0	—	—
(B)	37·5	12·5	—	—
(C)	40·7	8·5	23·7	—
(D)	37·2	7·8	—	55·0

4. Find the molecular formulæ of the following compounds from the percentage compositions and relative vapour densities provided:

	Per cent carbon	Per cent hydrogen	Per cent nitrogen	Per cent bromine	Relative vapour density
(X)	54·54	9·09	—	—	44·0
(Y)	65·5	9·1	25·4	—	27·5
(Z)	12·71	2·13	—	85·11	94·0

5. 0.0319 g of a compound containing carbon, hydrogen, and oxygen gave on oxidation 0.0722 g of carbon dioxide and 0.0303 g of water. The relative vapour density of the compound was 29.1. Find the molecular formula of the compound.

6. 25 cm³ of a mixture of ethane and ethene required 77.5 cm³ of oxygen for complete combustion. Calculate the percentage composition of the mixture.

7. Calculate the relative molecular masses of the organic acids in the following examples:

Mass of silver(I) salt/g	Mass of silver residue/g	Basicity of acid
(i) 0.334	0.214	1
(ii) 0.401	0.235	2
(iii) 0.3806	0.2377	2
(iv) 1.870	1.188	3

8. The combustion of a monobasic organic acid (containing carbon, hydrogen, and oxygen only) gave the following results:

Mass of acid taken $= 0.542$ g
Mass of carbon dioxide formed $= 1.084$ g
Mass of water formed $= 0.443$ g

1.344 g of the silver salt of this acid gave on ignition 0.744 g of metallic silver. What is the molecular formula of the acid? Suggest alternative constitutional formulæ for it.   (*W.J.E.C.*)

9. 16 cm³ of a gaseous aliphatic compound A, $C_nH_{3n}O_m$, was mixed with 60 cm³ of oxygen at room temperature and sparked. At the original temperature again, the final gas mixture occupied 44 cm³. After treatment with potassium hydroxide solution the volume of gas remaining was 12 cm³. Deduce the molecular and structural formulæ of A, and name it.

Give the name and structural formula of a compound B isomeric with A, and state briefly how A and B react separately with (*a*) sodium, (*b*) hydrogen iodide, (*c*) phosphorus trichloride.

(If there is no reaction in any one case, make this clear.)

Outline a reaction scheme, stating reagents, by which A might be prepared from **B**.   (*S.U.*)

10. An aromatic nitro-compound was found to contain $C = 61.31$, $H = 5.10$, $N = 10.2$ per cent, and on oxidation it gave a monobasic acid, the silver salt of which contained $Ag = 39.4$ per cent. What are the possible structural formulæ of the nitro-compound, and how would it behave on reduction?

(*J.M.B.*)

11. Outline the *principles* involved in **one** method used for the quantitative determination of nitrogen in an organic compound.

An organic compound of relative molecular mass 107 gave on analysis $C = 78.5$ per cent, $H = 8.5$ per cent, $N = 13.0$ per cent. The compound burned with a smoky flame and could be acetylated. Write down the possible structural formulæ for the compound.   (*O. and C.*)

12. Describe briefly **one** method by which benzene can be converted into phenol.

How would you separate a mixture of benzoic acid and phenol?

Bromine is added slowly to phenol (0·282 g) dissolved in water until a slight excess of bromine is present. Calculate the *mass* of bromine used.

(O. and C.)

More Difficult Problems

13. A neutral organic liquid **A** gave the following results on analysis: C = 54·5 per cent, H = 9·2 per cent, the remainder being oxygen. The relative vapour density of the liquid was found to be 43. The liquid (1·76 g) was boiled for an hour with M sodium hydroxide solution (25 cm³), and this solution was then titrated with 0·2 M hydrochloric acid; 25 cm³ of acid were required. When a small quantity of mercury(II) chloride was added to this neutral solution a white precipitate appeared which slowly turned grey. Suggest structures for **A**, and state how you might differentiate between them. *(O. and C.)*

14. An aliphatic compound, **A**, contained C = 21·2 per cent, H = 1·8 per cent, Cl = 62·8 per cent. On being treated with water in the cold, it gave a compound, **B**, containing C = 25·4 per cent, H = 3·15 per cent, Cl = 37·6 per cent. When **B** was heated with sodium carbonate solution and subsequently acidified an acid, **C**, of molecular formula $C_2H_4O_3$ was isolated. What were **A**, **B**, and **C**? Account for the above reactions.

(O. and C.)

15. A compound **A** contains C, 66·40, H, 5·54 per cent., the only other element present being chlorine. When **A** is treated with aqueous potassium hydroxide solution it affords a primary alcohol which, by oxidation, is ultimately converted into a monobasic acid. This, when heated with soda lime, affords benzene. From these data deduce the structural formula of **A** and give the structural formulæ and names of *three* other compounds which are isomeric with it. *(J.M.B.)*

16. Two isomeric benzene derivatives, **A** and **B**, have the composition C, 53·68, H, 3·20, Cl, 22·69 per cent., the only other element present being oxygen. Both **A** and **B** dissolve in aqueous sodium carbonate solution with effervescence, but on acidification of the resulting solutions **A** is reprecipitated unchanged while the solution of **B** gives a new compound, **C**, which contains no chlorine. With concentrated ammonia solution **B** gives a solution which contains chloride ion, but **A** does not.

Explain all these reactions, and deduce probable structural formulæ for **A**, **B**, and **C**.

Suggest any other reactions by means of which the structures assigned to **A** and **B** could be confirmed. *(J.M.B.)*

17. Three isomeric compounds, **A**, **B** and **C**, have the molecular formula $C_3H_6N_2$.

(a) When **A** is boiled for some time with dilute hydrochloric acid, it yields an organic acid and ammonium chloride. When the sodium salt of this acid is mixed with soda-lime and the mixture is heated, dimethylamine is evolved.

(b) When **B** is heated with water, or with dilute acid, nitrogen is evolved. When **B** is reduced it gives propylhydrazine, $CH_3.CH_2.CH_2.NH.NH_2$.

(c) When **C** is boiled with aqueous sodium hydroxide, both ammonia and ethylamine are evolved, leaving sodium carbonate.

In each of the above, draw what conclusions you can about the groups present in **A**, **B** and **C**. Suggest possible structures for these three compounds. *(C.L.)*

18. Three isomeric compounds, **A, B** and **C,** have the molecular formula $C_8H_8O_2$.

(*a*) When heated for a long time with soda-lime **A** gave benzene, while both **B** and **C** gave toluene.

(*b*) **A** was a neutral compound, whereas both **B** and **C** were monobasic acids.

(*c*) **B** and **C** were both readily chlorinated in sunlight; **B** gave a compound containing three chlorine atoms per molecule which was easily hydrolysed to a dibasic acid **D** which in turn gave an anhydride on heating. When **D** was heated with soda-lime it gave benzene. The chlorinated product from **C** contained two chlorine atoms per molecule and was a strong monobasic acid.

Identify **A, B, C** and **D,** write equations for the reactions, and indicate how you would prepare **A** from benzoic acid. (*C.L.*)

Answers
to Exercises

Exercise 1

1. (*a*) and (*c*).
2. (*d*).
3. (*a*) and (*c*).
4. $n = 10$.
5. $CH_3 - CH = CH_2$.
6. (*a*) and (*d*).
7. (*a*) Branched-chain, aliphatic; (*b*) closed-chain, alicyclic; (*c*) straight-chain, aliphatic.

Exercise 2

1. Ethanol.
2. (*c*).
3. (*d*).
4. (*b*) and (*d*).
5. (*a*) Above 65°C; (*b*) more; (*c*) less; (*d*) increases; (*e*) water.
6. (*b*) and (*d*).
7. (*c*) and (*d*)

Exercise 3

1. 'One molecule of.'
2. (*a*) Nitrogen; (*b*) hydrogen and oxygen; (*c*) sulphur; (*d*) iodine.
3. 8·8 g of carbon dioxide, 5·4 g of water.
4. $C_6H_3Cl_3$.
5. C_6H_{12}.
6. $CH_3—O—C_2H_5$.
7. (*a*) Mass spectrometry; (*b*) infrared spectroscopy; (*c*) electron diffraction; (*d*) X-ray diffraction.
8. (*a*) 0·154 nm (1·54 Å) and 0·120 nm (1·20 Å); (*b*) 0·154 nm (1·54 Å) and 0·133 nm (1·33 Å).

Exercise 4

1. (i) $a = 180°$; (ii) $b = 109°$; (iii) $c = 120°$; (iv) $d = 109°$, $e = 120°$.

2. (b) and (c).

3. (a), (c), (d) and (e).

4. (a) and (d).

5. (c), (d) and (e).

6. (a) Substitution; (b) re-arrangement; (c) condensation; (d) acid-base reaction; (e) elimination.

7. (a) Oxidized; (b) reduced; (c) neither; (d) reduced; (e) oxidized.

8. (a) Heterolytic fission; (b) nucleophilic reagent (nucleophile); (c) activation energy; (d) carbonium ion; (e) initiation, propagation, termination.

Exercise 5

1. (a) Alkene; (b) alkane; (c) alkene; (d) alkyne; (e) alkane.

2. (b), (c), and (d).

3. (a) Ethane; (b) propane; (c) propane; (d) hexane.

4. (e).

5. (a), (b), and (d).

6. (a) 2-Methylbutane; (b) 2,2-dimethylbutane; (c) 3-ethyl-2-methylpentane; (d) 2,3-dichloropentane.

7. (a) $C_2H_5.CH(CH_3).CH_2.C_2H_5$; (b) $(CH_3)_3 C.CH_2.C_2H_5$; (c) $CH_2Cl.CHCl.CH_3$; (d) $CH_2Cl.CH_2.CH(CH_3)_2$.

8. (a), (c), and (e).

9. (a) Synthesis gas; (b) cracking; (c) aromatization; (d) isomerization; (e) petrochemicals.

Exercise 6

1. (a) Concentrated; (b) excess of acid; (c) 180°C; (d) carbon dioxide and sulphur dioxide; (e) aqueous sodium(I) hydroxide.

(2) (a) Aluminium(III) oxide or porous pot; (b) 360°C; (c) propene; (d) $CH_3.CH_2.CH_2OH \rightarrow CH_3.CH=CH_2 + H_2O$; (e) dehydration.

3. (a) 1,2-dibromoethane (ethylene dibromide); (b) 1,2-dibromo-ethane and 2-bromoethanol (ethylene bromohydrin); (c) bromoethane (ethyl bromide); (d) ethyl hydrogen sulphate(VI); (e) ethane-1,2-diol (ethylene glycol).

4. (a) (1) Propene; (2) but-2-ene; (3) 2-methylbut-1-ene. (b) $(CH_3)_2 C=CH.CH_3$, 2-methylbut-2-ene.

5. (a) $CH_3—\overset{\displaystyle |}{\underset{\displaystyle CH_3}{C}}=CH_2$; (b) $C_2H_5—CH=CH—C_2H_5$;

(c) $C_2H_5—\overset{\displaystyle |}{\underset{\displaystyle CH_3}{CH}}—CH=CH_2$

6. (a) Hydrogenation; (b) polymerization; (c) unsaturated; (d) π (pi) bond; (e) σ (sigma) bond.

7. (a) $\overset{\delta+}{Br}—\overset{\delta-}{Br}$, $\overset{\delta+}{CH_2}=\overset{\delta-}{CH_2}$; (b) bromonium ion, Br^+; (c) $\overset{+}{CH_2}—CH_2Br + Br^- \rightarrow CH_2Br—CH_2Br$; (d) because the bromine molecule attacks the carbon atom which has the higher electron density; (e) $CH_2Br—CH_2Br$, $CH_2(OH)—CH_2Br$, $CH_2Cl—CH_2Br$.

Exercise 7

1. Calcium(II) dicarbide, ethanolic (alcoholic), potassium(I) hydroxide.

2. $C_2H_5OH \xrightarrow{H_2SO_4} C_2H_4 \xrightarrow{Br_2} C_2H_4Br_2 \xrightarrow{Alc.KOH} C_2H_2$.

3. (a) Both; (b) ethyne; (c) both; (d) ethyne; (e) ethyne.

4. (a) $CHCl=CHCl$ and $CHCl_2—CHCl_2$; (b) $CH_2=CHBr$ and $CH_3—CHBr_2$.

5. (a) 1,2-dichloroethene and 1,1,2,2-tetrachloroethane; (b) bromoethene and 1,1-dibromoethane.

6. (a) and (c).

7. (b), (c), and (d).

Exercise 8

1. (a) Both; (b) neither; (c) both; (d) neither.
2. (a) Rectified spirit; (b) absolute ethanol; (c) azeotropic mixture; (d) enzymes.
3. (a) Both; (b) ethanol; (c) both; (d) ethanol.
4. (a) Propan-2-ol; (b) butan-2-ol; (c) propan-1-ol; (d) 2-methyl-propan-1-ol.
5. (a) $C_2H_5.CH_2.CH_2OH$; (b) $C_2H_5.CH_2.CH(OH).CH_3$; (c) $C_2H_5.CH_2.CH_2.CH_2.CH_2OH$; (d) $(CH_3)_3 C(OH)$.
6. (a) Sodium propoxide; (b) propyl hydrogen sulphate(VI); (c) propene; (d) propyl ethanoate; (e) 1-chloropropane.

7. (a)
$$C_2H_5{-}OH + H^+ \rightarrow C_2H_5{-}\overset{+}{O}H_2$$
$$Br^- + C_2H_5{-}\overset{+}{O}H_2 \rightarrow Br{-}C_2H_5 + H_2O$$

(b) nucleophile.
8. (a), (b), and (c).

Exercise 9

1. (a) 1-Chloropropane; (b) 2-bromopropane; (c) 2-bromo-2-methylpropane.
2. (a) and (d).
3. (a) HO^- ion; (b) HO^- ion; (c) because it loses a half-share in two electrons; (d) slower (because the $CH_3{-}{-}$ radical increases the electron density on the carbon atom attacked).
4. (a) Ethoxyethane (diethyl ether); (b) ethyl ethanoate; (c) ethylamine and ethylammonium bromide.
5. (a) $C_2H_5O^-$, ethoxide ion; (b) 2-ethoxypropane (ethyl iso-propyl ether); (c) propene and ethanol.
6. (1) (a) Ethane-1,2-diol (ethylene glycol), (b) potassium(I) formate; (2) (c) ethyne, (d) isocyanobenzene (phenyl isocyanide).
7. (a) $C_2H_5.CH_2OH$, propan-1-ol; (b) $C_2H_5.CH(OH).CH_3$, butan-2-ol; (c) $C_2H_5.COOH$, propanoic acid.

Exercise 10

1. (a) Propanal; (b) butanal; (c) 2-methylpropanal.
2. (a) Propan-1-ol (n-propyl alcohol); (b) butan-1-ol (n-butyl alcohol); (c) 2-methylpropan-1-ol (isobutyl alcohol).
3. (b), (d), and (e).
4. (a) A purple colour; (b) a yellow precipitate (which turns red); (c) the purple solution is decolorized; (d) nothing; (e) a yellow precipitate.
5. (b) Ethanoic, or acetic, acid; (c) Ethanoic, or acetic, acid; (d) Sodium(I) 1-hydroxyethylsulphate(IV) (sodium acetaldehyde-bisulphite); (e) propanone 2,4-dinitrophenylhydrazone.
6. (b), (c).
7. (a).
8. (a) 2-hydroxy-2-methylpropanonitrile (acetone, or propanone, cyanohydrin; (b) sodium(I) 1-hydroxy-1-methylethylsulphate(IV); (c) propanone oxime; (d) bromopropanone.

Exercise 11

1. (a), (b), and (c).
2. (a) $C_2H_5OH \rightarrow C_2H_5Br \rightarrow C_2H_5CN \rightarrow C_2H_5.COOH$. (b) $C_2H_4 \rightarrow C_2H_5HSO_4 \rightarrow C_2H_5OH \rightarrow CH_3.COOH$.
3. (a), (b), and (e).
4. (a) Ammonium propanoate, $C_2H_5.COONH_4$; (b) propanoyl chloride, $C_2H_5.COCl$; (c) methyl propanoate, $C_2H_5.COOCH_3$; (d) propan-1-ol, $C_2H_5.CH_2OH$.
5. (b), (c), (d), and (e).
6. (a) Sodium(I) methanoate and carbon dioxide; (b) carbon dioxide; (c) carbon dioxide; (d) carbon monoxide; (e) carbon monoxide.
7. (d), (c), (e), (a), (b).
8. (a) $(CH_3)_2 CH.COOH$; (b) $CH_3.CCl_2.COOH$; (c) $CH_2Cl.COCl$; (d) $CH_2(OH).COOH$; (e) $CH_3.COOCH_3$.

Exercise 12

1. (a) Acyl; (b) ethanoyl, or acetyl; (c) propanoyl.

2. (a) $3C_2H_5.COOH + PCl_3 \rightarrow 3C_2H_5.COCl + H_3PO_3$.
(b) $C_2H_5.COOH + SOCl_2 \rightarrow C_2H_5.COCl + SO_2 + HCl$.

3. (a) Propanoic acid; (b) ethyl propanoate; (c) propanamide; (d) propan-1-ol; (e) propanoic anhydride.

4.

(a) $C_2H_5.CH_2.\overset{\overset{\textstyle O}{\|}}{C}{-}Cl$; 　　(b) $CH_3.\overset{\overset{\textstyle O}{\|}}{C}{-}NH_2$;

(c) $CH_3.\overset{\overset{\textstyle O}{\|}}{C}{-}NHC_2H_5$; 　　(d) $CH_3.\overset{\overset{\textstyle O}{\|}}{C}{-}O{-}\overset{\overset{\textstyle O}{\|}}{C}.CH_3$.

5.

$$R{-}\overset{\overset{\textstyle O}{\|}}{\underset{\underset{\textstyle Cl}{|}}{C}} + :OH^- \rightleftharpoons R{-}\overset{\overset{\textstyle \bar{O}}{|}}{\underset{\underset{\textstyle Cl}{|}}{C}}{-}OH \overset{-Cl^-}{\rightleftharpoons} R{-}\overset{\overset{\textstyle O}{\|}}{C}{-}OH.$$

6. (a) $(RCO)_2O + C_2H_5OH \rightarrow RCOOC_2H_5 + RCOOH$.
(b) $(RCO)_2O + 2NH_3 \rightarrow RCONH_2 + RCOONH_4$.

7. (b) and (c).

8. $CH_2Cl.COCl > CH_3.COCl > CH_3.COBr > C_2H_5.COBr$.

Exercise 13

1. (a), (b), and (e).

2. (a) Methyl ethanoate; (b) ethyl propanoate, (c) propyl ethanoate; (d) 1-methylethyl methanoate.

3.

$$H{-}\overset{\overset{\textstyle O}{\|}}{\underset{\underset{\textstyle OCH_3}{|}}{C}} + \ ^-OH \rightleftharpoons H{-}\overset{\overset{\textstyle \bar{O}}{|}}{\underset{\underset{\textstyle OCH_3}{|}}{C}}{-}OH \rightleftharpoons CH_3O^- + H{-}\overset{\overset{\textstyle O}{\|}}{C}{-}OH$$

$$CH_3O^- + H_2O \rightleftharpoons CH_3OH + OH^-$$

4. $CH_2Cl.COOCH_3 > HCOOCH_3 > CH_3COOCH_3$
　　　　　　　　　　　　　　　$> C_2H_5.COOCH_3$

5. (a) Methanol and propanoic acid; (b) methanol and sodium propanoate; (c) methanol and butyl propanoate; (d) methanol and propanamide; (e) methanol and propan-1-ol.

6. Propyl methanoate, $HCOOCH_2.C_2H_5$; 1-methylethyl methanoate, $HCOOCH(CH_3)_2$; ethyl ethanoate, $CH_3.COOC_2H_5$; methyl propanoate, $C_2H_5.COOCH_3$.

7. Soap (the only one not composed chiefly of esters).

8. (a) $C_{12}H_{25}SO_4^-Na^+$; (b) $-SO_4^-$; (c) $C_{12}H_{25}-$; (d) it does not form a precipitate, or scum.

Exercise 14

1. (b), (c), (d), (e).

2. (a) $R.CONH_2 + NaOH \rightarrow R.COONa + NH_3$
(b) $R.CONH_2 + HNO_2 \rightarrow R.COOH + N_2 + H_2O$
(c) $R.CONH_2 + P_2O_5 \rightarrow RCN + 2HPO_3$.

3. X = propanamide;
$C_2H_5.CONH_2 + H_2O + HCl \rightarrow C_2H_5.COOH + NH_4Cl$.

4. (a) Neither; (b) carbamide; (c) both; (d) carbamide; (e) both.

5. (a) Neither; (b) $-C{\equiv}N$; (c) $-C{\equiv}C-$; (d) both; (e) $-C{\equiv}C-$.

6. Acid amides, nitriles, ammonium salts of organic acids.

7. (a) $CH_3OH \rightarrow CH_3I \rightarrow CH_3CN \rightarrow CH_3.CH_2NH_2$.
(b) $C_2H_5OH \rightarrow C_2H_5I \rightarrow C_2H_5CN \rightarrow C_2H_5COOH$.

8. (a) Amino; (b) amido; (c) cyano; (d) peptide.

Exercise 15

1. (a) $C_2H_5.NH.CH_2.C_2H_5$; (b) $(CH_3)_2CHNH_2$;
(c) $(C_2H_5)_2N(CH_3)$; (d) $(CH_3)_2CH.CH_2NH_2$.

2. (a) Secondary; (b) primary; (c) tertiary; (d) primary.

3. (a) $C_2H_5.CONH_2 + Br_2 \rightarrow C_2H_5.CONHBr + HBr$
$C_2H_5.CONHBr + NaOH \rightarrow C_2H_5NCO + NaBr + H_2O$
$C_2H_5NCO + 2NaOH \rightarrow C_2H_5NH_2 + Na_2CO_3$
(b) N-bromopropanamide, ethyl isocyanate.

4. (b) and (c).

5. (a) Ethylammonium bromide; (b) ethylmethylamine;
(c) N-ethylethanamide; (d) isocyanoethane.

6. $CH_3.NH.CH_2.C_2H_5$, methylpropylamine.
$CH_3.NH.CH(CH_3)_2$, (I-methylethyl) methylamine.
$C_2H_5.NH.C_2H_5$, diethylamine.
$(CH_3)_2N(C_2H_5)$, ethyldimethylamine.

7. (a), (b), and (d).

Exercise 16

1. (a) Ethanedioic acid; (b) both; (c) methanoic acid; (d) both.

2. (a) Sodium ethanedioate; (b) ethanediamide; (c) ethanoic acid
and carbon dioxide; (d) butanedioic anhydride.

3. (a) $(COOH)_2.2H_2O$; (b) $(CONH_2)_2$;
(c) $(COCl)_2$; (d) $CH_2(COOCH_3)_2$.

4. (a) $C_2H_5OH \rightarrow CH_3.CHO \rightarrow CH_3.CH(OH)CN$
$\rightarrow CH_3.(CHOH).COOH$;

(b) $CH_2 \parallel CH_2 \rightarrow CH_2OH | CH_2Cl \rightarrow CH_2OH | CH_2CN \rightarrow CH_2OH | CH_2.COOH$

5. (a), (c), (d).

6. (a) Copper(II) aminoethanoate; (b) sodium(I) aminoethanoate;
(c) hydroxyethanoic acid; (d) ethyl aminoethanoate.

7. (a) Tautomerism (dynamic isomerism); (b) zwitterion; (c) polypeptide; (d) fibrous proteins; (e) denaturation.

8. (a) Yes; (b) no; (c) yes; (d) no.

9. Hydrogen bonds, disulphide (—S—S—) bonds, and ionic bonds between basic groups and acidic groups.

Exercise 17

1. (a), (c), and (d).

2. (a) (+)2-aminopropanoic acid, (−)2-aminopropanoic acid, and (±)2-aminopropanoic acid.

(b)
$$H-\overset{CH_3}{\underset{COOH}{C}}-NH_2 \qquad H_2N-\overset{CH_3}{\underset{COOH}{C}}-H$$

3. (a) Asymmetric; (b) plane-polarized; (c) enantiomorphs; (d) a racemic compound (or racemate).

4. (a), (b), and (c).

5. (a) Two;

(b)
$$H-\overset{COOH}{\underset{COOH}{\overset{|}{C}}}-Br \qquad Br-\overset{COOH}{\underset{COOH}{\overset{|}{C}}}-H$$
$$Br-\overset{|}{\underset{}{C}}-H \qquad H-\overset{|}{\underset{}{C}}-Br$$

6. The racemic compound (externally compensated), the *meso*-compound (internally compensated).

7. (*a*), (*b*), and (*d*).

8. (*a*), (*b*), and (*d*).

Exercise 18

1. (*a*) Disaccharide; (*b*) monosaccharide; (*c*) polysaccharide; (*d*) disaccharide; (*e*) monosaccharide.

2. X = —CH$_2$OH; Y = —CHO.

3. (*a*) Both; (*b*) glucose; (*c*) both; (*d*) glucose; (*e*) glucose.

4. (*a*) Aldohexose; (*b*) inversion of sucrose; (*c*) tautomerism; (*d*) Benedict's solution; (*e*) osazone.

5. (*b*), (*d*), (*e*).

6. (*a*) α-glucose; (*b*) α-glucose and γ-fructose; (*c*) α-glucose; (*d*) β-glucose; (*e*) α-glucose.

7. (*a*), (*c*), (*d*), (*e*).

8. (*a*) $C_{12}H_{22}O_{11} + H_2O \rightarrow 2C_6H_{12}O_6$.

(*c*) $C_{12}H_{22}O_{11} + H_2O \rightarrow C_6H_{12}O_6 + C_6H_{12}O_6$.

(*d*) $(C_6H_{10}O_5)_n + nH_2O \rightarrow nC_6H_{12}O_6$;

(*e*) $(C_6H_{10}O_5)_n + nH_2O \rightarrow nC_6H_{12}O_6$.

Exercise 19

1. (*a*) Methylbenzene (toluene); (*b*) chlorobenzene; (*c*) phenol; (*d*) nitrobenzene; (*e*) benzenesulphonic acid.

2. (*a*) Light oil; (*b*) benzol; (*c*) catalytic reforming; (*d*) (*i*) nitration, (*ii*) sulphonation.

3. Cyclohexane (the only non-aromatic compound).

4. (*a*) Both; (*b*) both; (*c*) hex-1-ene; (*d*) hex-1-ene; (*e*) both.

5. (*a*) C_6H_{12}, cyclohexane; (*b*) $C_6H_6Br_6$, 1,2,3,4,5,6-hexabromo-cyclohexane, (*c*) $C_6H_6(O_3)_3$, benzene triozonide.

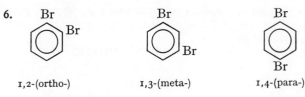

6.

1,2-(ortho-)   1,3-(meta-)   1,4-(para-)

7. (*a*), (*d*), and (*e*).

Exercise 20

1. (*c*), (*d*).

2. (*a*) Ethylbenzene ($C_6H_5.C_2H_5$); (*b*) ethylbenzene ($C_6H_5.C_2H_5$); (*c*) (1-methylethyl)benzene, $C_6H_5.CH(CH_3)_2$; (*d*) hexylbenzene ($C_6H_5.C_6H_{13}$); (*e*) dimethylbenzene, $C_6H_4(CH_3)_2$.

3. (*a*) Methyl-2-nitrobenzene and methyl-4-nitrobenzene; (*b*) (bromomethyl)benzene; (*c*) bromo-2-methylbenzene and bromo-4-methylbenzene; (*d*) 2- and 4-methylbenzenesulphonic acid.

4. (*a*)   (*b*)

CH$_2$Cl   Br   CH$_3$

(*c*)   (*d*)

CH$_3$ SO$_3$H   CH$_3$ CH$_3$

5. $C_6H_5.CH_3 \xrightarrow[\text{HNO}_3]{\text{Dil.}} C_6H_5.COOH \xrightarrow{\text{Soda-lime}} C_6H_5.COONa \xrightarrow{\text{Soda-lime}} C_6H_6.$

6. (*a*) 1,3-; (*b*) 1,2- and 1,4-; (*c*) 1,2- and 1,4-; (*d*) 1,3-; (*e*) 1,2- and 1,4-.
7. (*b*), (*d*), and (*e*).
8. (*a*) Faster; (*b*) slower; (*c*) slower; (*d*) slower; (*e*) faster.

Exercise 21

1. (*a*) and (*b*).
2. (*a*) Excess of bromine; (*b*) some of the bromine is wasted by (*i*) evaporation, (*ii*) formation of dibromobenzene.
3. (*a*) Both; (*b*) bromoethane; (*c*) bromoethane; (*d*) bromoethane; (*e*) both.
4. (*a*) $C_6H_5Br + HNO_3 \rightarrow C_6H_4\begin{smallmatrix}\diagup Br\\ \diagdown NO_2\end{smallmatrix} + H_2O$;

   (*b*) No reaction;

   (*c*) $C_6H_4\begin{smallmatrix}\diagup NO_2\\ \diagdown Cl\end{smallmatrix} + \overset{-}{O}H \rightarrow C_6H_4\begin{smallmatrix}\diagup NO_2\\ \diagdown OH\end{smallmatrix} + \overset{-}{Cl}$;

   (*d*) $C_6H_4\begin{smallmatrix}\diagup CH_3\\ \diagdown Br\end{smallmatrix} + 3O \rightarrow C_6H_4\begin{smallmatrix}\diagup COOH\\ \diagdown Br\end{smallmatrix} + H_2O$.

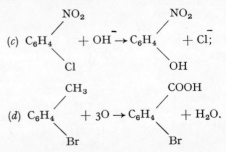

5. (*a*) Bromo-2-nitrobenzene and bromo-4-nitrobenzene; (*c*) 4-nitrophenol; (*d*) 2-bromobenzenecarboxylic acid.
6. (*a*), (*c*), and (*d*).

7. (*a*) Chloronium ion; (*b*) (trichloromethyl)benzene; (*c*) phenylmethyl ethanoate, or acetate; (*d*) (phenylmethyl)amine; (*e*) phenylmethanol.

Exercise 22

1. (*a*) NO_2^+; (*b*) $HNO_3 + 2H_2SO_4 \rightleftharpoons NO_2^+ + H_3\dot{O}^+ + 2HSO_4^-$.
2. $C_6H_6 + NO_2^+ \rightarrow \overset{+}{C_6}H_5\begin{smallmatrix}\diagup H\\ \diagdown NO_2\end{smallmatrix} \xrightarrow{HSO_4^-} C_6H_5NO_2 + H_2SO_4$.
3. (*a*), (*b*), and (*e*).
4. (*a*) Phenylamine (aniline); (*b*) benzene-1,3-diamine; (*c*) 4-methylphenylamine (*p*-toluidine).
5. (*a*) $\begin{smallmatrix}HO\diagdown\\ \bar{O}\diagup\end{smallmatrix}S\rightarrow O$; (*b*) $\begin{smallmatrix}HO\diagdown\\ \diagup\end{smallmatrix}S\begin{smallmatrix}\nearrow O\\ \searrow O\end{smallmatrix}$; (*c*) $\begin{smallmatrix}C_6H_5\diagdown\\ \bar{O}\diagup\end{smallmatrix}S\begin{smallmatrix}\nearrow O\\ \searrow O\end{smallmatrix}$.
6. (*b*), (*c*), and (*e*).
7. (*a*) Methylbenzenesulphonic acid (toluenesulphonic acid)

   $$C_6H_4\begin{smallmatrix}\diagup CH_3\\ \diagdown SO_3H\end{smallmatrix}$$

   (*b*) Sodium(I) phenoxide, C_6H_5ONa; (*c*) benzenesulphonyl chloride $C_6H_5SO_2Cl$.

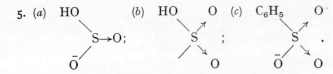

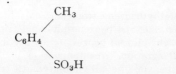

Exercise 23

1. (a) Exothermic; (b) phenylammonium hexachlorostannate(IV); (c) aqueous sodium(I) hydroxide; (d) ethoxyethane; (e) potassium(I) hydroxide.

2. (a) Below 100°C; (b) remains constant; (c) mass of X: mass of water = 1:4.

3. (a) Phenylammonium chloride ($C_6H_5NH_3{}^+Cl^-$); (b) benzene-diazonium chloride ($C_6H_5N_2{}^+Cl^-$); (c) N-phenylethanamide or N-phenylacetamide ($C_6H_5NH.COCH_3$); (d) 2,4,6-tribromophenyl-amine ($C_6H_2(NH_2)Br_3$); (e) 4-aminobenzenesulphonic acid.

$$\left(\begin{array}{c} NH_2 \\ C_6H_4 \\ SO_3H \end{array} \right)$$

4. (a) A white precipitate; (b) oily drops and a milky emulsion (due to liberated phenylamine); (c) a white precipitate; (d) a violet colour; (e) a red precipitate.

5. A = N-phenylethanamide ($C_6H_5NH.COCH_3$); B = 2- and 4-nitro-N-phenylethanamide

$$\left(\begin{array}{c} NH.COCH_3 \\ C_6H_4 \\ NO_2 \end{array} \right)$$

6. 2-, 3-, and 4-Methylphenylamine, N-methylphenylamine, (phenylmethyl)amine.

7. (a) Both; (b) both; (c) both; (d) neither; (e) both.

8.

$$C_6H_5NH_3{}^+Cl^- \xrightarrow[5°C]{HNO_2} C_6H_5N_2{}^+ \; Cl^- \xrightarrow[(KCN)]{CuCN} C_6H_5CN \xrightarrow{\substack{Na \; and \\ ethanol}}$$
$$C_6H_5CH_2NH_2$$

Exercise 24

1. (a) $C_6H_6 \rightarrow C_6H_5NO_2 \rightarrow C_6H_5NH_2 \rightarrow C_6H_5N_2Cl \rightarrow C_6H_5OH$,
(b) $C_6H_6 \rightarrow C_6H_5SO_3H \rightarrow C_6H_5SO_3Na \rightarrow C_6H_5ONa \rightarrow C_6H_5OH$.

2. (a) Propene; (b) anhydrous aluminium(III) chloride (or phosphoric(V) acid);

$$(c) \; C_6H_5.CH\Big\langle\begin{array}{c}CH_3\\CH_3\end{array} \quad ; \; (d) \; C_6H_5.C\Big\langle\begin{array}{c}O-O-H\\CH_3\end{array}$$

(e) propanone.

3. (c) and (e).

4. (a) Sodium(I) phenoxide; (b) no reaction; (c) phenyl ethanoate, or acetate; (d) 2- and 4-hydroxybenzenesulphonic acid; (e) phenyl benzenecarboxylate, or benzoate.

5. (a) No change; (b) an oily emulsion (due to liberated phenol); (c) a white precipitate; (d) a violet colour; (e) a yellow precipitate.

6. (a) and (e).

7. (a) Phenylmethyl ethanoate; (b) benzenecarbaldehyde, or benzaldehyde; (c) methylbenzene; (d) (chloromethyl)benzene; (e) sodium phenylmethoxide.

8. Phenylmethanol; 2-, 3-, and 4-methylphenol; methoxybenzene.

9. (a) $C_6H_6 \rightarrow C_6H_5.CH_3 \rightarrow C_6H_5.CH_2Cl \rightarrow C_6H_5.CH_2OH$;
(b) $C_6H_5.CH_2OH \rightarrow C_6H_5.COOH \rightarrow C_6H_5.COONa \rightarrow C_6H_6$.

Exercise 25

1. (a) Copper(II), or lead(II), nitrate(V) solution; (b) sodium(I) hydrogensulphate(IV); (c) a mixture of carbon monoxide and hydrogen chloride (AlCl$_3$ and CuCl catalyst); (d) air (vanadium compound as catalyst).

2. (a), (c), and (e).

3. (a) Sodium(I) hydroxyphenylmethylsulphate(IV)

$$(C_6H_5.CH(OH).SO_3^-Na^+);$$

(b) phenylmethanol (($C_6H_5.CH_2OH$) and sodium(I) benzenecarboxylate ($C_6H_5.COONa$); (c) methylbenzene ($C_6H_5.CH_3$); (e) benzenecarbaldehyde oxime or benzaldehyde oxime ($C_6H_5.CH:NOH$).

4. (a) Phenylethanone ($C_6H_5.CO.CH_3$); (b) diphenylmethanonephenylhydrazone; (c) 1-phenylethanol; (d) benzenecarboxylic acid and carbon dioxide.

5. (a) Benzene; (b) iron(III) benzenecarboxylate; (c) benzenecarbonyl chloride; (d) 3-chlorobenzenecarboxylic acid.

6. (a) $C_6H_5COO.CH_2.C_2H_5$; (b) $C_6H_5.COOC_6H_5$; (c) $(C_6H_5CO)_2O$; (d) $C_6H_5.CONH_2$.

7. (a), (c), and (d).

8. (a) Benzenecarbonitrile, or benzonitrile; (b) phenylmethyl benzenecarboxylate; (c) methylbenzenecarboxylic acid; (d) hydroxybenzenecarboxylic acid; (e) benzenedicarboxylic acid.

Exercise 26

1. (a) Unbranched linear polymer; (b) random linear polymer; (c) linear block polymer; (d) branched linear polymer.

2. (a) An initiator; (b) a Ziegler catalyst; (c) an isotactic polymer; (d) polyester, polyamide.

3. (a) and (e).

4. (a) Elastomers; (b) methylbuta-1,3-diene; (c) vulcanization; (d) anti-oxidants; (e) ($—CH_2—CH=CCl—CH_2$)$_n$.

5. (a) B; (b) B; (c) A; (d) B; (e) B.

6. (a) **A** = poly(phenylethene); (b) **B** = poly(tetrafluoroethene); (c) **C** = Bakelite; (d) **D** = poly(ethene).

7. (a) Benzene-1,4-dicarboxylic acid (HOOC—C_6H_4—COOH) and ethane-1,2-diol (HO—C_2H_4—OH).
(b) ($—OC—C_6H_4—COO—C_2H_4—O—$).

8. (a) Hexanedioic acid (HOOC—$(CH_2)_4$—COOH) and hexane-1,6-diamine ($H_2N—(CH_2)_6—NH_2$).
(b) ($—OC—(CH_2)_4—CO—NH—(CH_2)_6—NH—$).

9. (a) Cellulose diethanoate (approximately); (b) poly(propenonitrile); (c) polyester of benzene-1,4-dicarboxylic acid and ethane-1,2-diol; (d) poly(propene).

Answers to Miscellaneous Problems

Advanced Level

1. (i) 92·3 per cent carbon, 7·7 per cent hydrogen.
(ii) 52·2 per cent carbon, 13·0 per cent hydrogen, 34·8 per cent oxygen.
(iii) 40·0 per cent carbon, 6·7 per cent hydrogen, 53·3 per cent oxygen.

2. 31·6 per cent chlorine.

3. $A = CH_4$, $B = CH_4O$, $C = C_2H_5NO$, $D = C_2H_5Cl$.

4. $X = C_4H_8O_2$, $Y = C_3H_5N$, $Z = C_2H_4Br_2$.

5. C_3H_6O.

6. 20 per cent ethane, 80 per cent ethene.

7. (i) 61·5; (ii) 154·4; (iii) 131·8; (iv) 188·8.

8. $C_4H_8O_2$.

9. C_2H_6O; $CH_3—O—CH_3$; methoxymethane.

10.

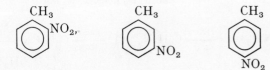

11. $C_6H_5.CH_2NH_2$, $C_6H_5.NH.CH_3$, $C_6H_4(CH_3)NH_2$ (three isomers).

12. 1·44 g bromine.

More Difficult Problems

13. $A = HCOOCH_2.C_2H_5$ or $HCOOCH(CH_3)_2$.

14. A = chloroethanoyl chloride, B = chloroethanoic acid, C = hydroxyethanoic acid.

15. $A = C_6H_5.CH_2Cl$.

16.

$$A = C_6H_4 \overset{\displaystyle COOH}{\underset{\displaystyle Cl}{\Big\langle}} \ , \ B = C_6H_4 \overset{\displaystyle OH}{\underset{\displaystyle COCl}{\Big\langle}} \ , \ C = C_6H_4 \overset{\displaystyle OH}{\underset{\displaystyle COOH}{\Big\langle}}$$

17. $A = CH_3.NH.CH_2CN$, $B = CH_3.CH_2.CHN_2$, $C = CH_3.CH_2.NH.CN$.

18. $A = C_6H_5.COOCH_3$, $B = CH_3.C_6H_4.COOH$, $C = C_6H_5.CH_2.COOH$, $D = C_6H_4(COOH)_2(1,2)$.

Index